Fundamentals of Environmental Engineering

James R. Mihelcic

Contributors

Martin T. Auer

David W. Hand

Richard E. Honrath, Jr.

Judith A. Perlinger

Noel R. Urban

Department of Civil and Environmental Engineering
Michigan Technological University

Michael R. Penn

Department of Civil and Environmental Engineering
University of Wisconsin-Platteville

John Wiley & Sons, Inc.

New York / Chichester / Weinheim / Brisbane / Singapore / Toronto

EDITOR *Wayne Anderson*
MARKETING MANAGER *Katherine Hepburn*
SENIOR PRODUCTION MANAGER *Lucille Buonocore*
SENIOR PRODUCTION EDITOR *Monique Calello*
DESIGN DIRECTOR *Madelyn Lesure*
COVER DESIGNER *Michael Jung*
ILLUSTRATION EDITOR *Sigmund Malinowski*
ILLUSTRATION STUDIO *JAK Graphics*

This book was set in Times Roman by UG Division of GGS Information Services and printed and bound by Courier Westford, Inc. The cover was printed by Phoenix Color Corporation.

This book is printed on acid-free paper. ♾

The paper in this book was manufactured by a mill whose forest management programs include sustained yield harvesting of its timberlands. Sustained yield harvesting principles ensure that the numbers of trees cut each year does not exceed the amount of new growth.

Library of Congress Cataloging in Publication Data:
Mihelcic, James R.
Fundamentals of environmental engineering / James R. Mihelcic & contributors, Martin T. Auer . . . [et al.].
p. cm.
Includes bibliographical references and index.
ISBN 0-471-24313-2 (cloth : alk. paper)
1. Ecological engineering. 2. Environmental sciences. I. Auer, Martin T. II. Title.
GE350.M54 1998
628—dc21 98-39323
CIP

Printed in the United States of America

10 9 8 7 6 5 4 3 2 1

Preface

In our local community many grade-school and high-school classes "adopt" a local stream. Working with their teachers and personnel from Michigan Tech's Environmental Outreach Center, students first learn about water quality and land issues in the classroom, then go into the field to measure the stream's flow and perform some field measurements of selected chemical and biological water quality parameters. One day I read these comments on the "Adopt a Stream" program provided by a third grader from the town of Bessemer. The third grader wrote:

> *"During "Adopt a Stream" we asked a ton of questions. Most were hard ones to answer. The teachers asked two tons of questions. Even I asked a ton of questions. I wanted to mostly know how long it takes nonpoint source pollution to go underground and travel into the water and pollute it."*

How insightful this third grader was! I seem to remember that I did not learn about nonpoint source pollution and how surface waters can be hydraulically connected to groundwater until late in my college career. The fact is that today, engineers and scientists who work on what seem to be increasingly complex environmental problems require a vast knowledge of the fundamentals of chemistry, biology, and physical processes.

Fundamentals of Environmental Engineering is the outgrowth of a team-teaching effort for a course required of all civil and environmental engineering undergraduates at Michigan Tech. Other students taking this course major in fields such as chemical and geological engineering, chemistry, and biology. The intention of the course is to provide a bridge for a student to move from the fundamentals studied in first- and second-year math and basic science courses to their introductory and upper level engineering courses, which apply those fundamentals.

Fundamentals of Environmental Engineering provides coverage of the basics required for design, operation, analysis, and modeling of both natural and engineered systems. Also, the fundamentals presented here are a necessity for solving both small-town and global environmental problems. Chapters include (1) reporting concentrations in air, water, and soil; (2) chemistry (kinetics, thermodynamics, equilibrium processes in air, soil, and water, and photochemistry); (3) physical processes (mass and energy balances, reactor engineering, mass transport, Stokes' law, and Darcy's law); and (4) biology (ecosystem structure and function, energy and material flow, population dynamics and modeling growth, oxygen demand and oxygen sag in surface water, wastewater treatment, eutrophication of lakes, ecosystem health, and public welfare). *Fundamentals of Environmental Engineering* also emphasizes the connections among the different specialty areas of environmental engineering—the contributors to this effort have expertise in drinking water and wastewater treatment, air quality engineering, groundwater engineering, solid and hazardous waste management and remediation, surface water quality, environmental chemistry, ecology, and assessing environmental risk. Thus the book presents those required fundamentals along with close to one hundred applications for a diverse set of relevant environmental situations including multimedia issues encompassing engineered treatment and chemical fate and transport in air, water, and soil. Our intent is not to replace the many excellent books that cover a particular topic or chapter of this text. Rather it is to present all the scientific fundamentals in one easy-to-understand format for an undergraduate audience.

Our student reviewers went well beyond pointing out typographical errors. They provided many insightful comments that make this text more "student friendly." One request was to note critical equations. We have done this by placing a next to those equations we felt were of major importance.

Web Site

Additional resources for instructors and students to support this text can be found at www.wiley.com/college/mihelcic.

Acknowledgments

The following Michigan Tech students majoring in civil engineering, environmental engineering, chemistry, and biology provided invaluable comments to improve the technical content and readability of this book. They are Janine Arnold, Evan Berglund, Elly Bunzendahl, Jessica Jirgl, Thomas McDowell, Katie Patterson, Raghuraman Venkatapathy, and Susan Weycker. Susan Bagley of Michigan Tech's Department of Biological Sciences provided helpful comments on the biology chapter and Kimberly Elenbaas assisted in assembling many of the figures in that chapter. I would also like to thank our other environmental engineering colleagues at Michigan Tech for their comments, support, and free exchange of ideas, many of which were incorporated into the text. They are Bob Baillod, John Crittenden, John Gierke, Neil Hutzler, Alex Mayer, and Kurt Paterson. Finally, I would like to thank Mark Milke, David Wareham, and other members of the Department of Civil Engineering, University of Canterbury, Christchurch, New Zealand, who made my sabbatical leave so pleasant and productive.

I would also like to thank the following reviewers for their help in the development of this text: William D. Burgos, *Pennsylvania State University*; John T. Coates, *Clemson University*; and Simeon J. Kosier, *Rensselaer Polytechnic Institute*.

James R. Mihelcic
South Shore of Lake Superior

About the Authors

Dr. James R. Mihelcic is Associate Professor of Civil and Environmental Engineering at Michigan Technological University. He teaches introductory courses in environmental engineering and advanced courses in environmental biochemistry, solid and hazardous waste management, biological processes, and water chemistry. He has been awarded the Department of Civil and Environmental Engineering Howard E. Hill Outstanding Faculty of the Year Award and has also been a finalist for the Michigan Tech Teacher of the Year Award. Dr. Mihelcic's research interests include environmental microbiology; optimizing biological treatment processes and understanding biological transformations in wastewater, soil, and groundwater; studying bioavailability of hydrophobic organic chemicals discharged to soil and aquatic environments; and development of estimation methods that predict environmental properties such as sediment–water partition coefficients, bioconcentration, and biodegradability from knowledge of chemical structure.

Dr. Martin T. Auer is Professor of Civil and Environmental Engineering at Michigan Technological University with 18 years of service on the faculty. He teaches introductory courses in environmental engineering and advanced coursework in surface water quality engineering and mathematical modeling of lakes, reservoirs, and rivers. Dr. Auer's present research activity includes studies of pollution problems in Onondaga Lake, New York and the New York City reservoir system and an investigation of cross-marginal transport processes in Lake Superior.

Dr. David W. Hand is Associate Professor of Civil and Environmental Engineering at Michigan Technological University. He teaches senior-level courses in drinking water treatment, wastewater treatment engineering, and air stripping/adsorption processes. Dr. Hand's present research focus is on the development of a mathematical model that describes the performance of the International Space Station's portable water treatment system, the development and application of photocatalytic oxidation processes for the removal of pollutants from water and air, and the development of environmental engineering software tools for pollution prevention practices.

Dr. Richard E. Honrath, Jr. is Associate Professor of Civil and Environmental Engineering at Michigan Technological University. He teaches courses on the fundamentals of environmental engineering science, current topics in air quality engineering, and advanced atmospheric chemistry. Dr. Honrath's research interests include the photochemistry of nitrogen oxides and ozone in the troposphere, with a focus on measurement and modeling studies of the hemispheric scale impacts of human activities on tropospheric composition and chemistry.

Dr. Michael R. Penn is Assistant Professor of Civil and Environmental Engineering at the University of Wisconsin-Platteville with teaching responsibilities for undergraduate courses in introductory environmental engineering, wastewater treatment and water quality, groundwater hydrology, solid and hazardous waste management, and hydrology. He taught comparable courses in a previous position at Wilkes University in Wilkes-Barre, Pennsylvania. Dr. Penn received his Ph.D. from Michigan Technological University on completion of research on nutrient cycling in lakes and sediments. His current research interests focus on involving undergraduates in surface and groundwater contaminant fate and transport investigations.

Dr. Judith A. Perlinger is Assistant Professor of Civil and Environmental Engineering at Michigan Technological University. She teaches courses on fundamentals of environmental engineering science, environmental and water chemistry, and environmental organic chemistry. Dr. Perlinger's research interests include the transport and transformation of xenobiotic organic chemicals in the environment. Most recently her research has focused on kinetic and mechanistic aspects of reductive transformations of these compounds in anaerobic environments such as groundwater aquifers and lake sediments.

Dr. Noel R. Urban is Assistant Professor of Civil and Environmental Engineering at Michigan Technological University. His research focuses on the biogeochemistry of nutrients and minor metals in wetlands and lakes. Dr. Urban's teaching activity centers around environmental chemistry.

Front Row (L-R): Judith Perlinger, Martin Auer, David Hand
Back Row (L-R): Mike Penn, Richard Honrath, Jim Mihelcic, Noel Urban

Contents

Chapter 5 Biology 209

Chapter 1

Student Preface: Why Do I Need Fundamentals?

James R. Mihelcic
Judith A. Perlinger

In this chapter a reader will be presented with examples of three environmental situations that require solutions by engineers and scientists. After each situation, a series of questions are posed with the purpose of demonstrating how an individual who has a strong command of chemical, biological, and physical process fundamentals will be able to more effectively solve environmental problems. As students read through *Fundamentals of Environmental Engineering* they should return to these questions to find how the fundamentals that they have learned can assist in answering a particular question(s). For example, to simply design and evaluate the performance of a physical/chemical treatment system, an individual must understand mass balances, reactor design and behavior, reaction order and rates, and chemical and physical treatment processes.

Today, the profession of environmental engineering draws upon many disciplines, including civil, environmental, chemical, mechanical, and geological engineering; geology; chemistry; microbiology; toxicology; atmospheric sciences; meteorology; and ecology. The civil engineering profession encompasses many specialty areas (i.e., structural, geotechnical, water resources, transportation, construction management, environmental), all of which have an environmental component to them. For example, an engineer may be designing the concrete structure of a reactor that another engineer sized or may be involved in traffic planning in an urban area that has a goal of meeting stringent air-emission regulatory limits. Also, many practicing engineers find themselves specifically working on an environmental engineering problem at some point in their career.

Today's environmental problems are complex and are no longer confined to one particular medium. For example, no longer is it acceptable to treat contaminated water by stripping the contamination into the air and thereby causing an air-pollution problem. Furthermore, an individual must now understand how chemicals move across boundaries of air, water, and soil, and how entry across each of these boundaries can influence a chemical's transport, fate, potential risk, and treatment. Not only is the study of environmental engineering the study of the effects of humans on the environment, but to some extent the effects of the environment on human activities.

One definition of environmental engineering is that it is *a field in which one applies the basic fundamentals of mathematics, physics, chemistry, and biology to the protection of human health and the environment.* At the 1996 Environmental Engineering Education Conference (AAEE, 1997) representatives from industry and academia stressed the necessity for environmental engineers to have a strong background in the fundamentals of chemistry, biology, and physical processes. The purpose of *Fundamentals of Environmental Engineering* is to demonstrate how fundamentals learned in mathematics, physics, chemistry, and biology are applied to environmental problems.

SITUATION 1 MUNICIPAL WASTEWATER TREATMENT

Municipal wastewater contains relatively large amounts of organic matter, solids, and nutrients such as nitrogen and phosphorus. As discussed in later chapters, all of these items may adversely impact the chemical and biological balances of a river or lake if they are improperly discharged. In addition, untreated wastewater may spread disease among humans by exposure during drinking, bathing, or swimming. The purpose of municipal wastewater treatment is to first collect and transport the wastewater to a treatment plant, where it is treated to remove dissolved and particulate organic matter, solids, nutrients, and pathogens to levels that will not cause an adverse impact on human health and the environment. These goals are met by constructing a treatment plant that employs physical, chemical, and biological treatment processes. Figure 1-1 shows an overhead photo of a large treatment plant used by a major U.S. city. There are many tanks and buildings in the photo, and each of them has some specialized job for pumping or treating the wastewater and sludge generated during treatment. Each of these specialized unit processes employs some fundamental aspect(s) of chemical, biological, or physical processes. For example, solids are removed by the physical process of gravity settling; organic matter is removed by a physical process like gravity settling if it is in a particulate form and by a biological process if it is dissolved and readily assimilated by microorganisms; and phosphorus can be removed by chemical and/or biological methods. In addition, there may be a need to determine how much of the potential pollutants in the wastewater a lake or river can safely assimilate. This requires knowledge of the chemistry and biology of the receiving water as well as of how physical processes such as mixing and transport influence the pollutants.

Figure 1-1. Aerial photo of a wastewater treatment plant used to treat municipal wastewater. Each reactor incorporates some combination of chemical, biological, or physical processes to treat the wastewater. (John Edwards/Tony Stone Images/New York, Inc.)

Fundamentals of Environmental Engineering provides a reader with the tools to begin answering questions such as:

How do I design, construct, and operate a treatment plant to treat a specific wastewater, given that I must treat a certain volume of wastewater daily, and the influent wastewater has specific pollutant characteristics?

How large a reactor must I construct, and can I speed up a chemical or biological reaction in order to construct a smaller, and less expensive reactor?

If I have several types of reactors to choose from, which one works best for a given treatment objective?

How do daily and seasonal variations in the flow of wastewater supplied by households and industry influence the design and operation of a particular treatment process?

Are there special considerations of the receiving water body that influence the plant design? For example, there may be a more strict phosphorus discharge standard for a more-nutrient-sensitive water body.

Can I use the same biological treatment reactor to remove both organic carbon and inorganic nitrogen?

How much of a particular waste can I discharge to a receiving water body; how do mixing and transport influence the fate of the waste; and will this discharge adversely affect the chemistry or biology of the water body?

How do I optimize a particular treatment process to more efficiently treat a particular waste?

Should I biologically treat a particular waste in the presence of oxygen (aerobically) or in the absence of oxygen (anaerobically), and how do these methods differ?

What are the most efficient methods of supplying needed oxygen to an aerobic biological reactor?

How do temperature changes during the summer and winter affect chemical or biological treatment processes or naturally occurring recovery processes of a lake or river?

SITUATION 2 ACID RAIN

Acid rain came to the forefront as a major environmental issue in the late 1960s and early 1970s. Acid rain has a much lower pH than natural rainwater (e.g., pH 3.2–4.5), although, as we will investigate in Chapter 3, rainwater not impacted by anthropogenic emissions is already slightly acidic (pH 5.6). Acid rain is primarily the result of humans discharging excessive amounts of sulfur dioxide (SO_2) and nitrogen oxides (NO and NO_2, which can be combined together and de-

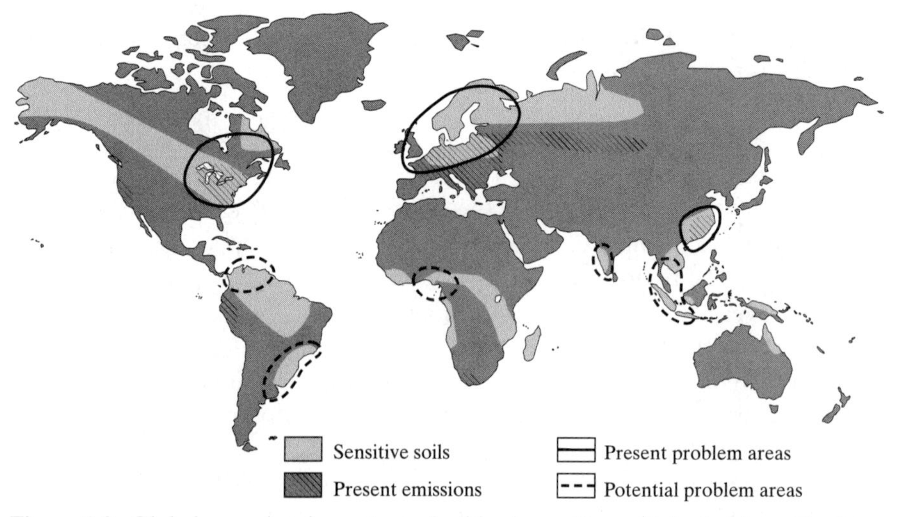

Figure 1-2. Global map showing extent of acid rain problem. (Adapted from Rodhue, 1989, "Acidification in a Global Perspective," with permission of the Royal Swedish Academy of Sciences.)

scribed as NO_x) into the atmosphere. After emission to the atmosphere, these oxides are chemically transformed into the strong acids, sulfuric acid (H_2SO_4) and nitric acid (HNO_3), that then return to the earth as either wet (i.e., rain, snow, fog) or dry deposition. These emissions result primarily from the combustion of fossil fuels (e.g., coal, oil) for electricity generation using coal, transportation, and home heating. Figure 1-2 shows the extent of the world's areas with current or potential acid-deposition problems. This figure shows that acid rain is a large-scale problem.

Figure 1-3 shows a simplified schematic of how sulfur and nitrogen oxide emissions may be transported to a terrestrial or aquatic system. The distances of transport can occur on a local or regional scale. Acid deposition may cause extensive damage to buildings and historical monuments and has been implicated in the decline of mountainous forests and the death of fish populations and other aquatic organisms. In the 1960s and 1970s, a popular method selected to control emissions at the local scale was the construction of tall stacks that released the gaseous emissions high in the air. This lessened high concentrations of pollutants at the ground level near the stack but resulted in transport of the problem to regions downwind. It is now clear that this method to control emissions was shortsighted. Because of political, technical, and economic reasons, little was done about reducing acid-deposition precursors until passage of the 1990 Clean Air Act Amendments. A section of this law (called Title IV) finally attempted to significantly reduce emissions of acid-rain precursors (especially SO_2 emis-

Figure 1-3. Simplified schematic that demonstrates how sulfur and nitrogen oxide emissions may be transported to terrestrial or aquatic ecosystems.

sions). This caused individuals to incorporate other solutions to the problems such as substitution of lower sulfur-content fossil fuels (some coal contains 1–5% sulfur!), removal of sulfur from fuels prior to combustion, reduction in demand for electricity and gasoline through energy conservation methods, and treatment of stack gases through removal of SO_2 and NO_x.

If you were an engineer or scientist, how would you solve the following aspects of the acid-rain problem? You might be developing a national policy to deal with the problem, or an employee at a coal-fired power plant attempting to reduce its emissions in order to meet new air regulations. *Fundamentals of Environmental Engineering* provides a reader some tools required to answer questions such as:

1. How do I design, construct, and operate a physical, chemical, or biological treatment method to scrub SO_2 and NO_x from stack gas?
2. If I scrub acid-rain precursors from the stack gas, do I create a water-pollution or solid-waste problem which I will then have to treat as well?
3. What is the atmospheric chemistry of acid-rain production, and how fast do the reactions take place?
4. How are air emissions mixed and transported downwind, and do such processes influence air quality on a local or regional scale?
5. What is the effect, if any, of acid deposition on a particular forest, lake, or agricultural area provided that I have some knowledge about the underlying geology, chemistry of a watershed or lake, and biotic community?
6. How does the design and implementation of a local or regional transportation system influence air quality on the local or regional scale?

SITUATION 3 FATE OF CHLORINATED ALIPHATIC HYDROCARBONS IN SOIL AND GROUNDWATER

The U.S. chemical industry manufactures approximately 20 million tons of chlorinated aliphatic hydrocarbons annually. A chlorinated aliphatic is a straight chain hydrocarbon consisting of hydrogen and carbon with chlorines substituted for some of the hydrogens. These chemicals are used as degreasing agents for metal parts and in the production of electronic parts, including the microchips in our computers. They are also used for dissolving rubber, extracting other chemicals (e.g., coffee production), dry cleaning, as intermediates in manufacturing other chemicals, and are also found in many household products, including paints, paint removers, and spill cleanup agents.

Table 1-1 lists the 20 most abundant organic constituents reported in groundwater at solid and hazardous waste disposal sites in the United States along with their common use and chemical formulae. Of these 20 chemicals, all but 7 are chlorinated aliphatic hydrocarbons. Some of the others include benzene, toluene, and ethylbenzene, which are components of fuel products like gasoline, diesel

Table 1-1. Twenty Most Abundant Organic Constituents Identified in Groundwater at 479 U.S. Waste Disposal Sites*

Rank	Chemical	Use	Chemical Formula
1	Dichloromethane	Paint stripping, solvent degreaser, blowing agent in foams	CH_2Cl_2
2	Trichloroethene	Dry cleaning agent, metal degreaser solvent	C_2Cl_3H
3	Tetrachloroethene	Dry cleaning, metal degreaser, solvent, paint remover	C_2Cl_4
4	*trans* 1,2-Dichloroethene	Solvent, additive to lacquer, low-temperature solvent for caffeine	$C_2H_2Cl_2$
5	Chloroform	Solvent, electronic circuit manufacturing	$CHCl_3$
6	1,1-Dichloroethane	Paint and varnish remover, metal degreaser, ore flotation	$C_2Cl_2H_4$
7	1,1-Dichloroethene	Paint and varnish remover, metal degreaser	$C_2Cl_2H_2$
8	1,1,1-Trichloroethane	Solvent	$C_2Cl_3H_3$
9	Toluene	Gasoline component, solvent thinner, adhesive solvent	C_7H_8
10	1,2-Dichloroethane	Paint and varnish remover, metal degreaser, fumigant	$C_2Cl_2H_4$
11	Benzene	Component of gasoline, used in chemical synthesis	C_6H_6
12	Ethylbenzene	Used in styrene manufacturing, solvent, asphalt construction	C_8H_{10}
13	Phenol	Disinfectant, pharmaceutical aid	C_6H_5OH
14	Chlorobenzene	Used in chemical synthesis	C_6H_5Cl
15	Vinyl chloride	Refrigerant, used in plastics industry	C_2ClH_3
16	Carbon tetrachloride	Dry cleaning, metal degreasing, veterinary medicine	CCl_4
17	Bis(2-ethylhexyl)phthalate	Used in vacuum pumps	$C_{24}H_{38}O_4$
18	Naphthalene	Used in manufacturing mothballs and motor fuel, component of coal tar	$C_{10}H_8$
19	1,1,2-Trichloroethane	Solvent	$C_2Cl_3H_3$
20	Chloroethane	Refrigerant, solvent, used to produce tetraethyl lead	C_2ClH_5

Adapted from Barbee, 1994.

*More information on uses and chemical properties can be found in Verschueren (1996) and Merck and Co., Inc. (1989).

Table 1-2. Properties of Selected Chlorinated Aliphatic Hydrocarbons*

Chemical	Vapor Pressure (mmHg)	Henry's Constant (atm-m^3/mole)	Water Solubility (mg/L)	Chemical Half-life (Years)
Carbon tetrachloride	90	0.0294	785	16–41
Chloroform	160	0.0040	8,200	742–3,000
Tetrachloroethene	14	0.0268	150	3.8×10^8–9.9×10^8
Trichloroethene	60	0.0117	1,100	4.9×10^5–1.3×10^6
Vinyl chloride	2,660	0.0224	2,700	>10

From Barbee, 1994.

*In later chapters, readers will learn about how these properties are used in evaluating and solving environmental problems.

fuel, and jet fuel. The chemicals benzene, toluene, ethylbenzene, and *ortho*-xylene/*meta*-xylene/*para*-xylene are found in gasoline and are commonly referred to as BTEX. Table 1-2 provides the vapor pressure, Henry's law constant, water solubility, and chemical degradation half-life for five of these chemicals. While at this point a reader may not understand the importance of these chemical properties, the properties do provide many important clues to the fate and treatability of the chemicals. They are also used in many design and modeling situations. In the remainder of this text, you will learn about how some of these properties are used to evaluate and solve environmental problems.

Think about a situation in which soil or groundwater is contaminated by one or more of the Table 1-1 chemicals from either the routine improper filling of a storage tank, a leaking underground storage tank, a large one-time spill, or a leaking landfill. *Fundamentals of Environmental Engineering* is designed to provide some of the tools so that an individual can begin to answer some of the following questions:

1. How do I design, construct, and operate a treatment system for air, water, or soil/sediment waste stream contaminated with these chemicals?
2. Should the treatment process I select use physical, chemical, or biological methods, or a combination?
3. If these chemicals were spilled on the ground's surface, how much of the chemical would partition to the air, water, and underlying soil after equilibrium is reached?
4. What is the speed at which these chemicals cycle between water, air, and soil, and what effect will variations in this speed have on the transport, fate, toxicity, and treatment of the chemical? How does dilution of the chemical as it spreads through the environment influence these parameters?
5. Will these chemicals degrade to nonhazardous endproducts by naturally occurring chemical or biological processes? Is it feasible to engineer a sys-

tem to degrade them to less hazardous end products? If I can engineer this system, what are the features of its design and operation?

6. If groundwater is contaminated, how long will it take before a downgradient drinking water source will be contaminated given that the chemical may undergo a chemical or biological reaction or have its movement slowed down by interacting with soil particles?
7. Will the degradation of a chemical like trichloroethene result in the production of vinyl chloride, a known human carcinogen, and more hazardous chemicals? And if it does, how fast will this reaction occur?

REFERENCES

Barbee, G. C. Vol. 14 Issue 1 1994. Fate of chlorinated aliphatic hydrocarbons in the vadose zone and ground water. Ground Water Monitoring and Remediation, Winter, 129–140.

Environmental Engineering Education: The Relationship to Engineering Practice. 1997. Proceedings of the 1996 Environmental Engineering Education Conference, Orono, Maine, August 3–6, published by the American Academy of Environmental Engineers (AAEE).

Merck and Co. 1989. The Merck Index, 11th Edition, Merck and Co., Rahway, N.J.

Rodhue, H. 1989. Acidification in a global perspective. Ambio. 18:155–160.

Verschueren, K. 1996. Handbook of Environmental Data on Organic Compounds. Van Nostrand Reinhold, New York.

Chapter 2

Units of Concentration

James R. Mihelcic
Richard E. Honrath, Jr.
Noel R. Urban

In this chapter a reader will become familiar with the different units used to measure pollutant levels in aqueous (i.e., water), soil/sediment, and atmospheric systems. In addition, the last section in the chapter will familiarize the reader with typical magnitudes and units of concentrations that are encountered in various engineered and natural systems.

Chemical concentration is one of the most important determinants in almost all aspects of chemical fate, transport, and treatment in both environmental and engineered systems. This is because concentration is the driving force that controls the movement of chemicals within and between different environmental media, as well as the rate of many chemical reactions. In addition, the severity of adverse effects, such as toxicity and bioconcentration, are often determined by concentration.

Concentrations of chemicals are routinely expressed in a variety of units. The choice of units to use in a given situation depends on the chemical, where it is located (e.g., air, water, or soil/sediments), and often on how the measurement will be used. It is therefore necessary to become familiar with the units used and methods for converting between different sets of units. Representation of concentration usually falls into one of the categories listed in Table 2-1. Important prefixes to know include pico (10^{-12}, abbreviated as p), nano (10^{-9}, abbreviated as n), micro (abbreviated as μ, 10^{-6}), milli (m, 10^{-3}), and kilo (k, 10^{+3}).

2.1 MASS CONCENTRATION UNITS

Concentration units based on chemical mass include mass chemical/total mass and mass chemical/total volume. Examples of these are shown below. In these descriptions, m_i is used to represent the mass of the chemical referred to as chemical i.

Table 2-1. Common Units of Concentration Used in Environmental Measurements

Representation	Example	Typical Units (Defined Below)
Mass chemical/total mass	mg/kg in soil	mg/kg, ppm_m
Mass chemical/total volume	mg/L in water or air	mg/L, $\mu g/m^3$
Volume chemical/total volume	volume fraction in air	ppm_V
Moles chemical/total volume	moles/L in water	M

2.1.1 Mass/Mass Units

Mass/mass concentrations are commonly expressed as parts per million, parts per billion, parts per trillion, and so on. For example, 1 mg of a solute placed in 1 kg of solvent equals 1 ppm_m. Parts per million by mass (referred to as ppm or ppm_m) is defined as the number of units of mass of chemical per million units of total mass. That is,

$$ppm_m = \text{g of } i \text{ in } 10^6 \text{ g total} \tag{2-1}$$

This definition is equivalent to the following general formula, which is used to calculate ppm_m concentration from measurements of chemical mass in a sample of total mass m_{total}:

$$ppm_m = \frac{m_i}{m_{total}} \times 10^6 \tag{2-2}$$

Note that the factor 10^6 in Equation 2-2 is really a conversion factor. It has the implicit units of ppm_m/mass fraction (mass fraction $= m_i/m_{total}$) as shown in Equation 2-3:

$$ppm_m = \frac{m_i}{m_{total}} \times 10^6 \frac{ppm_m}{\text{mass fraction}} \tag{2-3}$$

In Equation 2-3, m_i/m_{total} is defined as the mass fraction, and the conversion factor of 10^6 is similar to the conversion factor of 10^2 that is used to convert fractions to percentages. For example, the expression $0.25 = 25\%$ can be thought of as

$$0.25 = 0.25 \times 100^2\% = 25\% \tag{2-4}$$

Similar definitions are used for the units ppb_m, ppt_m, and % by mass. That is, 1 ppb_m equals 1 part per billion or 1 g chemical per billion (10^9) g total, so that the number of ppb_m in a sample is equal to $m_i/m_{total} \times 10^9$. And 1 ppt_m usually means 1 part per trillion. However, be cautious about interpreting ppt values because they may refer to either parts per thousand or parts per trillion (10^{12}). *Mass/mass concentrations can also be reported with the units explicitly shown (e.g., mg/kg, μg/kg).* In soils and sediments, 1 ppm_m equals 1 mg of pollutant per kg of solid (mg/kg) and 1 ppb_m equals 1 μg/kg. Percent by mass is analogously equal to the number of g pollutant per 100 g total.

EXAMPLE 2.1. CONCENTRATION IN SOIL

A one-kg sample of soil is analyzed for the chemical solvent trichloroethylene (TCE). The analysis indicates that the sample contains 5.0 mg of TCE. What is the TCE concentration in ppm_m and ppb_m?

SOLUTION

$$[\text{TCE}] = \frac{5 \text{ mg TCE}}{1.0 \text{ kg soil}} = \frac{0.005 \text{ g TCE}}{10^3 \text{ g soil}}$$

$$= \frac{5 \times 10^{-6} \text{ g TCE}}{\text{g soil}} \times 10^6 = 5 \text{ ppm}_\text{m} = 5{,}000 \text{ ppb}_\text{m}$$

Note that in soil and sediments, mg/kg equals ppm_m and μg/kg equals ppb_m.

2.1.2 Mass/Volume Units: mg/L and μg/m^3

In the atmosphere, it is common to use concentration units of mass/volume air such as mg/m^3 and μg/m^3. In water, mass/volume concentration units of mg/L and μg/L are common. *In most aqueous systems, ppm_m is equivalent to mg/L.* This is because the density of pure water is approximately 1,000 g/L. This is demonstrated in Example 2.2. The density of pure water is actually 1,000 g/L at 5°C. At 20°C the density has decreased slightly to 998.2 g/L. In addition, this equality is strictly true only for "dilute" solutions, in which any dissolved material does not contribute significantly to the mass of the water, and the total density remains approximately 1,000 g/L. Most wastewaters and natural waters can be considered dilute, except perhaps seawaters and brines.

EXAMPLE 2.2. CONCENTRATION IN WATER

One liter of water is analyzed and found to contain 5.0 mg TCE. What is the TCE concentration in mg/L and ppm_m?

SOLUTION

$$[\text{TCE}] = \frac{5.0 \text{ mg TCE}}{1.0 \text{ L H}_2\text{O}} = \frac{5.0 \text{ mg}}{\text{L}}$$

To convert to ppm_m, which is a mass/mass unit, it is necessary to convert the volume of water to mass of water, by dividing by the density of water, which is approximately 1,000 g/L:

$$[\text{TCE}] = \frac{5.0 \text{ mg TCE}}{1.0 \text{ L H}_2\text{O}} \times \frac{1.0 \text{ L H}_2\text{O}}{1{,}000 \text{ g H}_2\text{O}}$$

$$= \frac{5.0 \text{ mg TCE}}{1{,}000 \text{ g total}} = \frac{5.0 \times 10^{-6} \text{ g TCE}}{\text{g total}} \times \frac{10^6 \text{ ppm}_\text{m}}{\text{mass fraction}}$$

$$= 5.0 \text{ ppm}_\text{m}$$

See that *in most aqueous systems that are dilute, mg/L is equivalent to* ppm_m. Also, in this problem the TCE concentration is well above the allowable U.S. drinking water standard for TCE of 5 μg/L (or 5 ppb), which was set to protect human health. Five ppb is a small value. Think of it: if one assumes the Earth's human population is 5 billion people, this means that 25 individuals sitting in one of your classes constitute a human concentration of 5 ppb!

EXAMPLE 2.3. CONCENTRATION IN AIR

What is the carbon monoxide (CO) concentration expressed in $\mu g/m^3$ of a 10-L gas mixture that contains 10^{-6} mole of CO?

SOLUTION

In this case, the measured quantities are presented in units of moles chemical/total volume. To convert to mass of chemical/total volume, convert the moles of chemical to mass of chemical by multiplying by CO's molecular weight. Note that the molecular weight of CO (28 g/mole) is equal to 12 (atomic weight of C) plus 16 (atomic weight of O).

$$[\text{CO}] = \frac{1.0 \times 10^{-6} \text{ mole CO}}{10 \text{ L total}} \times \frac{28 \text{ g CO}}{\text{mole CO}}$$

$$= \frac{28 \times 10^{-6} \text{ g CO}}{10 \text{ L total}} \times \frac{10^6 \ \mu\text{g}}{\text{g}} \times \frac{10^3 \text{ L}}{\text{m}^3} = \frac{2{,}800 \ \mu\text{g}}{\text{m}^3}$$

2.2 VOLUME/VOLUME AND MOLE/MOLE UNITS

Units of volume fraction or mole fraction are frequently used for gas concentrations. The most common volume fraction units are ppm_V (parts per million by volume) (*referred to as ppm or* ppm_V), which is defined as:

$$\text{ppm}_\text{V} = \frac{V_i}{V_{\text{total}}} \times 10^6 \qquad \textbf{(2-5)}$$

In Equation 2-5, V_i/V_{total} is the volume fraction and the factor 10^6 is a conversion factor, with units of 10^6 ppm_V/(volume fraction).

Other common units for gaseous pollutants are ppb_V (parts per 10^9 by volume). *The advantage of volume/volume units is that gaseous concentrations reported in these units do not change as a gas is compressed or expanded.* Atmospheric concentrations expressed as mass/volume (e.g., $\mu g/m^3$) decrease as the gas expands, since the pollutant mass remains constant but the volume increases. Both mass/volume units, such as $\mu g/m^3$, and ppm_V units are frequently used to express gaseous concentrations (see Equation 2-9 for conversion between $\mu g/m^3$ and ppm_V).

2.2.1 Using the Ideal Gas Law to Convert ppm_V to $\mu g/m^3$

The Ideal Gas Law can be used to convert gaseous concentrations between mass/volume and volume/volume. The Ideal Gas Law states that *pressure* (P) times *volume occupied* (V) equals *the number of moles* (n) times the *gas constant* (R) times the *absolute temperature* (T) in degrees Kelvin or Rankine. This is written in the familiar form of

$$PV = nRT \quad \text{(2-6)}$$

Here R, the universal gas constant, may be expressed in many different sets of units. Some of the most common values are displayed below:

0.08205 L-atm/mole-K
8.205×10^{-5} m^3-atm/mole-K
82.05 cm^3-atm/mole-K
1.99×10^{-3} kcal/mole-K
8.314 J/mole-K
1.987 cal/mole-K
62,358 cm^3-torr/mole-K
62,358 cm^3-mmHg/mole-K

Because the gas constant may be expressed in a number of different units, always be careful of its units and cancel them out to ensure the use of the correct value of R.

The Ideal Gas Law states that the volume occupied by a given number of molecules of any gas is the same, no matter what the molecular weight or composition of the gas, as long as the pressure and temperature are constant. The Ideal Gas Law can be rearranged to show that the volume occupied by n moles of gas is equal to

$$V = n\frac{RT}{P} \quad \text{(2-7)}$$

At standard conditions (P = 1 atm, T = 273.15 K), one mole of any pure gas will occupy a volume of 22.4 L. This result can be derived by using the corresponding value of R (0.08205 L-atm/mole-K) and the form of the Ideal Gas Law provided in Equation 2-7. At other temperatures and pressures, this volume varies as determined by Equation 2-7.

EXAMPLE 2.4. GAS CONCENTRATION IN VOLUME FRACTION

A gas mixture contains 0.001 mole of sulfur dioxide (SO_2) and 0.999 mole of air. What is the SO_2 concentration, expressed in units of ppm_V?

SOLUTION

The concentration in ppm_V is determined using Equation 2-5.

$$[SO_2] = \frac{V_{SO_2}}{V_{total}} \times 10^6$$

To solve, convert the number of moles of SO_2 to volume using the Ideal Gas Law (Equation 2-6) and the total number of moles to volume. Then divide the two expressions:

$$V_{SO_2} = 0.001 \text{ mole } SO_2 \times \frac{RT}{P}$$

$$V_{total} = (0.999 + 0.001) \text{ mole total} \times \frac{RT}{P}$$

$$= (1.000) \text{ mole total} \times \frac{RT}{P}$$

Substitute these volume terms for ppm_V, to obtain

$$= \frac{0.001 \text{ mole } SO_2 \times \frac{RT}{P}}{1.000 \text{ mole total} \times \frac{RT}{P}} \times 10^6$$

$$= \frac{0.001 \text{ L } SO_2}{1.000 \text{ L total}} \times 10^6 = 1{,}000 \text{ ppm}_V$$

Note that in Example 2.4, the terms (RT/P) cancel out. This demonstrates an important point that is useful in calculating volume fraction or mole fraction concentrations. *For gases, volume ratios and mole ratios are equivalent.* This is clear from the Ideal Gas Law, because at constant temperature and pressure the volume occupied by a gas is proportional to the number of moles. Therefore, Equation 2-5 is equivalent to Equation 2-8:

$$\text{ppm}_V = \frac{\text{moles } i}{\text{moles total}} \times 10^6 \tag{2-8}$$

See that the solution to Example 2.4 could have simply been found by using Equation 2-8 and determining the mole ratio. Therefore, in any given problem, either units of volume or units of moles can be used to calculate ppm_V. Being aware of this fact will save unnecessary conversions between moles and volume. The mole ratio (moles i/moles total) is sometimes referred to as the *mole fraction, X*.

Example 2.5 and Equation 2-9 show how to use the Ideal Gas Law to convert concentrations between $\mu g/m^3$ and ppm_V.

EXAMPLE 2.5. CONVERT GAS CONCENTRATION BETWEEN ppb AND $\mu g/m^3$

The concentration of SO_2 is measured in air to be 100 ppb_V. What is this concentration in units of $\mu g/m^3$? Assume the temperature is 28°C and pressure is 1 atm. Remember that T(K) is equal to T(°C) plus 273.15.

SOLUTION

To accomplish this conversion, use the Ideal Gas Law to convert the volume of SO_2 to moles of SO_2, resulting in units of moles/L. This can be converted to $\mu g/m^3$ using the molecular weight of SO_2 (MW = 64). This method will be used to develop a general formula for converting between ppm_V and $\mu g/m^3$.

First, use the definition of ppb_V to obtain a volume ratio for SO_2:

$$100\ \text{ppb}_V = \frac{100\ \text{m}^3\ \text{SO}_2}{10^9\ \text{m}^3\ \text{air solution}}$$

Now convert the volume of SO_2 in the numerator to units of mass. This is done in two steps. First, convert the volume to a number of moles, using a rearranged format of the Ideal Gas Law (Equation 2-6) ($n/V = P/RT$) and the given temperature and pressure:

$$\frac{100\ \text{m}^3\ \text{SO}_2}{10^9\ \text{m}^3\ \text{air solution}} \times \frac{P}{RT}$$

$$= \frac{100\ \text{m}^3\ \text{SO}_2}{10^9\ \text{m}^3\ \text{air solution}} \times \frac{1\ \text{atm}}{8.205 \times 10^{-5}\left(\frac{\text{m}^3\ \text{atm}}{\text{mole K}}\right)(301\ \text{K})}$$

$$= \frac{4.05 \times 10^{-6}\ \text{mole SO}_2}{\text{m}^3\ \text{air}}$$

In the second step, convert the moles of SO_2 to mass of SO_2 using the molecular weight of SO_2.

$$\frac{4.05 \times 10^{-6} \text{ mole } SO_2}{\text{m}^3 \text{ air}} \times \frac{64 \text{ g } SO_2}{\text{mole } SO_2} \times \frac{10^6 \ \mu\text{g}}{\text{g}} = \frac{260 \ \mu\text{g}}{\text{m}^3}$$

Example 2.5 demonstrates that *a useful conversion for converting air concentrations between units of $\mu g/m^3$ and ppm_V can be written as*

$$\frac{\mu\text{g}}{\text{m}^3} = \text{ppm}_V \times \text{MW} \times \frac{1{,}000\ P}{RT} \tag{2-9}$$

where MW is the chemical species molecular weight; R equals 0.08205 L-atm/mole-K; T is the temperature in degrees K; and the 1,000 is a conversion factor (1,000 L = m^3). *Note that for 0°C, RT has a value of 22.4 L-atm/mole, while at 20°C RT has a value of 24.2 L-atm/mole.*

2.3 PARTIAL-PRESSURE UNITS

In the atmosphere, concentrations of chemicals in the gas and particulate phases may be determined separately. A substance will exist in the gas phase if the atmospheric temperature is above the substance's boiling (or sublimation) point or if its concentration is below the saturated vapor pressure of the chemical at a specified temperature (vapor pressure is defined in Section 3.4.1). The major and minor gaseous constituents of the atmosphere all have boiling points well below atmospheric temperatures. Concentrations of these species typically are expressed either as volume fractions (e.g., %, ppm_V, or ppb_V) or as partial pressures (units of atmospheres, atm). Table 2-2 summarizes the concentrations of the most abundant atmospheric gaseous constituents.

The total pressure exerted by a gas mixture may be considered as the sum of the partial pressures exerted by each component of the mixture. The partial pressure of each component is equal to the pressure that would be exerted if all of the other components of the mixture were suddenly removed. Partial pressure is commonly written as P_i, where i refers to the particular gas being considered. For example, the partial pressure of oxygen in the atmosphere P_{O_2} is 0.21 atm.

Remember that the Ideal Gas Law states that, at a given temperature and volume, pressure is directly proportional to the number of moles of gas present; therefore, pressure fractions are identical to mole fractions (and volume fractions). For this reason, *partial pressure can be calculated as the product of the mole or volume fraction and the total pressure.* For example,

$$\begin{aligned} P_i &= [\text{volume fraction}_i \text{ or mole fraction}_i \times P_{\text{total}}] \\ &= [(\text{ppm}_V)_i \times 10^{-6} \times P_{\text{total}}] \end{aligned} \tag{2-10}$$

Table 2-2. Composition of the Atmosphere*

Compound	Concentration (% volume or moles)	Concentration (ppm_V)
Nitrogen (N_2)	78.1	781,000
Oxygen (O_2)	20.9	209,000
Argon (Ar)	0.93	9,300
Carbon dioxide (CO_2)	0.035	350
Neon (Ne)	0.0018	18
Helium (He)	0.0005	5
Methane (CH_4)	0.00017	1.7
Krypton (Kr)	0.00011	1.1
Hydrogen (H_2)	0.00005	0.500
Nitrous oxide (N_2O)	0.000032	0.316
Ozone (O_3)	0.000002	0.020

Data from Graedel and Crutzen, 1993.

*Values represent concentrations in dry air at remote locations.

In addition, rearranging Equation 2-10 shows that ppm_V values can be calculated from partial pressures as follows:

$$ppm_V = \frac{P_i}{P_{total}} \times 10^6 \qquad \textbf{(2-11)}$$

Thus, partial pressure can be added to the list of unit types that can be used to calculate ppm_V. *That is, either volume (Equation 2-5), moles (Equation 2-8), or partial pressures (Equation 2-11) can be used in* ppm_V *calculations.*

EXAMPLE 2.6. CONCENTRATION AS PARTIAL PRESSURE

The concentration of gas-phase polychlorinated biphenyls (PCBs) in the air above Lake Superior was measured to be 450 picograms per cubic meter (pg/m^3). What is the partial pressure (in atm) of PCBs? Assume the temperature is 0°C, the atmospheric pressure is 1 atm, and the average molecular weight of PCBs is 325.

SOLUTION

The chemical structure of PCBs along with some general information is provided in Figure 2-1. The partial pressure is defined as the mole or volume fraction times the total gas pressure. First, find the number of moles of PCBs in a liter of air. Then use the Ideal Gas Law (Equation 2-7) to calculate that one mole of gas at

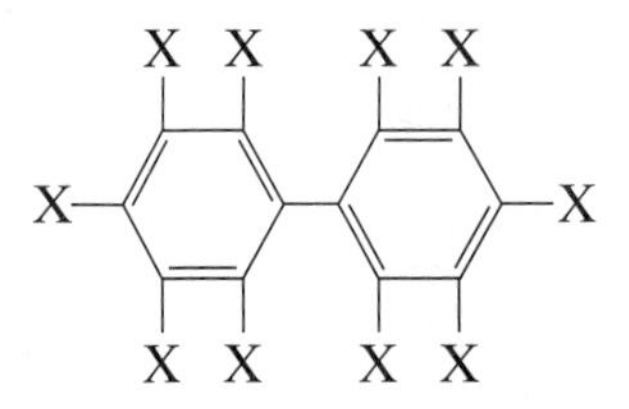

X = chlorine or hydrogen

Figure 2-1. Chemical Structure of Polychlorinated Biphenyls (PCBs). PCBs are a family of compounds produced commercially by chlorinating biphenyl. Chlorine atoms can be placed at any or all of ten available sites, with 209 possible PCB congeners. The great stability of PCBs caused them to have a wide range of uses, including serving as coolants in transformers and as hydraulic fluids and solvents. However, the chemical properties that resulted in this stability also resulted in a chemical that did not degrade easily, bioaccumulated in the food chain, and was also hazardous to humans and wildlife. Accordingly, in 1976 Congress passed the Toxic Substances Control Act (TSCA) that banned the manufacture of PCBs and PCB-containing products. TSCA also established strict regulations regarding the future use and sale of PCBs. PCBs were typically sold as mixtures that are commonly referred to as Arochlors. For example, the Arochlor 1260 mixture consists of 60% chlorine by weight, which meant the individual PCBs in the mixture are primarily substituted with 6–9 chlorines per biphenyl molecule. In contrast, Arochlor 1242 consists of 42% chlorine by weight; thus, it primarily consists of PCBs with 1–6 substituted chlorines per biphenyl molecule.

0°C and 1 atm occupies 22.4 L and substitute this value into the first expression to determine the mole fraction of PCBs:

$$450 \frac{\text{pg}}{\text{m}^3 \text{ air}} \times \frac{\text{mole}}{325 \text{ g}} \times 10^{-12} \frac{\text{g}}{\text{pg}} \times 10^{-3} \frac{\text{m}^3}{\text{L}} = 1.38 \times 10^{-15} \frac{\text{mole PCB}}{\text{L air}}$$

$$1.38 \times 10^{-15} \frac{\text{mole PCB}}{\text{L}} \times \frac{22.4 \text{ L}}{\text{mole air}} = 3.1 \times 10^{-14} \frac{\text{mole PCB}}{\text{mole air}}$$

Multiplying the mole fraction by the total pressure (1 atm) (see Equation 2-10) yields the PCB partial pressure of 3.1×10^{-14} atm.

EXAMPLE 2.7. CONCENTRATION AS PARTIAL PRESSURE CORRECTED FOR MOISTURE

What would be the partial pressure (atm) of carbon dioxide (CO_2) when the barometer reads 29.0 inches of Hg, the relative humidity is 80%, and the temperature is 70°F? Use Table 2-2 to obtain the concentration of CO_2 in dry air.

SOLUTION

The partial-pressure concentration units in Table 2-2 are for dry air, so the partial pressure must first be corrected for the moisture present in the air. In dry air, the

CO_2 concentration is 350 ppm_V. The partial pressure will be this volume fraction times the total pressure of dry air. The total pressure of dry air is the total atmospheric pressure (29.0 in Hg) minus the contribution of water vapor. The vapor pressure of water can be looked up at 70°F to be 0.36 lb/in^2. Thus the total pressure of dry air is

$$P_{total} - P_{water} = 29.0 \text{ inHg} - \left[0.36 \frac{\text{lb}}{\text{in}^2} \times \frac{29.9 \text{ inHg}}{\frac{14.7 \text{ lb}}{\text{in}^2}} \times 0.8 \right] = 28.4 \text{ inHg}$$

The partial pressure of CO_2 would be

$$\text{Vol fraction} \times P_{total} = 350 \text{ ppm}_V \times \frac{10^{-6} \text{ volfrac}}{\text{ppm}_V} \times \left[28.4 \text{ inHg} \times \frac{1 \text{ atm}}{29.9 \text{ inHg}} \right]$$

$$= 3.3 \times 10^{-4} \text{ atm}$$

2.4 MOLE/VOLUME UNITS

Units of moles per liter (molarity, M) are often used to report concentrations of compounds dissolved in water. Molarity is defined as the number of moles of compound per liter of solution. Thus a 10^{-4} M solution of copper contains 10^{-4} moles of copper per liter of solution. Concentrations expressed in these units are read as *molar*. Thus, the copper solution would be described as being 10^{-4} molar.

Molarity, M, should not be confused with molality, m. Molarity is usually used in equilibrium calculations and throughout the remainder of this book. Molality is the number of moles of a solute added to exactly one liter of solvent. Thus, the actual volume of a molal solution is slightly larger than one liter. Molality is more likely to be used when properties of the solvent, such as boiling and freezing point, are a concern. Therefore, it is rarely used in environmental situations.

EXAMPLE 2.8. CONCENTRATION AS MOLARITY

Convert the concentration of trichloroethene (TCE) (5 ppm) to units of molarity. The molecular weight of TCE is 131.5 g/mole.

SOLUTION

Remember, in water, ppm_m is equivalent to mg/L, so the concentration of TCE is 5.0 mg/L. Conversion to molarity units requires only the molecular weight:

$$5.0 \frac{\text{mg TCE}}{\text{L}} \times \frac{1 \text{ g}}{10^3 \text{ mg}} \times \frac{1 \text{ mole}}{131.5 \text{ g}} = \frac{3.8 \times 10^{-5} \text{ moles}}{\text{L}} \text{ or M}$$

Often, concentrations below 1 M are expressed in units of millimoles per liter, or millimolar (1 mM = 10^{-3} moles/L) or micromoles per liter or micromolar (1 μM = 10^{-6} moles/L). Thus, the concentration of TCE could be expressed as 0.038 mM or 38 μM.

EXAMPLE 2.9. CONCENTRATION AS MOLARITY

The concentration of alachlor, a common herbicide, in the Mississippi River was found to range from 0.04 to 0.1 μg/L. What is the concentration range in nmole/L? The molecular formula for alachlor is $C_{14}H_{20}O_2NCl$ (MW = 270).

SOLUTION

The lowest concentration range in nmole/L can be found as follows:

$$\frac{0.04\ \mu g}{L} \times \frac{mole}{270\ g} \times \frac{10^{-6}\ g}{\mu g} \times \frac{10^{9}\ nmole}{mole} = \frac{0.15\ nmol}{L}$$

Similarly, the upper limit (0.1 μg/L) can be calculated to be 0.37 nmol/L. Hence the concentration range of alachlor is 0.15–0.37 nmol/L.

2.5 OTHER TYPES OF UNITS

Sometimes concentrations are expressed as "normality," as a "common constituent," or represented by "effect." The following three sections describe these methods to express concentration in further detail.

2.5.1 Normality

Normality (equivalents/L) is typically used in defining the chemistry of water, especially in instances where acid/base and oxidation/reduction reactions are taking place. It is also used in determining the accuracy of a water analysis, as demonstrated in Example 2.12, and in calculating dosages of chemicals during some water treatment processes (e.g., softening hard water). Normality is also used frequently in the laboratory during the analytical measurement of water constituents. For example, "Standard Methods for the Examination of Water and Wastewater" (American Public Health Association (APHA), 1992) has many examples where concentrations of chemical reagents are prepared and reported in units of normality and not molarity.

Reporting concentration on an equivalent basis is useful because *if two chemical species react and the two species reacting have the same strength on an equiv-*

alent basis, a 1-mL volume of reactant number 1 will react with a 1-mL volume of reactant number 2. In acid/base chemistry the number of equivalents per mole of acid equals the number of moles of H^+ the acid can potentially donate. For example, HCl has 1 equivalent/mole, H_2SO_4 has 2 equivalents/mole, and H_3PO_4 has 3 equivalents/mole. Likewise, the number of equivalents per mole of a base equals the number of moles of H^+ that will react with one mole of the base. Thus, NaOH has 1 equivalent/mole, $CaCO_3$ has 2 equivalents/mole, and PO_4^{3-} has 3 equivalents/mole.

In oxidation/reduction reactions the number of equivalents is related to how many electrons a species donates or accepts. For example, the number of equivalents of Na^+ is 1 (where e^- equals an electron) because: $Na \rightarrow Na^+ + e^-$. Likewise, the number of equivalents for Ca^{2+} is 2 because: $Ca \rightarrow Ca^{2+} + 2e^-$. The equivalent weight (units = g/eqv) of a species is defined as *the molecular weight of the species divided by the number of equivalents in the species (g/mole/eqv/mole = g/eqv).*

EXAMPLE 2.10. CALCULATION OF EQUIVALENT WEIGHT

What is the equivalent weight of HCl, H_2SO_4, NaOH, $CaCO_3$, and aqueous CO_2?

SOLUTION

The equivalent weight is found by dividing the molecular weight by the number of equivalents.

$$\text{eqv wt of HCl} = \frac{1 + 35.5 \text{ g}}{\text{mole}} \div \frac{1 \text{ eqv}}{\text{mole}} = \frac{36.5 \text{ g}}{\text{eqv}}$$

$$\text{eqv wt of } H_2SO_4 = \frac{(2 \times 1) + 32 + (4 \times 16) \text{ g}}{\text{mole}} \div \frac{2 \text{ eqv}}{\text{mole}} = \frac{49 \text{ g}}{\text{eqv}}$$

$$\text{eqv wt of NaOH} = \frac{23 + 16 + 1 \text{ g}}{\text{mole}} \div \frac{1 \text{ eqv}}{\text{mole}} = \frac{40 \text{ g}}{\text{eqv}}$$

$$\text{eqv wt of } CaCO_3 = \frac{40 + 12 + (3 \times 16) \text{ g}}{\text{mole}} \div \frac{2 \text{ eqv}}{\text{mole}} = \frac{50 \text{ g}}{\text{eqv}}$$

Determining the equivalent weight of aqueous CO_2 requires a bit of thinking and some new information. Aqueous carbon dioxide is not an acid until it hydrates in water and forms carbonic acid ($CO_2 + H_2O \rightarrow H_2CO_3$). So aqueous CO_2 really has 2 eqv/mole. Thus one can see that the equivalent weight of aqueous carbon dioxide is

$$\frac{12 + (2 \times 16) \text{ g}}{\text{mole}} \div \frac{2 \text{ eqv}}{\text{mole}} = \frac{22 \text{ g}}{\text{eqv}}$$

EXAMPLE 2.11. CALCULATION OF NORMALITY

What is the normality of 1 M solutions of HCl and H_2SO_4?

SOLUTION

$$1 \text{ M HCl} = \frac{1 \text{ mole HCl}}{\text{L}} \times \frac{1 \text{ eqv}}{\text{mole}} = \frac{1 \text{ eqv}}{\text{L}} = 1 \text{ N}$$

$$1 \text{ M } H_2SO_4 = \frac{1 \text{ mole } H_2SO_4}{\text{L}} \times \frac{2 \text{ eqv}}{\text{mole}} = \frac{2 \text{ eqv}}{\text{L}} = 2 \text{ N}$$

Note that on an equivalent basis, a 1-M solution of sulfuric acid is twice as strong as a 1-M solution of HCl.

An example of how normality is used in analytical measurements can be seen when determining the chemical oxygen demand (COD) of a water sample. After performing a titration to determine the concentration of oxidized chromium (Cr^{6+}) remaining in solution, a conversion factor of 8,000 is employed in the expression to determine the COD based upon the amount of titrant used. This conversion factor results because an oxidation/reduction reaction occurs during the titration, the equivalent weight of oxygen is 8, and there are 1,000 mL per liter of water. Oxygen's equivalent weight can be found from dividing the molecular weight of O_2 (32 g/mole) by 4 eqv/mole.

EXAMPLE 2.12. USE OF EQUIVALENTS IN DETERMINING THE ACCURACY OF A WATER ANALYSIS

All aqueous solutions must maintain charge neutrality. Another way to state this is that the sum of all cations on an equivalent basis must equal the sum of all anions on an equivalent basis. Thus, water samples can be checked to determine if something is either incorrect in the analyses or a constituent(s) is missing. The following example shows how this is done. The label on a bottle of New Zealand mineral water purchased in the city of Dunedin stated that a chemical analysis of the mineral water resulted in the following cations and anions being identified with corresponding concentrations (in mg/L):

$$[Ca^{2+}] = 2.9; \quad [Mg^{2+}] = 2.0; \quad [Na^{+}] = 11.5; \quad [K^{+}] = 3.3;$$
$$[SO_4^{2-}] = 4.7; \quad [Fl^{-}] = 0.09; \quad [Cl^{-}] = 7.7$$

Is the analysis correct?

SOLUTION

First convert all concentrations of major ions to an equivalent basis. This is done by first multiplying the concentration in mg/L by a unit conversion (g/1,000 mg) and then dividing by the equivalent weight of each substance (#g/eqv). The concentrations of all cations and anions are then summed up on an equivalent basis. A solution with less than 5% error is generally considered acceptable.

Cations	**Anions**
$[Ca^{2+}] = \frac{1.45 \times 10^{-4} \text{ eqv}}{\text{L}}$	$[SO_4^{2-}] = \frac{9.75 \times 10^{-5} \text{ eqv}}{\text{L}}$
$[Mg^{2+}] = \frac{1.67 \times 10^{-4} \text{ eqv}}{\text{L}}$	$[Fl^-] = \frac{4.73 \times 10^{-6} \text{ eqv}}{\text{L}}$
$[Na^+] = \frac{5 \times 10^{-4} \text{ eqv}}{\text{L}}$	$[Cl^-] = \frac{2.17 \times 10^{-4} \text{ eqv}}{\text{L}}$
$[K^+] = \frac{8.5 \times 10^{-5} \text{ eqv}}{\text{L}}$	

The total amount of cations equals 9.87×10^{-4} eqv/L, and the total amount of anions equals 3.2×10^{-4} eqv/L.

The analysis is not within 5%. The analysis resulted in over 3 times more cations than anions on an equivalent basis. Therefore, it can be concluded that either (a) one or more of the reported concentrations are incorrect (assuming all major cations and anions are accounted for), or (b) one or more important anions were not accounted for by the chemical analysis (e.g., bicarbonate, HCO_3^-, would be a good guess for the missing anion, as it is a common anion in most natural waters).

2.5.2 Concentration as a Common Constituent

Concentrations can be reported as a common constituent, and can therefore include contributions from a number of different chemical compounds. Nitrogen and phosphorus are chemicals that have their concentration typically reported as a common constituent. For example, the phosphorus in a lake or wastewater may be present in inorganic forms called orthophosphates (i.e., H_3PO_4, $H_2PO_4^-$, HPO_4^{2-}, PO_4^{3-} and PO_4^{3-} and HPO_4^{-2} complexes), polyphosphates (e.g., $H_4P_2O_7$, $H_3P_3O_{10}^{2-}$), metaphosphates (e.g., $HP_3O_9^{2-}$), and/or organic phosphates. Because phosphorus can be chemically converted between these forms and can thus be found in several of these forms, it makes sense at some times to report the total P concentration, without specifying which form(s) are present. Thus each concentration for every individual form of phosphorus is converted to mg P/L using the MW of the individual species, the MW of P (32), and simple stoichiometry. These converted concentrations of each individual species can then be added to determine the total phosphorus concentration. The concentration is then reported in units of mg/L as phosphorus (written as mg P/L, mg/L as P, or mg/L P).

Nitrogen can also exist in many different chemical forms in aqueous systems. These forms include ammonia (NH_3 and NH_4^+), nitrate (NO_3^-), nitrite (NO_2^-), and organic nitrogen (e.g., amino acids like alanine (CH_3CHNH_2COOH) or glycine (H_2NCH_2COOH), amines, etc.). To express the concentration of total nitrogen (all forms of nitrogen combined), the concentration of each individual species of nitrogen is converted to mg N/L (MW = 14), as is done in Example 2.13. Thus the concentration could be reported as mg NO_3^-/L or mg NO_3^--N/L (mg of NO_3^- as nitrogen/L).

EXAMPLE 2.13. CONCENTRATIONS AS A COMMON CONSTITUENT

A water contains two nitrogen species. The concentration of NH_3 is 30 mg/L NH_3 and the concentration of NO_3^- is 5 mg/L NO_3^-. What is the total nitrogen concentration in units of mg N/L?

SOLUTION

Use the appropriate molecular weight and stoichiometry to convert each individual species to the requested units of mg N/L, then add the contribution of each species.

$$\frac{30 \text{ mg NH}_3}{\text{L}} \times \frac{\text{mole NH}_3}{17 \text{ g}} \times \frac{\text{mole N}}{\text{mole NH}_3} \times \frac{14 \text{ g}}{\text{mole N}} = \frac{24.7 \text{ mg NH}_3\text{-N}}{\text{L}}$$

$$\frac{5 \text{ mg NO}_3^-}{\text{L}} \times \frac{\text{mole NO}_3^-}{62 \text{ g}} \times \frac{\text{mole N}}{\text{mole NO}_3^-} \times \frac{14 \text{ g}}{\text{mole N}} = \frac{1.1 \text{ mg NO}_3^-\text{-N}}{\text{L}}$$

$$\text{Total nitrogen concentration} = 24.7 + 1.1 = \frac{25.8 \text{ mg N}}{\text{L}}$$

The "alkalinity" and "hardness" of a water are typically reported by determining all of the individual species that contribute to either alkalinity or hardness, then converting each of these species to units of mg $CaCO_3$/L, and finally summing up the contribution of each species. Thus, hardness is normally expressed as mg/L as $CaCO_3$.

The hardness of a water is caused by the presence of divalent cations in water. Ca^{2+} and Mg^{2+} are by far the most abundant divalent cations in natural waters, though Fe^{2+}, Mn^{2+}, and Sr^{2+} may contribute as well. These cations are released

Table 2-3. Scale to Quantify the Hardness of Water

Hardness (as mg/L $CaCO_3$)	Hardness Evaluation
<50	Soft
50–150	Moderately hard
151–300	Hard
>300	Very hard

From Hammer and Hammer, 1996.

from the dissolution of minerals. For example, calcium carbonate can react with the natural acidity found in rain water to release hardness (shown as Ca^{2+}) and alkalinity (here in the form of HCO_3^-) according to the following reaction:

$$CaCO_3 + H_2CO_3 \rightarrow Ca^{2+} + 2HCO_3^- \tag{2-12}$$

There is no adverse health effect from drinking hard waters; however, they can hinder soap formation and produce scale in boilers and piping. This increases the cost to society because of increased soap usage and plugging up of boilers and pipes with scale. Furthermore, hard waters can leave a "slimy" feeling on your body after bathing, so they are not aesthetically pleasing to some consumers.

The total hardness of a water can be found by summing the contributions of *all* divalent cations after converting their concentrations to a common constituent. Then the resulting value can be compared to the scale shown in Table 2-3 to evaluate the hardness of the water. In Michigan, Wisconsin, and Minnesota untreated waters usually have a hardness of 121–180 mg/L as $CaCO_3$. In Illinois and Iowa water is harder with many values greater than 180 mg/L as $CaCO_3$.

The conversion of concentration of specific cations (from mg/L) to hardness (as mg/L $CaCO_3$) can be accomplished by the following expression, where M^{2+} represents a divalent cation.

$$\frac{M^{2+} \text{ in mg}}{L} \times \frac{50}{\text{eqv wt of } M^{2+} \text{ in g/eqv}} = \frac{mg}{L} \text{ as } CaCO_3 \tag{2-13}$$

The origin of the 50 in Equation 2-13 comes from the fact that the equivalent weight of calcium carbonate is 50 (100 grams $CaCO_3$/2 equivalents). The equivalent weights (in units of g/eqv) of the following divalent cations are Mg = 24/2; Ca, 40/2; Mn, 55/2; Fe, 56/2; Sr, 88/2.

EXAMPLE 2.14. DETERMINATION OF A WATER'S HARDNESS

Water has the following chemical composition. $[Ca^{2+}] = 15$ mg/L; $[Mg^{2+}] = 10$ mg/L; $[SO_4^{2-}] = 30$ mg/L. What is the total hardness in units of mg/L as $CaCO_3$?

SOLUTION

Find the contribution of hardness from each divalent cation. Anions and all non-divalent cations are not included in the calculation.

$$\frac{15 \text{ mg Ca}^{2+}}{\text{L}} \times \left(\frac{\frac{50 \text{ g CaCO}_3}{\text{eqv}}}{\frac{40 \text{ g Ca}^{2+}}{2 \text{ eqv}}} \right) = \frac{38 \text{ mg}}{\text{L}} \text{ as CaCO}_3$$

$$\frac{10 \text{ mg Mg}^{2+}}{\text{L}} \times \left(\frac{\frac{50 \text{ g CaCO}_3}{\text{eqv}}}{\frac{24 \text{ g Mg}^{2+}}{2 \text{ eqv}}} \right) = \frac{42 \text{ mg}}{\text{L}} \text{ as CaCO}_3$$

Therefore, the total hardness is 38 + 42 = 80 mg/L as $CaCO_3$. Note this water is moderately hard. Also, note that if reduced iron (Fe^{2+}) or manganese (Mn^{2+}) were present, they would be included in the hardness calculation.

2.5.3 Reporting Particle Concentrations in Air and Water

Particles present in the air may reduce visibility (even in some of our remote National Parks!), blacken, corrode, or erode buildings and historical monuments, and adversely affect the health of humans and animals. The concentration of particles in an air sample is determined by pulling a known volume (e.g., several thousand m^3) of air through a filter. The increase in weight of the filter due to collection of particles on the filter can be determined and if this value is divided by the volume of air passed through the filter, the total suspended-particulate (TSP) concentration can be determined in units of g/m^3 or $\mu g/m^3$. In the United States, TSP concentrations average 20 $\mu g/m^3$ in clean areas and 60 to 200 $\mu g/m^3$ in urban areas. In order to penetrate deep into a human lung, particles must have a diameter between 0.1 and about 2.5 μm. Particulate concentrations determined after removing all particles larger than 2.5 μm are termed $PM_{2.5}$. Thus, $PM_{2.5}$ refers to particles less than 2.5 μm in size.

In aquatic systems and in the analytical determination of metals, the solid phase is distinguished by filtration using a filter opening of 0.45 μm. This size typically determines the cutoff between the "dissolved" and "particulate" phases. Therefore, in analytical chemistry, the 0.45-μm cutoff can be used to separate metals into a "dissolved" and "particulate" phase.

In the areas of drinking water, wastewater, and landfill leachate, solids are first divided into a "dissolved" or "suspended" fraction. This is done by a combination of filtration and evaporation procedures. Each of these two types of solids can be further broken down into a "fixed" and "volatile" fraction. Figure 2-2 shows the analytical difference between total solids, total suspended solids, total dissolved solids, and volatile suspended solids.

Total solids (TS) are determined by placing a well-mixed water sample of known volume in a drying dish and evaporating the water at 103–105°C. The

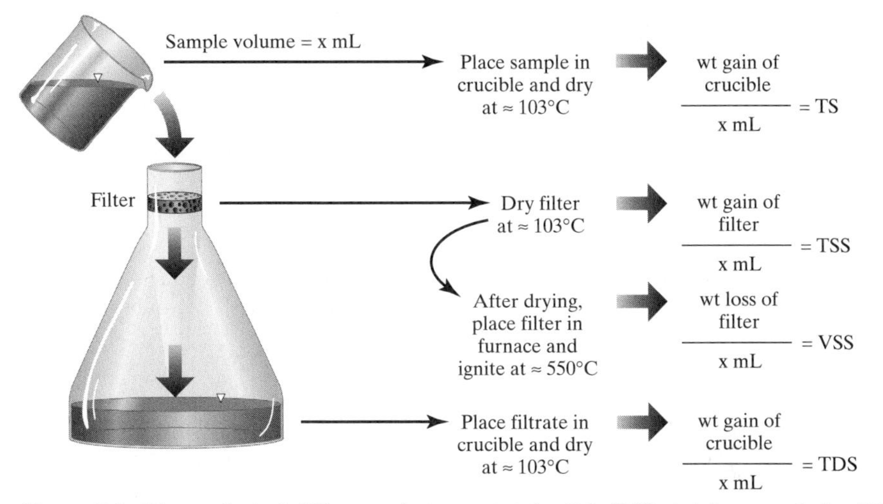

Figure 2-2. The analytical difference between total solids (TS), total suspended solids (TSS), volatile suspended solids (VSS), and total dissolved solids (TDS).

increase in the weight of the drying dish is due to the total solids. Total solids can thus be determined by dividing the increase in weight gain of the drying dish by the sample volume. Concentrations are typically reported in mg/L.

Total dissolved solids (TDS) and total suspended solids (TSS) are determined by first filtering a well-mixed sample of known volume through a glass-fiber filter (2-μm size opening). The suspended solids are the particles caught on the filter. The concentration of TSS can be determined by drying the filter at 103–105°C, determining the weight increase in the filter, and then dividing this weight gain by the sample volume (results in mg/L). Suspended solids (collected on the filter) may adversely impact aquatic ecosystems by impairing light penetration or act as a source of nutrients or oxygen-depleting organic matter. Also, a water high in suspended solids may be unsuited for human consumption or swimming. The TDS are determined by collecting the sample that passes through the filter, drying this filtrate at 180°C, and then determining the weight gain of the drying dish. This weight gain divided by the sample volume is the concentration of TDS (results in mg/L). Dissolved solids may adversely affect the taste of a water and can also lead to scale formation. This is because dissolved solids tend to be less organic in composition and consist of dissolved cations and anions. For example, one would expect that hard waters would also be high in dissolved solids.

TS, TDS, and TSS can be further broken down into a "fixed" and "volatile" fraction. For example, the volatile portion of the TSS is termed the volatile suspended solids (VSS) and the fixed portion is termed the fixed suspended solids (FSS). Determining the volatile fraction of a sample is done by taking each sample just discussed and igniting it in a furnace at 500°C (±50°C). The weight loss due to this high-temperature ignition provides the volatile fraction, and the fixed fraction is what sample remains after ignition. Figure 2-3 shows how to relate the various solid determinations used in water treatment, wastewater, and leachate situations. In wastewater treatment plants, the volatile fraction is a good approx-

TS	=	TDS	+	TSS
		=		=
TVS	=	VDS	+	VSS
		+		+
TFS	=	FDS	+	FSS

Figure 2-3. Matrix showing how the various measurements of solids in aqueous samples can be related. For example, if the TSS and VSS are measured, the FSS can be determined by difference.

imation of the organic matter content of the solids. Thus, determination of volatile solids in a biological aeration tank (e.g., 3,000 mg VSS/L) can be related to the number of microorganisms in the tank.

EXAMPLE 2.15. DETERMINING TSP AND $PM_{2.5}$ CONCENTRATIONS IN AN AIR SAMPLE

An air-sampling program sampled 100,000 L of air for particles. The following mass of particles were collected for particular size fractions: 12 mg retained with particle size > 2.5 μm and 6 mg retained with particle size < 2.5 μm. What are the $PM_{2.5}$ and total suspended particulate (TSP) concentrations of this air sample?

SOLUTION

By definition, $PM_{2.5}$ is the concentration of particles between 0.1 and 2.5 μm, or in this case the mass of particles retained with size < 2.5 μm. Therefore,

$$PM_{2.5} = \frac{6 \text{ mg}}{100{,}000 \text{ L}} \times \frac{10^3 \text{ L}}{\text{m}^3} \times \frac{10^3\ \mu\text{g}}{\text{mg}} = \frac{60\ \mu\text{g}}{\text{m}^3}$$

$$TSP = \frac{12 \text{ mg} + 6 \text{ mg}}{100{,}000 \text{ L}} \times \frac{10^3 \text{ L}}{\text{m}^3} \times \frac{10^3\ \mu\text{g}}{\text{mg}} = \frac{180\ \mu\text{g}}{\text{m}^3}$$

Note that these particulate concentrations are relatively high. The sample was most likely collected from an urban area where TSP concentrations range from 60 $\mu g/m^3$ to 200 $\mu g/m^3$.

EXAMPLE 2.16. DETERMINING CONCENTRATIONS OF SOLIDS IN A WATER SAMPLE

A laboratory provides the following analysis obtained from a 50-mL sample of wastewater. Total solids = 200 mg/L, total suspended solids = 160 mg/L, fixed suspended solids = 40 mg/L, and volatile suspended solids = 120 mg/L. (a) What is the concentration of total dissolved solids of this sample? (b) If this sample was filtered through a glass-fiber filter, then the filter was placed in a muffle furnace at 550°C overnight, what would be the weight of the solids (in mg) remaining on the filter after the night in the furnace? (c) Is this water sample turbid, and approximately what percent of the solids are composed of organic matter?

SOLUTION

(a) Refer to Figure 2-3 to see the relationship between the various forms of solids. TDS equals TS − TSS; thus,

$$\text{TDS} = \frac{200 \text{ mg}}{\text{L}} - \frac{160 \text{ mg}}{\text{L}} = \frac{40 \text{ mg}}{\text{L}}$$

(b) The solids remaining on the filter are suspended solids (dissolved solids would pass through the filter). Because the filter was subjected to a temperature of 550°C, the measurement was being made for the volatile and fixed fraction of the suspended solids, that is, the VSS and FSS. However, during the ignition phase, the volatile fraction is burned off, while what remains on the filter is the inert or fixed fraction of the suspended solids. Thus, this problem is requesting the fixed fraction of the suspended solids. Accordingly, the 50-mL sample had FSS = 40 mg/L. Therefore,

$$\text{FSS} = \frac{40 \text{ mg}}{\text{L}} = \frac{\text{wt of SS remaining on filter after ignition}}{\text{mL sample}} = \frac{x}{50 \text{ mL}}$$

The unknown, x, can be solved for and equals 2 mg.

(c) The sample is turbid. This is because the suspended matter, which is measured as TSS, causes the sample to appear turbid. Of course, if one allows the sample to sit for some time period, the suspended solids would settle and the sample might appear to be nonturbid. The solids found in this sample contain at least 60% organic matter. The total solids concentration is 200 mg/L and of this, 120 mg/L are volatile suspended solids. Therefore, because volatile solids consist primarily of organic matter (e.g., organic carbon, nitrogen, phosphorus, etc.), it can be concluded that approximately 60% (120/200) of the solids are organic.

2.5.4 Representation by Effect

In some cases, the actual concentration of a specific substance is not used at all, especially in instances where mixtures of ill-defined chemicals are present (e.g., raw sewage). Instead, the strength of the solution or mixture is defined by some common factor on which all the chemicals within the mixture depend (e.g., oxygen depletion from biological and chemical decomposition of the chemicals that make up the raw sewage). Thus, the strength of a wastewater is determined, not by measuring concentrations of a specific chemical(s), but by some direct effect the sample constituents have.

This method is used as a measure of the strength of municipal wastewater influent/effluent and other wastes. For many organic-bearing wastes, biological and chemical degradation of these wastes results in the depletion of oxygen from the stream that can result in fish kills if the dissolved oxygen concentration drops too low. Instead of identifying the hundreds of individual compounds that may be present, it is more convenient to report the effect, in units of the number of mg of oxygen that can be consumed per liter of water. This unit is referred to as biochemical oxygen demand (BOD). BOD has units of mg O_2/L and is used just like an ordinary mass/volume concentration. It is described in more detail in Chapter 5. Another measurement by effect is COD (chemical oxygen demand) also reported in mg O_2/L.

2.6 COMMON CONCENTRATIONS ENCOUNTERED IN WASTESTREAMS, SURFACE AND GROUNDWATERS, AND THE ATMOSPHERE

The purpose of this section is to familiarize the reader with the relative magnitude of concentrations that are typically encountered in the environmental science and engineering field, as well as the type of units commonly used in air, water, and soil measurements. In all of these examples, the chemical concentrations are very low. That is, the systems are "dilute" even though the chemical concentrations may still be of concern to human health and the environment.

Surface and Groundwater. Aqueous-phase (i.e., in water) concentrations typically are expressed either as mass or moles per volume. Concentrations of dissolved chemical species in water range from as low as pg/L (pg stands for picograms or 10^{-12} g) of contaminants to mg/L of major ions. The natural chemical composition of surface and groundwaters is strongly influenced by the chemical composition of the atmosphere and the chemical/biological weathering of rocks and minerals, which then release inorganic and organic chemical species into solution. There are other processes that can influence the chemistry of natural waters. For example, the chemistry of lakes in New Zealand is related to the lithology of the catchment area, the lake's proximity to the sea, and to a lesser degree, the origin of the lake (i.e., was the lake formed by volcanic activity, glacial activity, or constructed by humans) (Viner, 1987). In a second case, the chemistry of lakes situated in the Antarctic is profoundly influenced by the extreme cold and arid conditions of that region. Also lakes in the Adirondack Mountains of New York are influenced by the transport of air pollutants such as acid rain.

Table 2-4 summarizes the ranges of concentrations for some typical pollutants as well as for some natural constituents of surface waters. Major anions encountered in natural waters include Cl^-, SO_4^{2-}, HCO_3^-, and CO_3^{2-}, while major cations include Na^+, K^+, Ca^{2+}, and Mg^{2+}. Minor anions present in natural waters include

Table 2-4. Range of Concentrations Encountered in Natural Waters

Substance	Rain, Fog	Lakes, Rivers	Groundwater	Oceans
Trace metals (e.g., Pb, Cu, Hg, Zn)	0.01–100 μg/L	0.001–10 μg/L	0.1–10^6 ng/L	0.01–100 ng/L
Organic pollutants (e.g., PCBs, pesticides, solvents)	1–5,000 ng/L	0.1–500 ng/L	0.001–10^6 ng/L	0.001–10 pg/L
Major ions				
Ca^{2+}	0.1–20 mg/L	1–120 mg/L	1–120 mg/L	800 mg/L
Cl^-	0.05–10 mg/L	0.1–30 mg/L	0.1–50 mg/L	35,000 mg/L

OH^-, NO_3^-, and the various forms of orthophosphate, while minor cations in natural waters include H^+, Fe^{2+}, Fe^{3+}, Mn^{2+}, and Al^{3+}.

Organic chemicals that contaminate drinking water supplies and exhibit toxic effects have drinking water standards or guidelines ranging from 0.0005 to 10,000 μg/L, depending on the chemical and area of the country (McGeorge et al., 1992). Table 2-5 shows arsenic concentrations in Southern California's Central and West Basin groundwater supplies. The average arsenic concentration is 3.8 μg/L in the Central Basin and 0.8 μg/L in the West Basin. This example is typical in that pollutant concentrations in groundwaters are usually in the ppb to ppm range.

Municipal Wastewater and Landfill Leachate. Table 2-6 shows the concentration of dry-weather leachate from a landfill that has received hazardous and nonhazardous waste. Note that the concentrations of most constituents (e.g., nitrogen, COD, and total organic carbon (TOC)) are in the 10^2- to 10^3-mg/L range. Untreated municipal sewage may have concentrations of five-day biochemical oxygen demand (BOD_5) that range from 100 mg/L to 400 mg/L, TSS of 100 mg/L to 350 mg/L, total nitrogen concentrations that range from 20 mg N/L to 85 mg N/L, and total phosphorus concentrations that range from 4 mg P/L to 15 mg P/L. Treatment by activated sludge wastewater treatment process can

Table 2-5. Arsenic Concentrations in Southern California's Central and West Basin Groundwater Supplies

Groundwater Basin	Number Wells Tested	Number of Wells with Arsenic, Four Ranges			
		<0.5 μg/L	0.5–1.9 μg/L	2–5 μg/L	>5 μg/L
Central	227	13	58	119	37
West	35	14	19	1	1

Adapted from Ried, 1994.

Table 2-6. Dry Weather Leachate Concentration from the Goff Mountain Landfill, West Virginia

Parameter	Range (mg/L)	Average (mg/L)
Chemical oxygen demand (COD)	4,500–8,310	7,090
Total organic carbon (TOC)	169–2,820	1,270
Total suspended solids (TSS)	130–189	160
Volatile suspended solids (VSS)	108–149	120
NH_3-N	—	296
PO_4^{3-}-P	—	Below analytical detection
Alkalinity (as mg/L $CaCO_3$)	—	1,420

Adapted from Campbell et al., 1995.

reduce levels of BOD_5, TSS, and P by 85 to 95%. Though still small on a relative concentration basis, concentrations in the 10^2–10^3-mg/L range are high for most other environmental situations.

Atmosphere. Table 2-2 showed that the concentrations of the major atmospheric gaseous constituents were quite large. In fact, the concentration of oxygen in the atmosphere is approximately 210,000 ppm_V. However, the concentration of pollutants can be much smaller. Concentrations of particulates and of trace pollutants found in the atmosphere are frequently expressed in units of mass per unit volume on a ppm or ppb basis. Pollutant concentrations in the atmosphere are usually much lower than concentrations encountered in municipal, aqueous, and soil/sediment systems. Human exposure to air pollutants can be significant, however, as we breathe a large amount of air, approximately 20 m^3 per day. The following two paragraphs provide a brief description of typical concentrations of pollutants encountered in a wide variety of instances where air quality is important to humans and the environment.

Concentrations of total suspended particulate matter in the atmosphere may range from less than one to several hundred $\mu g/m^3$. Concentrations of airborne particles in Mexico City can average close to 600 $\mu g/m^3$ for total suspended particles and close to 300 $\mu g/m^3$ for particles that have an aerodynamic diameter of 10 μm or less (these smaller particles are termed PM_{10}) (Villalobos-Pietrini et al., 1995). Pollutant concentrations are only a small fraction of this total. For trace metals (e.g., Cu, Hg, Pb, Zn, Cd, etc.) as well as numerous trace organic pollutants, concentrations range from 0.001 ng/m^3 to 10 ng/m^3.

In Canada, ground-level ozone (O_3) levels have exceeded 190 ppb in some populated areas (also in some U.S. and European cities), while the acceptable Canadian air-quality objective of 82 ppb is exceeded up to 25% of the summer days in some areas. Rural and remote areas in Canada have ozone concentrations that peak in the 20–30 ppb range (Fuentese and Dann, 1994). In the 1960s, carbon monoxide (CO) levels averaged approximately 25–40 ppm in large U.S. cities like Washington, Chicago, Denver, and Los Angeles. Fortunately these values had been reduced to 9–15 ppm by the 1980s. However, one study determined

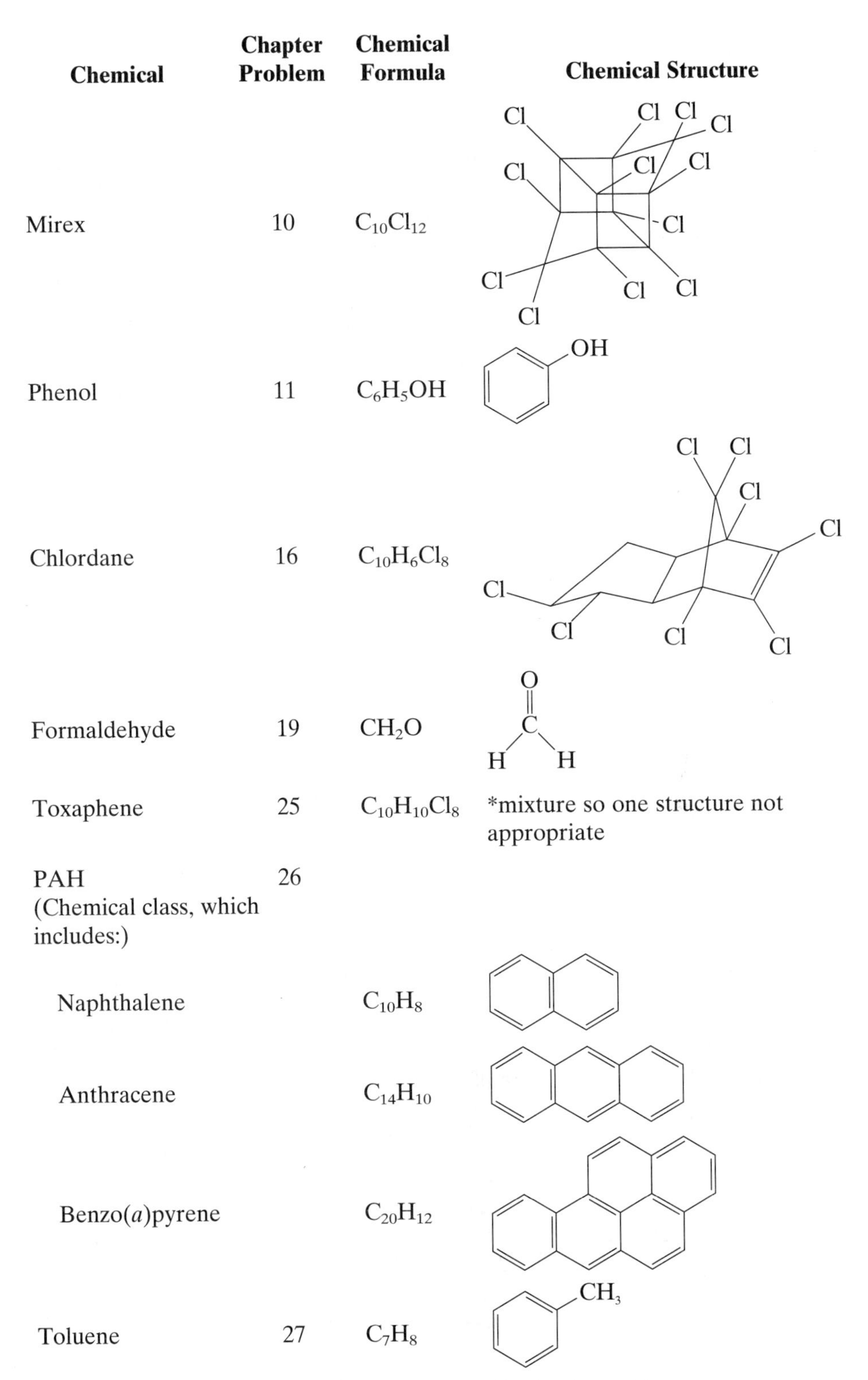

Chemical	Chapter Problem	Chemical Formula	Chemical Structure
Mirex	10	$C_{10}Cl_{12}$	
Phenol	11	C_6H_5OH	
Chlordane	16	$C_{10}H_6Cl_8$	
Formaldehyde	19	CH_2O	
Toxaphene	25	$C_{10}H_{10}Cl_8$	*mixture so one structure not appropriate
PAH (Chemical class, which includes:)	26		
Naphthalene		$C_{10}H_8$	
Anthracene		$C_{14}H_{10}$	
Benzo(*a*)pyrene		$C_{20}H_{12}$	
Toluene	27	C_7H_8	

Figure 2-4. Chemical structure of compounds used in chapter problems.

that Mexico City drivers were exposed to average CO levels of 56 ppm, with a maximum exposure to 80 ppm (Fernandez and Ashmore, 1995). In one study (Chan et al., 1991), in-vehicle commuters in Boston were exposed to average formaldehyde concentrations of 5.1 $\mu g/m^3$ and average toluene concentrations of 33.3 $\mu g/m^3$. That study also reported that cyclists were exposed to average formaldehyde concentrations of 6.3 $\mu g/m^3$ and average toluene concentrations of 16.3 $\mu g/m^3$.

Average mercury (Hg) concentrations in the atmosphere above Mace, Ireland, have been measured as 2.1 ng/m^3 and mercury concentrations of 8.2 ng/m^3 have been measured in the former German Democratic Republic. Concentrations of total mercury in rain range from 17 ng/L to 460 ng/L at the same sites (Ebinhaus et al., 1995). Global background levels of Hg° average 1 ng/m^3 and global particulate Hg averages 0.01 ng/m^3, which are quite low.

Finally radon (a colorless, odorless gas that is a potential indoor air-pollution problem) levels in California have been reported to range from 0.10 pCi/L to 16 pCi/L (picocuries/L) in one study (Liu et al., 1991). A curie is a unit of measuring radioactivity. A curie of a radioactive substance will release 3.7×10^{10} disintegrations per second.

CHAPTER PROBLEMS

2-1. Use your knowledge of what units of concentration are appropriate for a particular medium(s) to match the given concentration units with the appropriate pollutant/medium where those units are commonly used. For example, in part (a) mg/L O_2 corresponds to (i) and (ii).

Concentration Units: (a) mg/L O_2; (b) ppm_m; (c) ppm_V; (d) moles/L; (e) equivalents/L; (f) mg/L

Pollutant/medium: (i) BOD in river; (ii) oxygen dissolved in water; (iii) lead in water; (iv) benzene in soil; (v) nitrogen oxide in air; (vi) alkalinity

2-2. A typical loaf of bread contains 120 mg of sodium in each 1-ounce slice. (a) What is the concentration of sodium in the bread in ppm? (b) Is the answer to part (a) in units of ppm_m or ppm_V? Which makes sense and why?

2-3. (a) During drinking water treatment, 17 lb of chlorine are added daily to disinfect 5 million gallons of water. What is the aqueous concentration of chlorine in mg/L? (b) The chlorine demand is the concentration of chlorine used during disinfection. The chlorine residual is the concentration of chlorine that remains after treatment and is used to maintain a residual of chlorine in the water distribution system so that the water maintains its disinfecting power during the water's journey from the water-treatment plant to a household. If the residual concentration is 0.20 mg/L, what is the chlorine demand in mg/L?

2-4. A water sample contains 10 mg NO_3^-/L. What is the concentration in (a) ppm, (b) moles/L, (c) mg NO_3^--N/L, and (d) ppb?

2-5. A liquid sample has a concentration of iron (Fe) of 5.6 mg/L. The density of the liquid is 2,000 gm/L. What is the Fe concentration in ppm?

2-6. Coliform bacteria (e.g., *E. coli*) are excreted in large numbers in human and animal feces (approximately 50 million coliforms per gram of feces!). Untreated domestic sewage may contain greater than 3 million coliforms per 100 mL. Water that meets a standard of less than one coliform per 100 mL is considered safe for human consumption. Is a one-liter water sample that contains 9 coliforms safe for human consumption?

2-7. The treated effluent from a domestic wastewater-treatment plant contains ammonia at 9 mg N/L and nitrite at 0.5 mg N/L. Convert these concentrations to mg NH_3/L and mg NO_2^-/L.

2-8. A chemist reports that the concentration of nitrite (NO_2^-) plus nitrate (NO_3^-) in a groundwater sample from an agricultural region is 0.850 mM. Nitrite and nitrate are often elevated in groundwater in agricultural regions due to nitrogen fertilization. Regulations require that the total concentration ($[NO_2^-] + [NO_3^-]$) be below 10.0 mg/L as N to avoid methemoglobinemia, or blue-baby syndrome, which can be fatal. (a) What is the concentration of ($[NO_2^-] + [NO_3^-]$) expressed as {mg/L as N}? (b) What is the concentration expressed as ppm_m as N? (c) What is the concentration expressed as % by mass as N?

2-9. Concentrations of nitrate exceeding 44.3 mg NO_3^-/L are a concern in drinking water due to the infant disease, methemoglobinemia. Nitrate levels can be enhanced by improper use of synthetic and natural fertilizers, irrigation practices, livestock-handling operations, and industrial-waste handling. Due to the presence/absence of animal wastes, fertilizer application, and groundwater recharge patterns, nitrate concentrations near three rural wells were reported as 0.01 mg NO_3^- N/L, 1.3 NO_3^- N/L, and 20 NO_3^- N/L. Do any of these three wells exceed the 44.3 ppm level?

2-10. Mirex (MW = 540) is a fully chlorinated organic pesticide that was manufactured to control fire ants. It was also used as a fire retardant and in pyrotechnics. Due to its structure, Mirex is very unreactive; thus, it persists in the environment. Lake Erie water samples have had Mirex measured as high as 0.002 μg/L and lake trout samples with 0.002 μg/g (Journal of Great Lakes Research, 19:145–157, 1993). (a) What is the aqueous concentration of Mirex in units of (i) ppb, (ii) ppt, (iii) μmoles/L? (b) What is the concentration of Mirex in fish in (i) ppm, (ii) ppb?

2-11. Chlorophenols impart unpleasant taste and odor to drinking water at concentrations as low as 5 mg/m^3. They are formed when the chlorine disinfection process is applied to phenol-containing waters. What is the unpleasant taste and odor threshold in units of (a) mg/L, (b) μg/L, (c) ppm, (d) ppb?

2-12. The LC_{50} is used in determining toxicity to aquatic species. It is the aqueous concentration of a test chemical that results in 50% mortality in the test species (LC = lethal concentration) during a predetermined time period. The LC_{50} measured for fathead minnows, during a 96-h exposure test, is 151 mg/L for caffeine ($C_8H_{10}N_4O_2$) and 44.1 mg/L for trichloroethene (TCE) (C_2Cl_3H). Is the LC_{50} exceeded if the aqueous concentration of caffeine is 6.2×10^{-4} M and TCE is 4.5×10^{-4} M?

2-13. Chloroacetic acids are byproducts of the breakdown of hydrofluorochlorocarbons (HFCs), which have been implicated in the depletion of the ozone layer. HFCs are the current replacement products for freons. The concentration of monochloroacetic acid in rain water collected in Zurich was 7.8 nmol/L. Given that the formula for monochloroacetic acid is $CH_2ClCOOH$, calculate the concentration in μg/L.

2-14. In 1981, concentrations of Pb, Cu, and Mn in rainwater collected in Minneapolis were found to be 9.5, 2.0, and 8.6 μg/L, respectively. Express these concentrations as nmole/L, given that the atomic weights are 207, 63.5, and 55, respectively.

2-15. The dissolved oxygen (DO) profile in a biological reactor (i.e., aeration tank) at a domestic wastewater-treatment plant is 0.5 mg/L in the anoxic zone and 8 mg/L near the end of the 108-ft-long tank. What are these two DO concentrations in units of (a) ppm, (b) moles/L?

2-16. The average concentration of chlordane, a chlorinated pesticide now banned in the United States, in the atmosphere above the Arctic circle in Norway was found to be 0.6 pg/m^3 (Oehme et al., 1996). In this measurement, approximately 90% of this compound is present in the gas phase, the remainder is adsorbed to particles. For this problem, assume that all of the compound occurs in the gas phase, the humidity is negligibly low, and the average barometric pressure is 1 atm. Calculate the partial pressure of chlordane. The molecular formula for chlordane is $C_{10}Cl_8H_6$. The average air temperature through the period of measurement was −5°C.

2-17. What is the concentration in (a) ppm, and (b) percent by volume, of carbon monoxide (CO) that has a concentration of 103 mg/m^3?

2-18. Ice resurfacing and edging machines use internal combustion engines that give off exhaust containing CO and NO_x. The *Detroit News* (March 5, 1995) reported that average CO concentrations measured in local ice rinks were as high as 107 ppm and as low as 36 ppm. How do these concentrations compare to the U.S. outdoor-air-quality 1-h standard of 35 mg/m^3? Assume the temperature equals 20°C.

2-19. The concentration of formaldehyde in a mobile home was found to be 0.7 ppm. If the inside volume of the home is 800 m^3, what mass of formaldehyde vapor is inside the home, in units of grams? Assume $T = 298$ K, $P = 1$ atm. The molecular weight of formaldehyde is 30.

2-20. The concentration of ozone (O_3) in Los Angeles on a hot summer day ($T = 30°C$, $P = 1$ atm) is 125 ppb_V. What is the O_3 concentration in units of (a) $\mu g/m^3$, and (b) number of moles of O_3 per 10^6 moles of air?

2-21. An empty balloon is filled with exactly 10 g of nitrogen (N_2) and 2 g of oxygen (O_2). The pressure in the room is 1.0 atm and the temperature is 25°C. (a) What is the oxygen concentration in the balloon in units of percent by volume? (b) What is the volume of the balloon after it's blown up, in L?

2-22. A gas mixture contains 1.5×10^{-5} mole CO and has a total of 1 mole. What is the CO concentration in ppm?

2-23. "Clean" air might have a sulfur dioxide (SO_2) concentration of 0.01 ppm, while "polluted" air might have a concentration of 2 ppm. Exposure to SO_2 levels over 1 ppm may lead to breathing constrictions in the upper respiratory system. Convert these two concentrations to $\mu g/m^3$. Assume a temperature of 298 K.

2-24. Carbon monoxide (CO) affects the oxygen carrying capacity of your lungs. CO competes with oxygen for one of the four iron sites of your blood's hemoglobin molecules. In fact, CO has about a 210-times stronger affinity for these sites than oxygen. Exposure to 50 ppm CO for 90 min has been found to impair one's ability to time/interval discriminate; thus, motorists in heavily polluted areas may be more prone to accidents. Are motorists at a greater risk to accidents if the CO concentration is 65 mg/m^3? Assume a temperature of 298 K.

2-25. The State of Michigan's Department of Environmental Quality has determined that toxaphene concentrations in soil that exceed 60 $\mu g/kg$ can pose a threat to underlying groundwater. (a) If a 100-g sample of soil contains 10^{-5} g of toxaphene, what are the toxaphene soil and Michigan action level concentrations reported in units of ppb? (b) Does this sample pose a threat to groundwater in Michigan?

2-26. Polycyclic aromatic hydrocarbons (PAHs) are a class of organic chemicals that consist of two or more fused benzene rings placed in a linear, angular, or cluster arrangement. They are associated with the combustion of fossil fuels and some are considered carcinogenic or mutanogenic. Undeveloped areas may have total PAH concentrations in the soil of 5 $\mu g/kg$, while urban areas may have soil concentrations that range from 600 $\mu g/kg$ to 3,000 $\mu g/kg$. What is the concentration of PAHs in undeveloped areas in units of ppm?

2-27. The concentration of toluene (C_7H_8) in subsurface soil samples collected after an underground storage tank was removed indicated the toluene concentration was 5 mg/kg. What is the toluene concentration in ppm?

2-28. The following table contains a typical chemical analysis of a surface water. (a) Using the four most abundant cations and three most abundant anions found in this specific surface water, show whether the analysis is correct or not. (b) What is the hardness of this water in units of mg/L $CaCO_3$?

Alkalinity	108	Lead	ND
Alkylbenzene sulfonate (detergent)	0.1	Magnesium	9.9
Arsenic	ND	Nitrate	2.2
Barium	0	pH	7.6
Bicarbonate	131	Inorganic phosphorus	0.5
Cadmium	ND	Potassium	3.9
Calcium	35.8	Selenium	ND
Chloride	7.1	Silver	ND
Chromium	ND	Sodium	4.6
Copper	0.1	Sulfate	26.4
Cyanide	ND	Total dissolved solids	220
Fluoride	0.7	Zinc	ND

Taken from Hammer, 1975.

Note: All values are reported in mg/L. ND means the chemical was nondetectable.

2-29. While visiting Zagreb, Croatia, Mr. Arthur Van de Lay visits the Mimara Art Museum and then takes in the great architecture of the city. He stops at a café in the old town and orders a bottle of mineral water. The reported chemical concentration of this water is: $[Na^+] = 0.65$ mg/L, $[K^+] = 0.4$ mg/L, $[Mg^{2+}] = 19$ mg/L, $[Ca^{2+}] = 35$ mg/L, $[Cl^-] = 0.8$ mg/L, $[SO_4^{2-}] = 14.3$ mg/L, $[HCO_3^-] = 189$ mg/L, $[NO_3^-] = 3.8$ mg/L. The pH of the water is 7.3. (a) What is the hardness of this water in mg/L $CaCO_3$? (b) Is the chemical analysis correct?

2-30. A laboratory provides the following solids analysis for a wastewater sample: TS = 200 mg/L; TDS = 30 mg/L; FSS = 30 mg/L. (a) What is the total suspended solids concentration of this sample? (b) Does this sample have appreciable organic matter? (Why or why not?)

2-31. A 100-mL water sample is collected from a municipal wastewater treatment plant's biological aeration treatment reactor. The sample is placed in a drying dish (weight = 0.5000 g before the sample is added), and then placed in an oven at 104°C until all the moisture is evaporated. The weight of the dried dish is recorded as 0.5625 g. A similar 100-mL sample is filtered and the 100-mL liquid sample that passes through the filter is collected and placed in another drying dish (weight of dish before sample is added is also 0.5000 g). This sample is dried at 104°C and the dried dish's weight is recorded as 0.5325 g. What is the concentration (in mg/L) of (a) total solids? (b) total suspended solids? (c) total dissolved solids? and (d) volatile suspended solids? (Assume that in the aeration basin, VSS = 0.7*TSS.)

REFERENCES

American Public Health Association (APHA). 1992. Standard Methods for the Examination of Water and Wastewater, American Public Health Association, Washington, D.C.

Campbell, M. P., D. P. Smith, and A. D. Levine. 1995. Biological treatment of hazardous waste landfill leachate: A comparative study of fixed film reactors. *In* Innovative Technologies for Site Remediation and Hazardous Waste Management, Pittsburgh, Pa.

Chan, C. C., J. D. Spengler, H. Özkaynak, and M. Lefkopoulou. 1991. Commuter exposures to VOCs in Boston, Massachusetts. J. Air & Waste Management Association. 41:1594–1600.

Ebinghaus, R., H. H. Kock, S. G. Jennings, P. McCartin, and M. J. Orren. 1995. Measurements of atmospheric mercury concentrations in northwestern and central Europe—Comparison of experimental data and model results. Atmospheric Environment. 29(2):3333–3344.

Fernandes-Bremauntz, A. A., and M. R. Ashmore. 1995. Exposure of commuters to carbon monoxide in Mexico City—I. Measurement of in-vehicle concentrations. Atmospheric Environment. 29(4):525–532.

Fuentes, J. D., and T. F. Dann. 1994. Ground-level ozone in eastern Canada: Seasonal variations, trends, and occurrences of high concentrations. J. Air & Waste Management Association. 4:1019–1026.

Graedel, T. E., and P. J. Crutzen. 1993. Atmospheric Change: An Earth System Perspective. W. H. Freeman, New York.

Hammer, M. J. 1975. Water and Waste-Water Technology. John Wiley & Sons, New York.

Hammer, M. J., and M. J. Hammer, Jr. 1996. Water and Wastewater Technology. Prentice Hall, Englewood Cliffs, N.J.

Liu, K. S., S. B. Hayward, J. R. Girman, B. A. Moed, and F. Y. Huang. 1991. Annual average radon concentrations in California residences. J. Air & Waste Management Association. 41:1207–1212.

McGeorge, L. J., S. J. Krietzman, C. J. Dupuy, and B. Mintz. 1992. National Survey of drinking water standards and guidelines for chemical contaminants. J. American Water Works Association. 84(3):72–76.

Ried, J. 1994. Arsenic occurrence: USEPA seeks cleaner picture. J. American Water Works Association. 86(9):44–51.

Villalobos-Pietrini, R., S. Blanco, and S. Gomez-Arroyo. 1995. Mutagenicity assessment of airborne particles in Mexico City. Atmospheric Environment. 29(4):517–524.

Viner, A. B. 1987. Inland Waters of New Zealand. Department of Scientific and Industrial Research, Wellington, New Zealand.

Chapter 3

Chemistry

James R. Mihelcic
Noel R. Urban
Judith A. Perlinger
David W. Hand

In this chapter a reader will review several important chemical principles and apply them to understanding the chemistry of pollutants in a variety of engineered and natural systems. The chapter begins with a brief review of the difference between concentration and activity. This is followed by coverage of chemical kinetics; thermodynamics and its relationship to equilibrium processes; specific equilibrium processes between chemicals in the water, air, and solid phase; oxidation/reduction processes; and a discussion of photochemistry using the examples of the ozone hole and urban ozone.

Chemistry is important to environmental engineering and science because the ultimate fate of many pollutants discharged to air, water, soil, and treatment facilities is controlled by their "reactivity" and "chemical speciation." Likewise, cost-effective design, construction, and operation of many waste-treatment unit processes is dependent on the fundamental chemical processes that may take place in a particular treatment process. Also, individuals who predict (i.e., model) how pollutants move through groundwater, surface water, soil, the atmosphere, or a reactor are interested in whether a chemical degrades over time and how they can mathematically describe the rate of chemical disappearance or equilibrium in their overall model.

3.1 THE DIFFERENCE BETWEEN ACTIVITY AND CONCENTRATION

Activity can be thought of as the *effective* or *apparent* concentration in water or that portion of the true mole-based concentration of a species that participates in a chemical reaction. In most environmental situations, activity and concentration are used interchangeably. Places where they begin to differ greatly include seawater, some briny groundwaters, and some highly concentrated wastewaters. This section explains the difference between activity and concentration and shows how to relate them.

The remainder of Chapter 3 assumes that activity (designated by { } brackets) and concentration (designated by [ ] brackets) are equal and uses the notation for concentration ([ ] brackets).

In an ideal system, the molar free energy of a solute in water depends on the mole fraction (mole fraction defined in Chapter 2). However, this fraction does not include the effect of other dissolved species or the composition of the water, which also affect a solute's molar free energy. Chemical species interact by covalent bonding, van der Waals interactions, volume exclusion effects, and long-range electrostatic forces (i.e., repulsion and attraction between ions). In dilute aqueous systems, most interactions are caused by these long-range electrostatic forces. On a molecular scale these interactions can lead to local variations in the electron potential of the solution, which results in a decrease in the total free energy of the system.

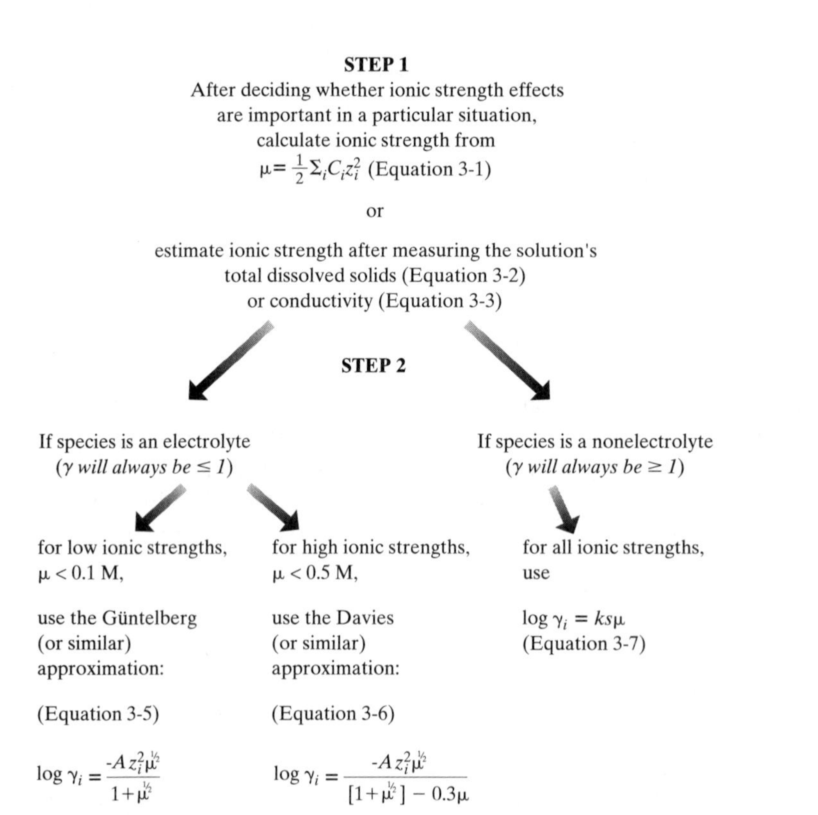

Figure 3-1. Two-step process to determine activity coefficients for electrolytes and nonelectrolytes.

All these complexities are then superimposed on solute–solvent interactions. *Use of activity instead of concentration accounts for these nonideal effects.* Activity is related to concentration by use of activity coefficients. Activity coefficients are determined by first finding the solution's ionic strength. Then one of several equations, developed specifically for either electrolytes (i.e., ions) or nonelectrolytes (uncharged species), are used to determine the activity coefficient of each individual species. Figure 3-1 summarizes this two-step process.

3.1.1 Ionic Strength

The ionic strength of a solution (referred to as I or μ) has units of moles/liter and can be determined from the following expression:

$$\mu = \frac{1}{2} \Sigma_i C_i z_i^2 \quad \textbf{(3-1)}$$

where C_i is the molar concentration of an ionic species i in solution, and z_i is the charge of the ion. In most natural waters the ionic strength is derived primarily from the major "background" cations and anions that make up natural waters. These latter were listed in Section 2.6. Freshwaters typically have an ionic strength of 0.001 to 0.01 M and seawaters have an ionic strength of approximately 0.7 M. The ionic strength of aqueous systems rarely exceeds 0.7 M. It can be correlated to easily measured water-quality parameters such as total dissolved solids (TDS) or specific conductance as shown below:

$$\mu = 2.5 \times 10^{-5} \times (\text{TDS}), \text{ where TDS is in mg/L} \quad \textbf{(3-2)}$$

or

$$\mu = 1.6 \times 10^{-5} \times (\text{specific conductance}) \quad \textbf{(3-3)}$$

where specific conductance is in μmho/cm and is measured with a conductivity meter.

3.1.2 Calculating Activity Coefficients for Electrolytes

When describing equilibrium relationships, aqueous chemical concentrations are reported as either [concentration] or {activity}. As mentioned previously, both have units of moles/liter when used in chemical-equilibrium relationships. Activity and concentration are related by an activity coefficient, γ, as follows:

$$\{C_i\} = \gamma_i \, [C_i] \quad \textbf{(3-4)}$$

where $\{C_i\}$ is the activity of species i in units of moles per liter, $[C_i]$ is the concentration of species i in moles per liter, and γ_i is the activity coefficient for species i and is unitless. Note that *activity* and *activity coefficients* are different. For electrolytes (species like Na^+, OH^-, $HgCl^+$, etc.) the activity coefficient is always equal to or less than 1. Figure 3-2 demonstrates through photos of an uncrowded and crowded dance floor how an increase in ionic strength reduces the activity of the reacting species. This can be thought of as reducing the "apparent" or "active" concentration of the species in solution.

For nonelectrolytes (such as O_2, CO_2, trichloroethene, benzene) the activity coefficient is always equal to or greater than 1.

Also, note that in an equilibrium reaction quotient (see Equation 3-39), activity coefficients are raised to the molar stoichiometric power just as the concentrations are. Activity coefficients for electrolytes can be determined by a variety of methods including the Güntelberg approximation, which is useful for solutions with $\mu < 0.1$ M.

$$\log \gamma_i = \frac{-Az_i^2\ \mu^{1/2}}{1 + \mu^{1/2}} \tag{3-5}$$

In Equation 3-5, A equals $1.82 \times 10^6\ (DT)^{-3/2}$, where T is the solution temperature in K and D is the dielectric constant for the solvent, which is usually water. For water, D is approximately 78.5. Thus, A is approximately 0.5 for water at 25°C.

The Davies approximation is useful for solutions with $\mu < 0.5$ M:

$$\log \gamma_i = \frac{-Az_i^2\ \mu^{1/2}}{[1 + \mu^{1/2}] - 0.3\ \mu} \tag{3-6}$$

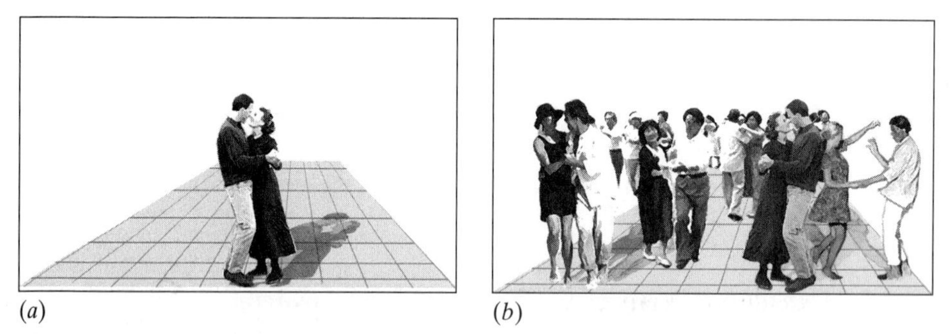

(*a*) (*b*)

Figure 3-2. Part (*a*) shows an uncrowded dance floor. Here a dancer can easily find their partner and this is not strongly influenced by the presence of other dancers. Part (*b*) shows a crowded dance floor. In this situation, a dancer who wants to find their dance partner may be adversely influenced by contact, attraction, and repulsion from other dancers who have crowded the dance floor. In an aqueous solution, an increased number of background ions (i.e., background electrolytes) also influences the activity of other ions attempting to react with one another. In solution, an ion's reactivity or apparent concentration is decreased in the presence of a large number of background ions (i.e., high ionic strength).

Other relationships include the Debye–Hückel and extended Debye–Hückel (see Stumm and Morgan, 1996, for a summary). Note that activity coefficients for electrolytes depend on the absolute value of the charge, as is discussed in Example 3-1.

EXAMPLE 3.1. CALCULATING IONIC STRENGTH AND ACTIVITY COEFFICIENTS FOR ELECTROLYTES

Calculate the ionic strength and all the individual activity coefficients for a 1-liter solution in which 0.01 mole of $FeCl_3$ and 0.02 mole of H_2SO_4 are dissolved.

SOLUTION

After the two compounds are placed in water they will completely dissociate to form: 0.01 M Fe^{3+}, 0.04 M H^+, 0.03 M Cl^-, and 0.02 M SO_4^{2-}. The ionic strength is given by Equation 3-1:

$$\mu = \tfrac{1}{2}[0.01(3+)^2 + 0.04(+1)^2 + 0.03(-1)^2 + 0.02(-2)^2] = 0.12\ \text{M}$$

Note that this ionic strength is relatively high but still below that of seawater. The Davies approximation (Equation 3-6) is useful for $\mu < 0.5$ M and can thus be used to determine each species's activity coefficient:

$$\gamma(H^+) = 0.78,\ \gamma(Cl^-) = 0.78,\ \gamma(SO_4^{2-}) = 0.36,\ \gamma(Fe^{3+}) = 0.10$$

As is apparent from this example, the activity coefficients of ions with higher valence deviate much more from 1.0 for a given ionic strength. That is, for electrolytes, use of activity coefficients is much more important for ions with a higher valence, because they are strongly influenced by the presence of other ions. Thus, while at a particular ionic strength it may not be important to calculate activity coefficients for monovalent ions, it may be very important for di, tri, and tetravelent ions.

3.1.3 Calculating Activity Coefficients for Nonelectrolytes

Activity coefficients for nonionic species are usually estimated from empirical expressions such as:

$$\log \gamma_i = ks\mu \tag{3-7}$$

where ks is referred to as the "salting out" coefficient and has units of L/mole. Table 3-1 lists some salting-out coefficients for several organic chemicals of environmental interest. Examples of salting-out coefficients are 0.208 for toluene, 0.140 for chloroform, and 0.132 for O_2.

Table 3-1. Salting-out Coefficients (at 20°C) for Several Organic Chemicals

Organic Chemical	Salting-out Coefficient (L/mole)	Reference
Tetrachloroethene	0.213	Gossett, 1987
Trichloroethene	0.186	Gossett, 1987
1,1,1-Trichloroethane	0.193	Gossett, 1987
1,1-Dichloroethane	0.145	Gossett, 1987
Chloroform	0.140	Gossett, 1987
Dichloromethane	0.107	Gossett, 1987
Benzene	0.195	Schwarzenbach et al., 1993
Toluene	0.208	Schwarzenbach et al., 1993
Naphthalene	0.220	Schwarzenbach et al., 1993
Oxygen	0.132	Snoeyink and Jenkins, 1980

EXAMPLE 3.2. CALCULATING ACTIVITY COEFFICIENTS FOR NONELECTROLYTES

An air stripper is being used to remove benzene (C_6H_6) from seawater and freshwater. Assume the ionic strength of seawater is 0.7 M and that of freshwater is 0.001 M. What is the activity coefficient for benzene in seawater and in freshwater? The salting-out coefficient for benzene is 0.195.

SOLUTION

Because benzene is a nonelectrolyte, use Equation 3-7 to determine the activity coefficients. This expression is used for solutions of both low and high ionic strength:

$$\log \gamma = ks\mu$$

$$\log \gamma = 0.195\ (0.001\ \text{M}) \text{ results in } \gamma \text{ (freshwater)} = 1$$

$$\log \gamma = 0.195\ (0.7\ \text{M}) \text{ results in } \gamma \text{ (seawater)} = 1.4$$

Note that if $\mu < 0.1$ M, the activity coefficient for a nonelectrolyte does not deviate much from a value of 1. Therefore, determination of the activity coefficients for nonelectrolytes becomes important for high ionic strengths solutions such as seawater. Needless to say that for most dilute environmental systems, activity coefficients are usually assumed equal to one. Places where they can gain importance are in the ocean, estuaries, briny groundwaters, or concentrated wastestreams.

This result of activity coefficients being greater than one can be explained by visualizing atmospheric oxygen dissolving in a water of high ionic strength. A water with a high ionic strength has many ions dissolved in it. The ions dissolved in water are covered with a shell of hydrated water and these hydration shells are very tightly packed together when compared to the water molecules in the bulk solution, which take up the remainder of the solution. Because of this tight packing, nonelectrolytes, such as oxygen in this case, dissolve less well in this matrix of hydrated water than the bulk volume of more loosely associated non-hydrated water. Therefore, a given total volume of water (hydrated plus non-hydrated water) can accept less nonelectrolyte. Thus, the concentration of the total volume of water is less than the activity.

3.2 CHEMICAL KINETICS

There are two very different approaches used in evaluating a chemical's fate and treatment: kinetics and equilibrium. *Kinetics* (Section 3.2) deals with the rates of reactions, and *equilibrium* (Sections 3.3 and 3.4) deals with the final result or stopping place of reactions. If reactions happen very rapidly relative to the time frame of our interest, the final conditions that result from the reaction are likely to be of more interest than are the rates at which the reaction occurs. In this case, an equilibrium approach is used. Examples of rapid reactions include acid–base reactions, complexation reactions, and some phase–transfer reactions (e.g., volatilization, dissolution).

Other reactions either occur relatively slowly or are prevented from reaching equilibrium (e.g., via subsequent reactions of the products of the first reaction). In such cases, the kinetic approach is used to understand how fast the reaction will occur. Biologically mediated reactions, light-driven (photochemical) reactions, redox reactions, and decay of radioactive elements all are examples of reactions that generally do not reach equilibrium quickly and thus require the kinetic approach. It is extremely important to be able to decide which approach to take in solving chemical problems.

Both the equilibrium and kinetic approaches are based on thermodynamics. As is discussed in Section 3.3, only reactions that result in thermodynamically favorable changes in their energy state can occur. This change in energy state is called Gibbs free energy change and denoted ΔG. It is this change in energy state that defines the equilibrium condition. However, not all reactions that occur would result in a favorable change in Gibbs free energy, and the magnitude of this energy change seldom is related to the rate of the reaction. For a reaction to occur, it generally is necessary that atoms collide, and that this collision has the right orientation and enough energy to overcome the "activation energy" required for the reaction. These energetic relationships are shown in Figure 3-3.

3.2.1 The Rate Law

Rate laws are used to predict the rates of chemical (abiotic) and biological (biotic) processes. The rate law expresses the dependence of the reaction rate on measurable, environmental parameters. Of particular interest is the dependence of

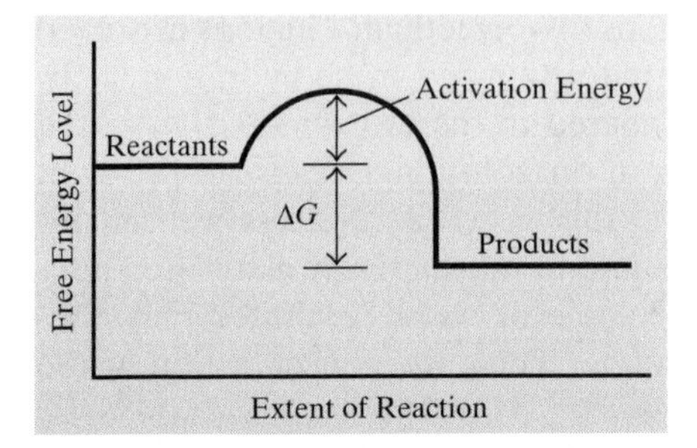

Figure 3-3. The energetic relationships required for a reaction to occur.

the rate on the concentrations of the reactants. Other parameters that may influence the reaction rate include temperature and the presence of catalysts (including microorganisms).

The rate of an irreversible reaction, and the exact form of the rate law, depends on the mechanism of the reaction. Consider the hydrolysis of dichloromethane (DCM). In this reaction one molecule of DCM reacts with a hydroxide ion (OH^-) to produce chloromethanol (CM) and chloride ion:

$$H-\overset{\displaystyle Cl}{\underset{\displaystyle Cl}{\overset{|}{\underset{|}{C}}}}-H + OH^- \longrightarrow H-\overset{\displaystyle Cl}{\underset{\displaystyle OH}{\overset{|}{\underset{|}{C}}}}-H + Cl^-$$

For the reaction depicted here to occur, one molecule of DCM must collide and react with one molecule of OH^-. This reaction is termed a *binary reaction* because it involves two molecules. Molecules in nature are constantly moving at a rate that is dependent on temperature. As a result of this motion, molecules are constantly colliding, but only a small fraction of collisions result in a reaction. The more molecules of DCM in a fixed volume (i.e., the concentration of DCM is high), the greater the probability that a collision between a molecule of DCM and OH^- will occur. Similarly, the more molecules of OH^- there are, the greater the probability of a collision between a molecule of DCM and OH^-. Although, as stated earlier, only a small fraction of the collisions actually result in a reaction, the rate of the reaction is still proportional to the number of collisions per unit time.

Based on this discussion, the rate of an irreversible binary reaction should be proportional to the concentration of each chemical species and can be written as:

$$\boldsymbol{R = k[\mathrm{DCM}][\mathrm{OH}^-]} = \frac{-d[\mathrm{DCM}]}{dt} = \frac{-d[\mathrm{OH}^-]}{dt} = \frac{d[\mathrm{CM}]}{dt} = \frac{d[\mathrm{Cl}^-]}{dt} \qquad \textbf{(3-8)}$$

where R is the rate of reaction, k is the rate constant for this particular reaction, [DCM] is the concentration of dichloromethane, $[OH^-]$ is the concentration of hydroxide ion, [CM] is the concentration of chloromethanol, and $[Cl^-]$ is the concentration of chloride ion. The minus signs in Equation 3-8 indicate that the products' concentrations are decreasing over time.

The bold portion of Equation 3-8 is referred to as the *rate law* for the reaction. The rate law expresses the dependence of the reaction rate on the concentrations of the reactants. The rate law in this case would be called first order with respect to DCM and first order with respect to OH^-. The term *first order* indicates that each species is raised to the first power. The rate law is second order overall because it involves the product of two species, each raised to the first power. Because the reaction was depicted as being irreversible, it was assumed that the concentration of products did not influence the rate of the forward reaction.

To generalize these terms, a hypothetical rate law can be constructed for a generic irreversible reaction of a moles of species A reacting with b moles of species B to yield products, P. The rate law is written

$$R = k[A]^a[B]^b \tag{3-9}$$

This reaction would be termed ath order with respect to A and bth order with respect to B. The overall order of the reaction would be $(a + b)$. This reaction is termed an *elementary* reaction because *the reaction order is controlled by the stoichiometry of the reaction*. That is, a equals the molar stoichiometric coefficient of species A, and b equals the molar stoichiometric coefficient for B.

The order of a reaction should be determined experimentally, however, because it often does not correspond to the reaction stoichiometry. This is because the mechanism or steps of the reaction do not always correspond to that shown in the reaction equation.

The collision-based reaction of the hydrolysis of dichloromethane can be contrasted with some biological transformations of organic chemicals that occur in either treatment plants or natural environments where soils and sediments are present. In some of these situations, zero-order transformations are observed. This can be due to several items, including the rate-limiting diffusion of oxygen from the air to the aqueous phase, which may be slower than the demand of oxygen by the microorganism biodegrading the chemical. Another explanation for observing zero-order kinetics is the slow, rate-limiting movement of a chemical (required by the microorganisms for energy and growth) that has a low water solubility (ppb and ppm range) from an oil or soil/sediment phase into the aqueous phase where the chemical is then available for the organism to utilize.

2,4-D is one such chemical that has been observed to have zero-order kinetics of biodegradation. It is a herbicide, commonly used by farmers and households (it is found in many dandelion weed killers). 2,4-D can be transported to a river or lake by horizontal runoff or vertical migration to groundwater that is then hydraulically connected to a lake or river. It has been found to disappear in lake water according to zero-order kinetics (Subba-Rao, 1982):

The rate law for this type of reaction can be written as

$$R = \frac{-d[2,4\text{-D}]}{dt} = k \tag{3-10}$$

This reaction is termed *zero-order* because *it does not depend on the concentration of the compound involved in the reaction*. The two cases just described illustrate how the form of the rate law depends on the mechanism of the reaction. Similarly, if the form of the rate law is known, it can provide insight into a reaction's actual mechanism.

3.2.2 Zero- and First-order Reactions

Many environmental situations can be described by zero- or first-order kinetics. Figure 3-4 summarizes the major differences between these two types of kinetics. These types of chemical and biological kinetic reaction terms will be incorporated into mass-balance expressions in Chapter 4. In this section, these two types of kinetic expressions are discussed in depth by first constructing a generic chemical reaction whereby a chemical "C" is converted to some unknown products:

$$C \rightarrow \text{products} \tag{3-11}$$

The rate law for describing the decrease in concentration of chemical C with time can be written as

$$\frac{d[C]}{dt} = -k[C]^n \tag{3-12}$$

Here $[C]$ is the concentration of "C," t is time, k is a rate constant that has units dependent on the order of the reaction, and the reaction order, n, is typically an integer (e.g., 0, 1, 2).

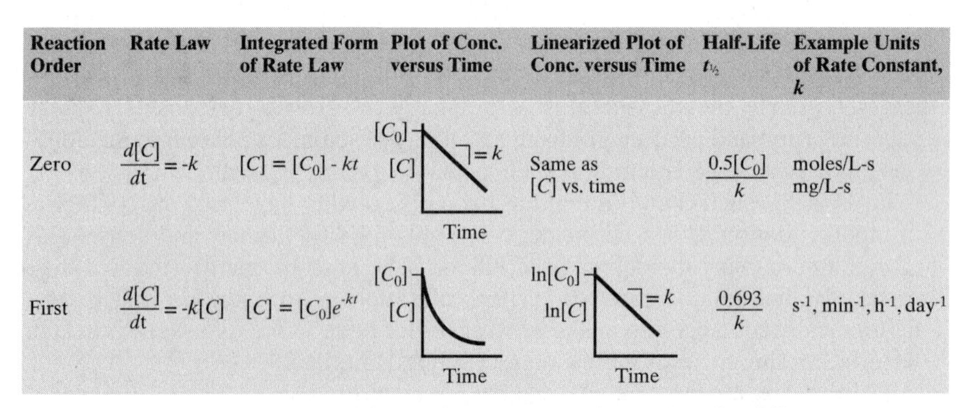

Reaction Order	Rate Law	Integrated Form of Rate Law	Plot of Conc. versus Time	Linearized Plot of Conc. versus Time	Half-Life $t_{1/2}$	Example Units of Rate Constant, k
Zero	$\frac{d[C]}{dt} = -k$	$[C] = [C_0] - kt$	$[C_0]$; $[C]$; $= k$; Time	Same as $[C]$ vs. time	$\frac{0.5[C_0]}{k}$	moles/L-s mg/L-s
First	$\frac{d[C]}{dt} = -k[C]$	$[C] = [C_0]e^{-kt}$	$[C_0]$; $[C]$; Time	$\ln[C_0]$; $\ln[C]$; $= k$; Time	$\frac{0.693}{k}$	s^{-1}, min^{-1}, h^{-1}, day^{-1}

Figure 3-4. Summary of zero and first-order rate expressions. Note the differences between each of these expressions.

Zero-order Reaction

If n *is zero, Equation 3-12 becomes*

$$\frac{d[C]}{dt} = -k \qquad \textbf{(3-13)}$$

This is the rate law describing a zero-order reaction. This is the rate-law expression used in Equation 3-10 to describe the transformation of 2,4-D. Here, the rate of disappearance of C with time is zero order with respect to "C" and the overall order of the reaction is zero order. Equation 3-13 can be rearranged and integrated for the following conditions; at time zero the concentration of "C" equals C_0, and at some future time, t, the concentration equals C:

$$\int_{C_0}^{C} d[C] = -k \int_0^t dt \qquad \textbf{(3-14)}$$

Integration of Equation 3-14 yields:

$$[C] = [C_0] - kt \qquad \textbf{(3-15)}$$

A reaction is zero order if concentration data are plotted versus time and the result is a straight line. This is illustrated in Figure 3-4. The slope of the resulting line is the zero-order rate constant k, which has units of concentration/time (e.g., moles/liter-day).

First-order Reaction

If n = 1, *Equation 3-12 becomes*

$$\frac{d[C]}{dt} = -k[C] \qquad \textbf{(3-16)}$$

This is the rate law for a first-order reaction. Here the rate of disappearance of C with time is first order with respect to $[C]$ and the overall order of the reaction is first order. Equation 3-16 can be rearranged and integrated for the same two conditions used in Equation 3-14 to obtain an expression that describes the concentration of C with time:

$$[C] = [C_0]e^{-kt} \qquad \textbf{(3-17)}$$

Here k is the first-order reaction rate constant and has units of $time^{-1}$ (e.g., hr^{-1}, day^{-1}). A reaction is first order when the natural logarithm of concentration data plotted versus time results is a straight line. The slope of this resulting straight line is the first-order rate constant, k, as illustrated in Figure 3-4.

Radioactive decay is an example of a first-order reaction. For example, the radioisotope cesium (^{137}Cs) was released into the atmosphere as a result of the Chernobyl nuclear accident. The nucleus of any radioactive atom disintegrates at a constant rate by emitting a particle from the nucleus, and in the process the atom is transformed to a different element. Although this reaction does not depend on the collisions of molecules, the rate of atomic decay is proportional to the amount of the radioisotope present. Therefore, the rate law for cesium decay would be written as

$$\frac{d[^{137}\mathrm{Cs}]}{dt} = -k[^{137}\mathrm{Cs}] \tag{3-18}$$

This rate law is first order with respect to ^{137}Cs and is first order overall.

There are some important things to note about first- and zero-order chemical reactions. First, when comparing the concentration over time in the two reactions (as shown in Figure 3-4), the first-order reaction appears to "slow down" over time, while the zero-order reaction decreases at a steady rate. To say this another way, the rate of the first-order reaction (slope of concentration data versus time) decreases over time, while in the zero-order reaction the slope remains constant over time. This suggests that the rate of a zero-order reaction is independent of the chemical concentration (see Equation 3-13), while the rate of a first-order reaction is dependent on the concentration of the chemical (see Equation 3-16).

This has important implications for chemical reactions that occur in either the environment or in a treatment plant. If a reaction follows first-order kinetics, the rate of the reaction decreases with time. Thus, a chemical whose disappearance follows concentration-dependent kinetics, like first order, will disappear more slowly as its concentration decreases. Most environmental pollutants typically exist in very low concentrations in nature and wastewater-treatment plants. For example, wastewater entering a municipal wastewater-treatment plant is over 99% water (it is relatively dilute), and, as was seen in Chapter 2, typical environmental concentrations of many pollutants are in the ppm and ppb range. Additionally, treatment objectives often are set at very low concentrations in order to protect human health and the environment. For example, the concentrations of many chemicals that have contaminated groundwater may have to be reduced down to ppb levels in order to protect drinking water or an aquatic ecosystem if the groundwater is hydraulically connected to surface water. This means that some rates of chemical and biological transformation reactions that occur either naturally or are engineered may be very slow.

EXAMPLE 3.3. USE OF A RATE LAW

How long will it take the carbon monoxide (CO) concentration in a room to decrease by 99% after the source of carbon monoxide is removed and the windows are opened? Assume the first-order rate constant for CO removal (due to dilution by incoming clean air) is 1.2 h^{-1}.

SOLUTION

This is a first-order reaction, so use Equation 3-17. Let $[CO_0]$ equal the initial CO concentration. When 99% of the CO goes away, $[CO] = 0.01[CO_0]$. Therefore,

$$0.01[CO_0] = [CO_0]e^{-kt}$$

where $k = 1.2\ h^{-1}$. Solve for t, which equals 3.8 h.

3.2.3 Pseudo-first-order Reactions

There are many circumstances in which the concentration of one participant in a reaction remains constant during the reaction. For example, if the concentration of one reactant initially is much higher than the concentration of another, it is impossible for the reaction to cause a significant change in the concentration of the substance with the high initial concentration. Alternatively, if the concentration of one substance is buffered at a constant value (e.g., pH in a lake does not change because it is buffered by the dissolution and precipitation of alkalinity-containing solid $CaCO_3$), then the concentration of the buffered species will not change even if the substance participates in a reaction. A pseudo-first-order reaction is used in these situations. It can be modeled as if it were a first-order reaction. Consider the irreversible elementary reaction:

$$aA + bB \rightarrow cC + dD \tag{3-19}$$

The rate law for this reaction is

$$R = k[A]^a[B]^b \tag{3-20}$$

If the concentration of A does not change significantly during the reaction for one of the reasons provided above (i.e., $[A_0] \gg [B_0]$ or $[A] \cong [A_0]$, the concentration of "A" may be assumed to remain constant and can be incorporated into the rate constant, k. The rate law then becomes:

$$R = k'[B]^b \tag{3-21}$$

where k' is the pseudo-first-order rate constant and equals $k[A_0]^a$. This manipulation greatly simplifies the rate law for the disappearance of substance B:

$$\frac{d[B]}{dt} = -k'[B]^b \tag{3-22}$$

If b is equal to 1, then the solution of Equation 3-22 is identical to that for Equation 3-17. In this case, the pseudo-first-order expression can be written as

$$[B] = [B_0]e^{-k't} \tag{3-23}$$

EXAMPLE 3.4. PSEUDO-FIRST-ORDER REACTION

Lake Silbersee is located in the German city of Nuernberg. The lake's water quality has been diminished because of high hydrogen sulfide concentrations (the rotten egg smell) that originates from a nearby leaking landfill. To combat the problem, the city decided to aerate the lake in an attempt to oxidize the odorous H_2S to nonodorous sulfate ion according to the following oxidation reaction: $H_2S + 2O_2 \rightarrow SO_4^{2-} + 2H^+$.

It has been determined experimentally that the reaction follows first-order kinetics with respect to both oxygen and hydrogen sulfide concentrations:

$$\frac{d[H_2S]}{dt} = -k[H_2S][O_2]$$

The present rate of aeration maintains the oxygen concentration in the lake at 2 mg/L. The rate constant (k) for the reaction was determined experimentally to be 1,000 L/mole-day. If the aeration completely inhibited anaerobic respiration and thus stopped the production of sulfide, how long would it take to reduce the H_2S concentration in the lake from 500 μM to 1 μM?

SOLUTION

The dissolved oxygen of the lake is maintained at a constant value and therefore is a constant. It can be combined with the rate constant to make a pseudo-first-order rate constant. Thus

$$[H_2S] = [H_2S_0]e^{-k't}$$

where $k' = k[O_2]$.

$$1\ \mu\text{m} = 500\ \mu\text{m}\ \exp\left\{-\frac{1{,}000\ \text{L}}{\text{mole-day}} \times \frac{2\ \text{mg}}{\text{L}} \times \frac{\text{g}}{1{,}000\ \text{mg}} \times \frac{\text{mole}}{32\ \text{g}} \times t\right\}$$

The time, t, can be calculated to equal 100 days.

3.2.4 Half-Life and Its Relationship to the Rate Constant

It is often useful to express a reaction in terms of the time required to react one-half of the concentration initially present. The "half-life" is used for this purpose. The *half-life*, $t_{1/2}$, is defined as *the time required for the concentration of a chemical to decrease by one-half* (e.g., $[C] = 0.5[C_0]$). The relationship between half-life and reaction rate constant depends on the order of the reaction as shown in Figure 3-4.

For zero-order reactions, the half-life can be related to the zero-order rate constant, k, by substituting $[C] = 0.5 \times [C_0]$ into Equation 3-15:

$$0.5[C_0] = [C_0] - kt_{1/2} \tag{3-24}$$

Equation 3-24 can be solved for the half-life:

$$t_{1/2} = \frac{0.5[C_0]}{k} \tag{3-25}$$

Likewise, for a first-order reaction, the half-life can be related to the first-order rate constant, k, by substituting $[C] = 0.5[C_0]$ into Equation 3-17:

$$0.5C_0 = C_0 e^{-kt} \tag{3-26}$$

The half-life for a first-order relationship then is given by:

$$t_{1/2} = \frac{0.693}{k} \tag{3-27}$$

EXAMPLE 3.5. CONVERSION OF RATE CONSTANT TO HALF-LIFE

An engineer is modeling the transport of a chemical contaminant in groundwater. The individual has a mathematical model that only accepts first-order degradation rate constants and a handbook of "subsurface chemical transformation half-lives." Subsurface half-lives for benzene, TCE, and toluene are listed as 69, 231, and 12 days, respectively. What are the first-order rate constants for all three chemicals?

SOLUTION

The model only accepts concentration-dependent, first-order rate constants. Thus, to solve the problem, convert half-life to a first-order rate constant with the use of Equation 3-27.

$$k_{\text{benzene}} = \frac{0.693}{t_{1/2}} = \frac{0.693}{69 \text{ days}} = 0.01 \text{ day}^{-1}$$

$$k_{\text{TCE}} = \frac{0.693}{t_{1/2}} = \frac{0.693}{231 \text{ days}} = 0.003 \text{ day}^{-1}$$

$$k_{\text{toluene}} = \frac{0.693}{t_{1/2}} = \frac{0.693}{12 \text{ days}} = 0.058 \text{ day}^{-1}$$

EXAMPLE 3.6. USE OF HALF-LIFE IN DETERMINING FIRST-ORDER DECAY

After the Chernobyl nuclear accident, the concentration of ^{137}Cs in milk was proportional to the concentration of ^{137}Cs in the grass that cows consumed. The concentration in the grass was, in turn, proportional to the concentration in the soil. Assume that the only reaction by which ^{137}Cs was lost from the soil was through radioactive decay and the half-life for this isotope is 30 years. Calculate the concentration of ^{137}Cs in cow's milk after 5 years if the concentration in milk shortly after the accident was 12,000 Bq/L. (*Note*: A Bequerel is a measure of radioactivity. One Bequerel equals one radioactive disintegration per second.)

SOLUTION

Since the half-life equals 30 years, the rate constant k can be determined from Equation 3-27:

$$k = \frac{0.693}{t_{1/2}} = 0.023 \text{ yr}^{-1}$$

Therefore:

$$[^{137}Cs]_{t=5} = [^{137}Cs]_{t=0} \exp(-kt) = \frac{12{,}000\ Bq}{L} \exp\left(-\frac{0.023}{\text{yr}} \times 5 \text{ yr}\right) = \frac{10{,}700\ Bq}{L}$$

3.2.5 Effect of Temperature on Rate Constants

Typically rate constants are determined and compiled for temperatures of either 20 or 25°C. However, groundwaters typically have temperatures around 8–12°C, and surface waters, wastewaters, and soils generally have temperatures that range from 0°C to 30°C. Thus, when a different temperature is encountered, an engineer or scientist must first determine if the effect of temperature is important, and secondly, if important, how to convert the rate constant for the new temperature.

The rates of most reactions are dependent on temperature. In Section 3.2.1 it was mentioned that the rate of molecular motion was a function of temperature. Thus, the higher the temperature, the faster the molecules move, which results in an increase in the number of collisions per unit time. A higher temperature also increases the energy of the collisions. As a result, a greater fraction of the collisions results in a chemical reaction. Consequently, the rate of chemical reactions that depend on collisions of two or more molecules will be increased by an increase in temperature. In contrast, reactions not dependent on molecular collisions (e.g., radioactive decay) will not be influenced in this manner. In addition, biological processes (enzyme catalyzed) have some optimal temperature range, above or below which the enzyme becomes destroyed (i.e., denatured). That is, biological processes will show an increase in reaction rate with increase in temperature up to a certain point, after which the rate will begin to decrease with further increases in temperature.

An example demonstrating the importance of temperature effects on a rate constant is the decomposition of peroxyacetyl nitrate (PAN). PAN is formed in the atmosphere by the oxidation of hydrocarbons in the presence of nitrogen oxides (NO_x). PAN is an eye irritant to humans, a plant toxicant, and can act as a reservoir of NO_x during urban-smog formation. The rate constant for the atmospheric thermal decomposition of PAN is strongly temperature dependent. The first-order rate constant increases from $5.0 \times 10^{-6}\ s^{-1}$ at 0°C to $1.6 \times 10^{-3}\ s^{-1}$ at 35°C (Finlayson-Pitts and Pitts, 1986). The result is that the half-life of PAN drops from 2,300 min at 0°C to only 7.2 min at 35°C.

The *Arrhenius equation* is used to adjust rate constants for changes in temperature. It is written as

$$k = Ae^{-(Ea/RT)} \tag{3-28}$$

where k is the rate constant of a particular order, A is termed the preexponential factor (same units as k), E_a is the activation energy (kcal/mole), R is the gas constant, and T is temperature (K). The preexponential factor is related to the number of collisions per time; therefore, the preexponential factor is different for gas- and liquid-phase reactions. The preexponential factor, A, also has a small dependence on temperature for many reactions; however, most environmental situations span a relatively small temperature range. The activation energy, E_a, is the energy required for the collision to result in a reaction. A plot of $\ln(k)$ versus $1/T$ can be used to determine E_a and A, as shown in Figure 3-5. After E_a and A are known for a particular reaction, Equation 3-28 can be used to adjust a rate constant for changes in temperature.

The Arrhenius equation is the basis for another commonly used relationship between rate constants and temperature used for biological processes over narrow temperature ranges. The carbonaceous biochemical oxygen demand (CBOD) rate constant, k_L, known at a particular temperature, is typically converted to other temperatures using the following expression:

$$k_{T2} = k_{T1} \times \Theta^{(T2-T1)} \tag{3-29}$$

where Θ is a dimensionless temperature coefficient. In fact, Θ equals $\exp\{Ea \div [R \times T1 \times T2]\}$ as can be seen from the Arrhenius equation. Θ is temperature dependent and has been found to range from 1.056 to 1.13 for biological decay of municipal sewage.

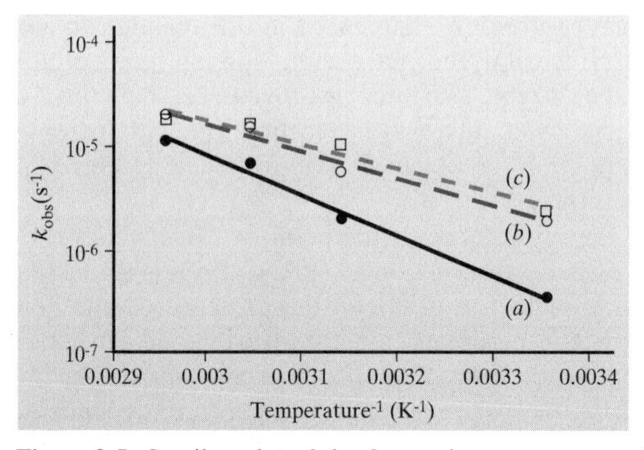

Figure 3-5. Semilog plot of the first-order rate constant (k) versus $1/T$, which can be used to determine E_a and A. The slope equals $-E_a/R$ and the intercept equals ln A. In order to obtain this plot, the rate constant was determined at various temperatures for the chemical reaction of hexachloroethane with hydrogen sulfide in the presence of a compound called juglone (see Figure 3-7 for a detailed description of the chemistry involved in this reaction). This chemical reaction may be an important process determining the natural disappearance of halogenated alkanes and alkenes that have contaminated subsurface soils and groundwaters due to improper land disposal of industrial, commercial, and household solvents. This semilog plot can be used to determine the activation energy, E_a, of a reaction and the pre-exponential factor, A, in the Arrhenius relation (Equation 3-28). The slope equals $-E_a/R$ and the y-intercept equals ln A. When (a) 1 mM of hydrogen sulfide reacts with hexachloroethane at pH 7.0, E_a of the reaction is 75 kJ/mole. In the presence of either (b) 20 μM juglone or (c) 200 μM juglone at pH 7.0 the activation energy decreases to about 50 kJ/mole and the reaction rate increases by approximately an order of magnitude in the temperature range tested (25°C to 65°C). (Reprinted with permission from Perlinger et al., Copyright, 1996, American Chemical Society).

EXAMPLE 3.7. EFFECT OF CBOD RATE CONSTANT ON TEMPERATURE

The rate constant for carbonaceous biochemical oxygen demand (CBOD) at 20°C is 0.1 day^{-1}. What is the rate constant at 30°C? Assume $\Theta = 1.072$.

SOLUTION

Using Equation 3-29,

$$k_{30} = 0.1 \text{ day}^{-1}[1.072^{(30C-20C)}] = 0.2 \text{ day}^{-1}$$

This example demonstrates that for *biological systems used in wastewater treatment, one would often observe a doubling in the biological reaction with every 10°C increase in the temperature*. This is a general rule for biological systems. In biology, the term Q_{10} is sometimes used to describe the change in enzyme activity with a 10°C rise in temperature. Thus, as expected, Q_{10} for many enzyme systems is approximately 2. Remember, though, as mentioned previously, biological systems have some optimal temperature above which there is an adverse effect on chemical removal.

3.2.6 Catalysts

Catalysts are substances that accelerate chemical reactions without being transformed in the reaction. They are used to enhance the rates of slow reactions. In brief, catalysts lower the activation energy required for a particular reaction to take place. Numerous examples of catalysts are available in natural and engineered environmental situations. The rest of this section discusses a few examples.

An excellent example of the importance of catalysts is provided in Section 3.6 where the formation of urban ozone (smog) and the destruction of the Earth's protective ozone layer are all shown to be catalyzed by many chemicals of anthropogenic origin.

Particle surfaces often act as catalysts and the oxidation of reduced manganese (Mn^{2+}) and iron (Fe^{2+}) is catalyzed by the presence of particles. Since the oxidation of these metals also produces particles (i.e., Fe^{3+} and Mn^{4+} precipitates), the reaction is said to be autocatalytic. Several common groundwater pollutants can be destroyed with the help of light-activated semiconductors such as titanium dioxide, and development of catalysts to promote these reactions has been a major area of research and engineering development. Another innovative method for the catalytic destruction of chlorinated solvents in groundwater focuses on the use of iron and natural organic compounds to catalyze the reduction of the halogenated pollutants.

Biological enzymes are used by microorganisms to catalyze many biochemical reactions. Many of these enzymes are currently being identified by scientists and engineers for their importance in the degradation of synthetic organic chemicals. For example, one of the most important classes of biological enzymes is the oxygenases. These enzymes are instrumental in the initial microbial transformation of many aromatic organic chemicals such as benzene and naphthalene. These compounds are found in fuel products such as gasoline and diesel fuel. Figure 3-6 shows an example of how dioxygenase enzymes incorporate oxygen into an aromatic ring, thus making the ring susceptible to ring cleavage, after which it may be utilized as an energy source.

Methane monooxygenase (MMO) is an enzyme recently identified in the cometabolic transformation of some chlorinated alkanes and alkenes such as trichloroethene (TCE) and tetrachloroethene (PCE). A cometabolic reaction is one in which bacterial enzymes react with a minor species that is similar in structure to the primary substrate with which the enzyme normally reacts. The primary substrate is the organism's primary source of energy for growth, movement, and maintenance. For instance, methanotrophic bacteria have an enzyme, MMO, that oxidizes methane to methanol in the presence of oxygen. In this case methane

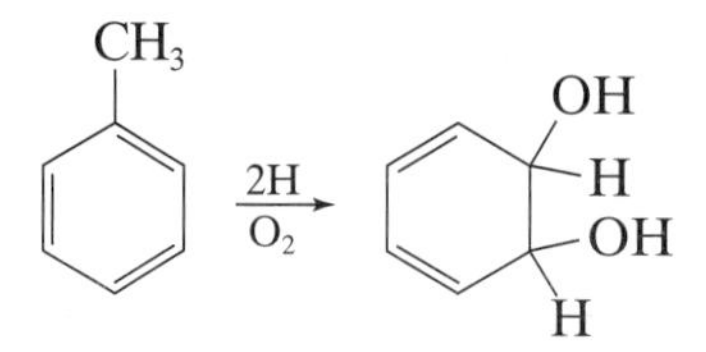

Figure 3-6. Dioxygenase enzymes assist the incorporation of molecular oxygen into the aromatic ring of toluene. This makes the ring susceptible to ring cleavage and useful as an energy source to bacteria.

is the primary substrate that provides the energy required by the methanotrophs for cell growth, maintenance, and mobility. However, MMO also will transform the organic pollutant TCE to something less hazardous, not because the microorganism has any desire to utilize TCE as a source of energy, but only because the enzyme, MMO, has broad specificity. This means that MMO cannot distinguish between the somewhat similar chemical structures of methane and TCE. Consequently, TCE is cometabolized to a less hazardous compound.

Alkaline phosphatases are used by activated sludge organisms and algae to catalyze the hydrolysis of unavailable phosphorus compounds (e.g., polyphosphates like $H_4P_2O_7$) to the biologically available orthophosphates. Production of these enzymes occurs when organisms are grown in a phosphorus-limited environment and need to utilize all available types of phosphorus in their environment.

Many catalysts are surface-active agents. That is, the reaction takes place on a surface that has been coated or "impregnated" with a catalyst. For example, catalytic converters in automobiles use trace metals bound on the surface of the converter to enhance the combustion of unburned hydrocarbons at temperatures lower than the engine temperature. Similarly, in order to meet air-emission limits set for particulates in the United States after July 1, 1990, many woodstove manufacturers installed catalytic converters in their stoves. These stoves typically contain a ceramic base in the shape of a honeycomb. The honeycomb's large surface area is coated with a noble metal such as platinum or palladium. When unburnt smoke (which contains small unburnt particles and gases) is passed through the honeycomb, the catalyst allows the secondary combustion of the smoke particles to occur at temperatures of 260–315°C versus the 540–650°C range required in the absence of the catalyst. Thus, this secondary combustion allows newer woodstoves to emit fewer particles as well as to get more heat out of a load of wood. Woodstove manufacturers recommend that an individual not burn garbage or colored newsprint in their woodstove because garbage and newsprint may contain trace elements that can "poison" or "mask" the catalyst, resulting in its inactivation. This is the same reason leaded gasoline should not be used in a car equipped with a catalytic converter.

Finally, Figure 3-7 depicts the situation where a chemical intermediate accelerates a chemical reaction but is not a catalyst.

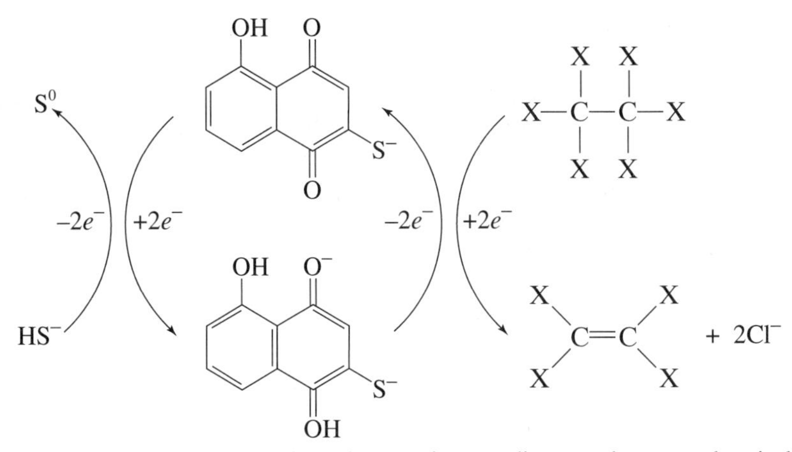

Figure 3-7. Chemical reaction where an intermediate accelerates a chemical reaction but is not a catalyst. The chemical reaction of a halogenated alkane, hexachloroethane, with hydrogen sulfide and an electron-transfer mediator, juglone (1,5-hydroxy-4-naphthoquinone) (Perlinger et al., 1996). The juglone mimics quinone functional groups found in natural organic matter that can also carry out electron-transfer reactions from a bulk reductant (hydrogen sulfide, in this case) to a pollutant (hexachloroethane, in this case). Natural organic matter imparts the yellow and brown color found in many bogs and wastewaters. This reaction may be important in controlling the natural disappearance of some environmental pollutants in subsurface environments. Electron transfer reactions such as this are also very common in biotic systems. In the absence of juglone, the hydrogen sulfide reacts directly with the hexachloroethane, but at a slower rate, as was demonstrated in the Arrhenius plot provided in Figure 3-5. In this case, the juglone accelerates the reaction, but is not considered to be a catalyst because it is continually oxidized and re-reduced in mediating the transformation of hexachloroethane to tetrachloroethene. In addition, catalysts usually accelerate reactions by a factor of 10^7 or even more, much more than the juglone in this case.

3.3.1 CHEMICAL THERMODYNAMICS AND EQUILIBRIUM

As mentioned in the previous section, kinetics and equilibrium represent two very different approaches in environmental chemistry. The kinetic approach is appropriate when the reaction is slow relative to our time frame, or when we are interested in the rate of change of concentration. The equilibrium approach is useful whenever reactions are very fast, whenever we want to know in which direction a reaction will go, or whenever we want to know the final, stable conditions that will exist at equilibrium. In this chapter, the equilibrium approach is demonstrated for a variety of types of reactions. This approach is identical for all types of chemical reactions.

The equilibrium approach is based upon fundamental principles of thermodynamics. This section begins with a review of this theoretical background. The thermodynamic principles are used to develop a general approach toward solving equilibrium problems. This approach is then applied to five types of reactions (volatilization, air–water partitioning, adsorption and sorption, acid–base, precipitation/dissolution) in Section 3.4. These reactions are involved in the fate, transport, and treatment of many chemicals.

In chemistry, equilibrium has a very precise definition that is based on thermodynamics. Thermodynamics provides a well-defined set of rules that determine first whether a reaction can occur, and secondly, if the reaction will proceed in the forward or reverse direction. Therefore, equilibrium, as defined by thermodynamics, is a condition that can be predicted.

As the roots of the word imply (thermo equals heat; dynamo equals change), thermodynamic deals with conversions of energy from one form to another. Physical chemists have defined a quantity, the Gibbs free energy (G), as that energy that is available to do work. The formal definition of Gibbs free energy is

$$G = H - T \times S \tag{3-30}$$

Equation 3-30 shows that the Gibbs free energy is related to the system's enthalpy (H), entropy (S), and temperature (T). The energy of inter- and intramolecular bonds that bind various atoms and molecules together is designated by the enthalpy, while entropy refers to the "disorder" of the system.

One can logically ask, "How does disorder contribute to work?" The answer to this question may be more obvious if the question is asked in reverse. It clearly requires work to put things in order. There is work involved in packing clothes into a suitcase, packing molecules closer together (as in converting a gas to a liquid), and arranging molecules in an intricate pattern (e.g., the effect of dissolving them in water). Equation 3-30 indicates that the more disorder there is (i.e., the larger the value of S), the less energy there is available to do work. Again, this may be easier to understand in reverse. That is, the more disorganized something is, the more work it will require to bring it into order.

Another way of visualizing this statement is that there is more potential to do work in liquid water where each of the molecules occupies about 4×10^{-20} cm^3 (i.e., a low-entropy system) versus water vapor where each molecule occupies 3×10^{-17} cm^3. If liquid water molecules were allowed to expand 1,000-fold so that they occupied the volume of the water vapor, a large amount of work would be accomplished. This has far-reaching implications. Any chemical reaction that results in the conversion of a liquid or solid to a gas will result in a large increase in entropy. An increase in entropy is therefore equivalent to a decrease in the free energy of the system or an accomplishment of work.

The second law of thermodynamics states that perpetual motion machines are impossible because heat may not be converted to work and back to heat isothermally. This law may be restated as: all systems tend to lose "useful" energy and approach a state of minimum free energy, or an equilibrium state. Thus *a process will proceed "spontaneously" (i.e., without energy being put into the system from the outside) only if the process leads to a decrease in the free energy of the system (i.e., $\Delta G < 0$).*

In Figure 3-8, a process could proceed if it reduced the free energy from its value at point A in the direction of point C, but it could not proceed if it raised

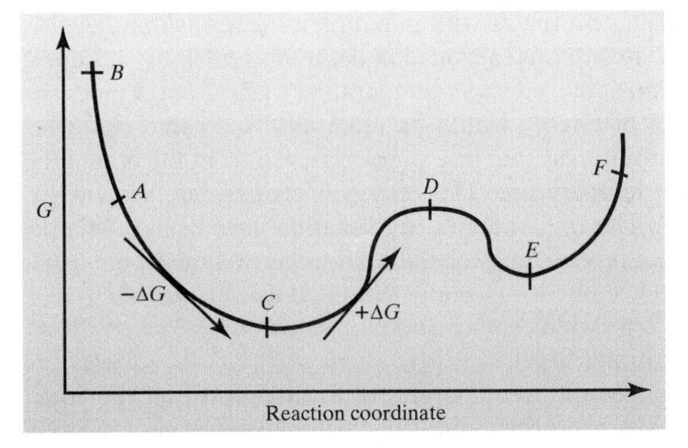

Figure 3-8. The change in free energy (G) during a reaction. If the change leads to a decrease in free energy (i.e., for the forward reaction, if the slope of a tangent to the curve is negative), then the reaction can proceed spontaneously. Points C and E represent possible equilibrium points because the slopes of tangents at these points would be zero.

the energy in the direction of point B. The process could proceed from A as far as point C, but it could not go further toward point D. A reaction could also proceed from point D toward point C or point E. This is because moving in either direction results in a decrease in free energy. Point E is called a *local equilibrium*. It is not the minimum-possible energy point of the system (point C is), but to leave point E requires an input of energy. Hence, if the free energy of a system under all conditions could be quantified, the changes that could occur spontaneously in that system could then be determined (i.e., any changes that would cause a decrease in the free energy).

3.3.1 Calculation of Change in Free Energy at Standard Conditions

Changes in the total free energy of a system cannot be measured. In fact, the free energy of a system cannot be measured without defining a reference state. In chemistry, the reference conditions are the most stable state of each element (i.e., elemental) at 1-atmosphere pressure and 25°C. Standard conditions also refers to a 1-molar concentration for all dissolved and precipitated reactants and 1-atmosphere concentration for all gaseous reactants. The free energy of each element under these conditions is defined to be zero.

Provided with this reference state, the free energy of formation at standard conditions (ΔG_f^0) of any compound can be measured by determining the energy required to combine the elements at 25°C and 1-atmosphere pressure to produce the given compound. For example, consider the reaction:

$$C_{(s)} + O_{2(g)} \rightarrow CO_{2(g)} \quad \textbf{(3-31)}$$

in which one mole of solid carbon reacts with one mole of gaseous dimolecular oxygen to produce one mole of gaseous carbon dioxide. If this reaction proceeded in a closed container maintained at a constant temperature of 25°C and a pressure of 1 atmosphere, 394.37 kJ of energy would be released. This value could be determined from the amount of heat that had to be removed from the container to maintain it at a constant temperature. This energy is termed the free energy of formation (termed ΔG_f^0). The free energies of formation have been tabulated for a vast number of compounds and are available in many standard chemistry handbooks (e.g., CRC, 1991). Values are usually tabulated in units of either kcal/mole or kJ/mole. Table 3-2 provides some common values of ΔG_f^0 as well as values for the heat of formation (ΔH_f^0).

If an individual had access to the free energies of formation for all the compounds that participate in any given reaction, the change in free energy (ΔG^0) could be determined for any reaction. If that free-energy change (ΔG^0) were negative, it could be concluded that the reaction leads to a decrease in the total free energy, and the reaction therefore can proceed spontaneously as written under standard conditions. If the free-energy change was positive, it could be concluded that the reaction cannot proceed as written, but could, in fact, proceed spontaneously in the opposite direction under standard conditions.

The change in free energy for a reaction at standard conditions, ΔG^0, can be determined as the sum of the free energies of formation of the products multiplied by their stoichiometric coefficients, minus the sum of the free energies of formation of the reactants multiplied by their stoichiometric coefficients:

$$\Delta G^0 = \Sigma n_i \Delta G^0_{fi(\text{products})} - \Sigma n_i \Delta G^0_{fi(\text{reactants})} \tag{3-32}$$

In Equation 3-32 n_i refers to the stoichiometric coefficient for each species, i, which is obtained from a balanced chemical reaction; ΔG^0_{fi} is obtained for individual species, i, from tabulated values as provided in Table 3-2. If ΔG^0 is positive, under standard conditions, the reaction could not proceed in the forward direction, but could proceed in the reverse direction. Likewise, if ΔG^0 is negative, under standard conditions, the reaction could proceed spontaneously in the direction the reaction was written. The change in heat of formation at standard conditions (ΔH^0) could be calculated in a similar manner. Note that predicting whether a reaction can proceed spontaneously says nothing about the rate at which the reaction will proceed.

EXAMPLE 3.8. CALCULATION OF THE CHANGE IN FREE ENERGY AT STANDARD CONDITIONS

Acid rain is the result of burning fossil fuels. The sulfur contained in fossil fuels is oxidized to sulfur dioxide, SO_2, during combustion. In the atmosphere, SO_2 dissolves in water droplets where it forms sulfurous acid, H_2SO_3. The sulfurous acid can then be oxidized by other substances in the cloud water (substances such as peroxide, iron, copper) to form sulfuric acid, H_2SO_4. Could the following two

Table 3-2. Values of the Change in Free Energy of Formation (ΔG_f^0) and Change Heat of Formation (ΔH_f^0) for Some Common Chemical Species at 25°C*

Species	Change in Free Energy of Formation (ΔG_f^0) (kJ/mole)	Change in Heat of Formation (ΔH_f^0) (kJ/mole)
Al^{3+} (aq)	−489	−531
$Al(OH)_3$ (s)	−1,140	−1,270
CO_2 (g)	−394	−393
H_2CO_3* (aq)	−623	−670
HCO_3^- (aq)	−587	−692
CO_3^{2-} (aq)	−528	−677
Ca^{2+} (aq)	−554	−543
CaF_2 (s)	−1,203	−1,107
$Ca(OH)_2$ (s)	−898	−986
Cd^{2+} (aq)	−77.6	—
$Cd(OH)_2$ (s)	−470	−561
Cr^{3+} (aq)	−216	−256
CrO_4^{2-} (aq)	−728	−881
F^- (aq)	−279	−333
Fe^{2+} (aq)	−78.9	−89.1
$Fe(OH)_3$ (s)	−699	−822
Fe^{3+} (aq)	−4.60	−48.5
H^+ (aq)	0	0
HgO (s)	−58.5	−90.80
Hg (g)	+31.8	+61.2
H_2S (g)	−33.6	−20.6
H_2S (aq)	−27.9	−39.8
HS^- (aq)	+12.05	−17.6
S^{2-} (aq)	85.8	33.0
Mn^{2+} (aq)	−228	−221
$Mn(OH)_2$ (s)	−616	—
MnO_2 (s)	−465	−502
$MnOOH$ (s)	−558	—
NH_3 (g)	−16.5	−46.1
NH_3 (aq)	−26.6	−80.3
NH_4^+ (aq)	−79.4	−132
NO_3^- (aq)	−111	−207
O_2 (g)	0	0
O_2 (aq)	16.3	−11.7
OH radical (g)	34.2	38.9
OH^- (aq)	−157	−230
H_2O (l)	−237	−286
H_2O (g)	−229	−242
H_3PO_4 (aq)	−1,140	1,290
$H_2PO_4^-$ (aq)	1,130	1,300
HPO_4^{2-} (aq)	−1,090	−1,290
PO_4^{3-} (aq)	−1,020	−1,280
SO_4^{2-} (aq)	−745	−909

Note: The parenthetical (s) refers to a species in a solid or precipitated state; the parenthetical (g) refers to a species in a gaseous state; the parenthetical (aq) refers to a species dissolved in an aqueous or liquid state; and the parenthetical (l) refers to a species existing as a liquid.

*In the literature, these values are tabulated as kJ/mole or kcal/mole. (Values were first compiled from Stumm and Morgan, 1996. If not available, CRC (1991) was consulted. All values have been reduced to three significant figures.)

reactions, the combustion of FeS_2 (pyrite) in the coal to yield SO_2, and the oxidation of SO_2 by Fe^{3+} in the clouds to yield H_2SO_4 proceed spontaneously at standard conditions?

$$2FeS_{2(s)} + 5.5O_{2(g)} \longleftrightarrow Fe_2O_{3(s)} + 4SO_{2(g)}$$

$$H_2SO_{3(aq)} + 2Fe^{3+}{}_{(aq)} + H_2O_{(l)} \longleftrightarrow H_2SO_{4(aq)} + 2Fe^{2+}{}_{(aq)} + 2H^+$$

SOLUTION

The relevant free energies of formation at standard conditions are tabulated below together with the stoichiometric coefficients from the above two equations.

Compound	G_f^0 **(kJ/mole)**	n **(no. moles)**
$FeS_{2(s)}$	160	2
$O_{2(g)}$	0	5.5
$Fe_2O_{3(s)}$	−743	1
$SO_{2(g)}$	−300	4
$H_2SO_{3(aq)}$	−534	1
$Fe^{3+}{}_{(aq)}$	−4.60	2
$H_2O_{(l)}$	−237	1
$H_2SO_{4(aq)}$	−745	1
$Fe^{2+}{}_{(aq)}$	−78.9	2
$H^+{}_{(aq)}$	0	2

Use Equation 3-32 to determine the change in free energy at standard conditions for both reactions. For pyrite combustion:

$$\Delta G^0 = [\Delta G^0_{f\mathrm{Fe_2O_3}} + 4\Delta G^0_{f\mathrm{SO_2}}] - [2\Delta G^0_{f\mathrm{FeS_2}} + 5.5\Delta G^0_{f\mathrm{O_2}}]$$

$$\Delta G^0 = [-743 + 4(-300)] - [2(-160) + 5.5(0)] = -1{,}623 \text{ kJ}$$

Similarly, for sulfuric acid production:

$$\Delta G^0 = [\Delta G^0_{f\mathrm{H_2SO_4}} + 2\Delta G^0_{f\mathrm{Fe^{2+}}} + 2\Delta G^0_{f\mathrm{H^+}}] - [\Delta G^0_{f\mathrm{H_2SO_3}} + 2\Delta G^0_{f\mathrm{Fe^{3+}}} + \Delta G^0_{f\mathrm{H_2O}}]$$

$$\Delta G^0 = [-745 + 2(-78.9) + 2(0)] - [(-534) + 2(-4.60) + (-237)] = -123 \text{ kJ}$$

Both reactions result in a decrease in free energy (i.e., ΔG^0 is negative). Therefore, both reactions can proceed spontaneously as written at standard conditions.

3.3.2 Calculating Changes in Free Energy under Nonstandard "Ambient" Conditions

Section 3.3.1 considered only the free-energy change for reactions occurring under standard conditions. Remember that standard conditions refers to 1-atmosphere pressure, 25°C, a 1-molar concentration for all aqueous and precipitated species, and a 1-atmosphere concentration for all gaseous species. The free-energy change under standard conditions was denoted by the symbol ΔG^0, where the superscript 0 denoted the standard conditions. Obviously, standard conditions do not prevail in most "ambient" or real-world situations. In any location there are many days when the pressure is not exactly 1 atmosphere and the temperature is not 25°C (77°F). In addition, it is extremely rare when the reactants and products for any process are present in equal concentrations, and those concentrations equal one. Hence, for the thermodynamic rules discussed earlier to be useful, they must be applicable under realistic, nonstandard, "real-world" conditions.

To adjust the calculated value of ΔG^0 to ambient conditions a simple correction factor must be applied. ΔG *(without the superscript* 0) *is the free-energy change under ambient conditions* (i.e., the prevailing environmental conditions). The value of ΔG is calculated according to the relationship:

$$\Delta G = \Delta G^0 + RT \ln(Q) \quad \textbf{(3-33)}$$

where R is the gas constant, T is the ambient temperature in K, and Q is the "reaction quotient." The reaction quotient is defined as the product of the activity (i.e., apparent concentration of the reaction products) raised to the power of their stoichiometric coefficients divided by the product of the activity (or concentration) of the reactants raised to the power of their stoichiometric coefficients. Thus for the generalized reaction

$$aA + bB \longleftrightarrow cC + dD \quad \textbf{(3-34)}$$

in which a moles of compound A react with b moles of compound B to form c moles of compound C and d moles of compound D, the reaction quotient (Q) is given by

$$Q = \frac{[C]^c[D]^d}{[A]^a[B]^b} \quad \textbf{(3-35)}$$

Four rules are used to determine what value to use for the concentration (i.e., activity), $[i]$.

Rule 1. Solvents (e.g., water): $[i]$ is equal to the mole fraction of the solvent. Thus, in aqueous solutions the mole fraction of water can be assumed to equal one. Thus, $[H_2O]$ always equals 1.

Rule 2. Pure solids in equilibrium with a solution (e.g., $CaCO_{3(s)}$, $Fe(OH)_{3(s)}$): $[i]$ always equals 1.

Rule 3. Gases in equilibrium with a solution (e.g., $CO_{2(g)}$, $O_{2(g)}$): $[i]$ equals the partial pressure of the gas (units of atmospheres, atm).

Rule 4. Compounds dissolved in water: $[i]$ is always reported in units of moles/liter (not mg/L or ppm!).

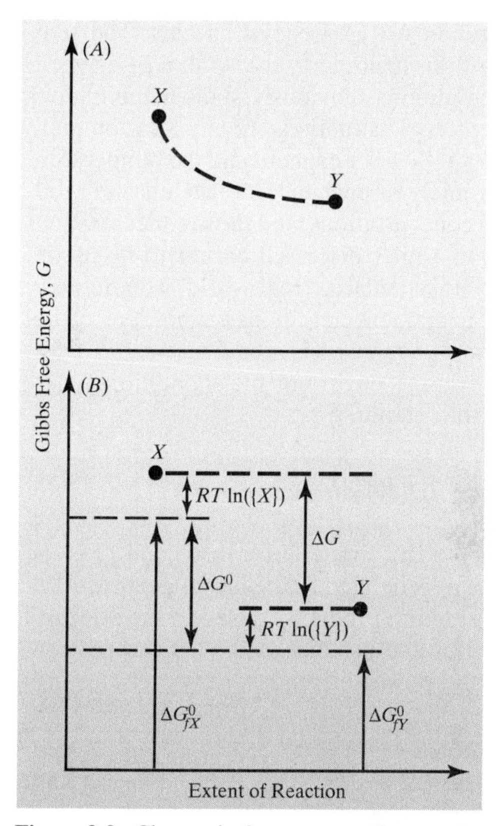

Figure 3-9. Change in free energy of a reaction. (*a*) Only if a decrease in free energy occurs can a reaction occur spontaneously. A reaction going from position X to position Y is thermodynamically possible. (*b*) The change in free energy (ΔG) consists of two components: $\Delta G = \Delta G^0 + RT\ln(Q)$ (Equation 3-33). The value of ΔG^0 can be calculated as the difference in free energies of formation (ΔG_f^0) of products minus free energies of formation of reactants (Equation 3-32). $RT\ln(Q)$ is equal to $RT\ln(\{Y\}/\{X\})$ or $RT\ln(\{Y\}) - RT\ln(\{X\})$.

Figure 3-9 summarizes how ΔG and ΔG^0 are related by $\ln Q$. Example 3.9 illustrates how to perform the adjustment from standard to ambient conditions.

EXAMPLE 3.9. ADJUSTING THE CHANGE IN FREE ENERGY FROM STANDARD TO AMBIENT CONDITIONS

Determine whether the oxidation of sulfurous acid to sulfuric acid could proceed in clouds under the following conditions. Assume that in clouds on an autumn day $T = 25°C$, $[H_2SO_3] = 10^{-6}$ M, $[H_2SO_4] = 10^{-3.7}$ M, $[Fe^{3+}] = 10^{-8}$ M, $[Fe^{2+}] = 10^{-10}$ M, and $[H^+] = 10^{-4}$ M.

$$H_2SO_{3(aq)} + 2Fe^{3+}{}_{(aq)} + H_2O_{(l)} \longleftrightarrow H_2SO_{4(aq)} + 2Fe^{2+}{}_{(aq)} + 2H^+$$

SOLUTION

Assume ideal conditions, so concentration equals activity. Example 3.8 provided a ΔG^0 equal to -123 kJ for the reaction of interest. Therefore,

$$\Delta G = \Delta G^0 + R \times T \times \ln \frac{[H_2SO_4][Fe^{2+}]^2[H^+]^2}{[H_2SO_3][Fe^{3+}]^2[H_2O]}$$

$$\Delta G = \frac{-123 \text{ kJ}}{\text{mole}} + \frac{8.314 \text{ J}}{\text{mole} - K} \times \frac{10^{-3} \text{ kJ}}{\text{J}} \times 298K \times \ln \frac{(10^{-3.7})(10^{-10})^2(10^{-4})^2}{(10^{-6})(10^{-8})^2(1)}$$

$$= \frac{-178 \text{ kJ}}{\text{mole}}$$

ΔG at ambient conditions is negative; therefore, the oxidation of sulfurous acid to sulfuric acid could proceed under the stated conditions.

Equation 3-33 corrects ΔG^0 for concentrations that are not equal to "standard conditions"; however, it does not correct it for temperature. Because most tabulated values for ΔG^0 are at 25°C, if the ambient temperature is not equal to 25°C, an additional correction is necessary to obtain a value for ΔG^0 at the new temperature. For the curious, the derivation of this temperature correction is presented in the Advanced Topic box below. An example that demonstrates the use of temperature correction of equilibrium constants is postponed until Section 3.3.4 because the material covered in that section allows a simplification to be made.

ADVANCED TOPIC: DERIVATION OF THE TEMPERATURE CORRECTION FOR ΔG^0

This temperature correction is based on the definition of Gibbs free energy:

$$G = H - T \times S \qquad \textbf{(a)}$$

From this definition, it is clear that the change in free energy per unit change in temperature (i.e., $\partial G/\partial T$) is equal to $-S$:

$$\frac{\partial G}{\partial T} = -S \qquad \textbf{(b)}$$

The derivative of this expression with respect to temperature again yields that $(\partial \Delta G^0/\partial T)$ equals $-\Delta S^0$. (Note that by measuring the change in standard free energy with change in temperature, one could determine the magnitude of the standard change in entropy.) The important implication of these relationships is that ΔG^0 is dependent on temperature. Further transformation of Equation b yields a useful generalization that allows one to correct ΔG^0 for changes in temperature. If

$$\frac{\partial \frac{\Delta G^0}{T}}{\partial T} = -\frac{1}{T^2} G^0 + \frac{1}{T}\left(\frac{\partial \Delta G^0}{dT}\right) = -\frac{1}{T^2}(\Delta G^0 + T\Delta S^0) \qquad \textbf{(c)}$$

and ΔH^0 can be substituted for $(\Delta G^0 + T\Delta S^0)$ to obtain

$$\frac{\partial \frac{\Delta G^0}{T}}{\partial T} = \frac{-\Delta H^0}{T^2} \qquad \textbf{(d)}$$

A simple substitution into Equation d yields the van't Hoff equation that is discussed later (Equation 3-41). Integration of Equation d yields the generalized correction in ΔG^0 for changes in temperature:

$$\int_{T_1}^{T_2} d\left(\frac{\Delta G}{T}\right) = \int_{T_1}^{T_2} \frac{\Delta H^0}{T^2}\, dT \rightarrow \frac{\Delta G_2^0}{T_2} - \frac{\Delta G_1^0}{T_1} = \Delta H^0\left(\frac{1}{T_1} - \frac{1}{T_2}\right) \qquad \textbf{(e)}$$

Here ΔG_1^0 and ΔG_2^0 are the values of ΔG^0 at temperatures T_1 and T_2. In practice, one must correct ΔG^0 for the change in temperature, and then use the corrected value of ΔG^0 in Equation 3.33.

3.3.3 Free Energy and Its Relationship to the Equilibrium Constant

In the previous sections equilibrium was defined as the state (or position) with the minimum possible free energy. This occurred at point C in Figure 3-8. The value of the slope at the point of equilibrium (point C) is zero. In other words, *the change in free energy is zero at equilibrium.* If the change in free energy (ΔG) is equal to zero at equilibrium, Equation 3-33 can be rewritten as follows in order to determine a reaction's equilibrium constant, K:

$$\Delta G = 0 = \Delta G^0 + RT \ln K \qquad \textbf{(3-36)}$$

or

$$\Delta G^0 = -RT \ln K \qquad \textbf{(3-37)}$$

At *equilibrium*, $Q = K$, so K was substituted for the reactant quotient, Q, in the two previous equations. According to Equation 3-37, there is a special relationship between the concentrations of all of the species at equilibrium and the value of ΔG^0. This relationship is

$$K = \exp \frac{-\Delta G^0}{RT} \tag{3-38}$$

At 25°C all of the terms on the right-hand side of Equation 3-38 are constant, and hence the value of the quotient Q at equilibrium (i.e., K) is a constant. Equation 3-38 is useful because it can be used to solve for unknown equilibrium constants as long as the reaction temperature, stoichiometry, and values for ΔG_f^0 are known. The equilibrium expression, or equilibrium quotient, is called the equilibrium constant and is usually denoted by the symbol K.

Do not confuse the equilibrium constant, K, with the reaction rate constant, k. Note that K is constant for a specific reaction (as long as temperature is constant) and its units are dependent on reaction stoichiometry. Later on in this chapter, equilibrium constants will be defined for reactions that describe acid–base chemistry (K_a and K_b); precipitation/dissolution reactions (K_{sp}); and air/water exchange (Henry's constant). The equilibrium constant for the reaction provided in Equation 3-34 is given by the reaction quotient, Q:

$$Q = \frac{[C]^c[D]^d}{[A]^a[B]^b} = K \tag{3-39}$$

The equilibrium constant is useful because it provides the ratio of the concentration (or activity) of individual reactants and products for any reaction at equilibrium.

EXAMPLE 3.10. USE OF THE EQUILIBRIUM CONSTANT

Reduced ferrous iron (Fe^{2+}) can be oxidized to ferric iron (Fe^{3+}) in the presence of oxygen according to the following reaction.

$$4Fe^{2+}{}_{(aq)} + O_{2(g)} + 10\ H_2O_{(l)} \longleftrightarrow 4Fe(OH)_{3(s)} + 8H^+{}_{(aq)}$$

This reaction produces the reddish-brown stain seen in many sinks and bathtubs where the water is aerated as it exits the faucet. This chemical reaction is also used during water treatment to remove iron from water. During water treatment, after the $Fe(OH)_3$ precipitate is formed from aerating the water, the precipitate is then removed by either settling tanks or filtration. The equilibrium constant for this reaction at 25°C is $10^{17.37}$. What equilibrium concentration of reduced iron is present in a system of water at 25°C and pH = 7 ($[H^+] = 10^{-7}$ M) that has air bubbled through it? Assume ideal conditions (i.e., activity and concentration are the same).

SOLUTION

Set up the equilibrium expression

$$10^{17.37} = \frac{[Fe(OH)_{3(s)}]^4[H^+]^8}{[Fe^{2+}]^4[O_{2(gas)}][H_2O]^{10}}$$

Remember that $[H_2O_{(l)}] = [Fe(OH)_{3(s)}] = 1$ and the reaction is written using gaseous oxygen, so use the partial pressure of oxygen in air, 0.21 atm, not the aqueous phase oxygen concentration. If the reaction was written with $O_{2(aq)}$ you would use the aqueous molar O_2 concentration:

$$10^{17.37} = \frac{[10^{-7}]^8}{[Fe^{2+}]^4[0.21 \text{ atm}]}$$

Solve for $[Fe^{2+}] = 6.7 \times 10^{-19}$ M.

Note that there is very, very little dissolved iron in this solution. Most of it precipitates out in the Fe^{3+} form. This "low solubility" of ferric iron explains why there is little dissolved Fe^{3+} available for plant and animal growth in most natural systems that are near neutral pH. In fact, some microorganisms produce a "chelating" agent that complexes iron in order to enhance the amount of apparent Fe^{3+} available for microbial use.

EXAMPLE 3.11. USE OF EQUILIBRIUM CONSTANT

From the reaction that produces acid rain in Example 3.8, determine the ratio of sulfurous acid to sulfuric acid at equilibrium using the measured concentrations of H^+, Fe^{2+}, and Fe^{3+} provided in Example 3.9.

SOLUTION

In Example 3.8 ΔG^0 was determined to equal −123 kJ for the reaction of interest. Hence the equilibrium constant is determined from Equation 3-38 because ΔG equals zero at equilibrium:

$$K = \exp\frac{(-\Delta G^0)}{RT} = \exp\frac{(123 \text{ kJ})}{} \times \frac{(\text{mole} - K)}{(8.314 \text{ J})} \times \frac{(10^3 \text{ J})}{(\text{kJ})} \times \frac{1}{298K} = 10^{21.6}$$

Remember that K is merely the reaction quotient, Q, at equilibrium. Hence

$$10^{21.6} = \frac{[H_2SO_4][Fe^{2+}]^2[H^+]^2}{[H_2SO_3[[Fe^{3+}]^2[H_2O]}$$

From this expression the ratio that the problem requested can be found is

$$\frac{[H_2SO_3]}{[H_2SO_4]} = \frac{[Fe^{2+}]^2[H^+]^2}{K[Fe^{3+}]^2[H_2O]}$$

Filling in the appropriate values the ratio is

$$\frac{[H_2SO_3]}{[H_2SO_4]} = \frac{[10^{-10}]^2[10^{-4}]^2}{10^{21.6}[10^{-8}]^2[1]} = 10^{-33.6}$$

Clearly this reaction proceeds extremely far to the right such that virtually all of the sulfur is oxidized to sulfuric acid under the assumed conditions inside of clouds.

3.3.4 Effect of Temperature on the Equilibrium Constant

Most tabulated equilibrium constants are recorded at 25°C. The van't Hoff relationship is used to convert equilibrium constants to temperatures other than those for which the tabulated values are provided. Van't Hoff discovered that the equilibrium constant (K) varied with absolute temperature and the enthalpy of a reaction (ΔH^0). Van't Hoff proposed the following expression to describe this:

$$\frac{d \ln K}{dT} = \frac{\Delta H^0}{RT^2} \tag{3-40}$$

Here ΔH^0 is found from the heat of formation (ΔH_f^0) for the reaction of interest determined at standard conditions (see Table 3-2 for some typical values). Most temperatures encountered in environmental problems are relatively small. Therefore, the temperature differences are not that large. If ΔH^0 is assumed to not change over the temperature range being investigated, Equation 3-40 can be integrated to yield:

$$\ln\left[\frac{K_2}{K_1}\right] = \frac{\Delta H^0}{R} \times \left(\frac{1}{T_1} - \frac{1}{T_2}\right) \tag{3-41}$$

Equation 3-41 can be used to calculate an equilibrium constant for any temperature (i.e., temperature two, T_2) if the equilibrium constant is known at another absolute temperature (temperature one, T_1, which is usually 20 or 25°C). Remember, though that, as stated previously, it is assumed that ΔH^0 (usually measured at 25°C) remains constant over the temperature range of interest. Later in this chapter, Example 3.25 demonstrates the use of Equation 3-41. Similarly, if ΔG^0 was known at a specific temperature, the equilibrium constant at that temperature could be calculated using Equation 3-38.

3.4 EQUILIBRIUM PROCESSES

In environmental systems, some important equilibrium processes include those between pure compounds and air (volatilization); chemicals dissolved in water and air (Henry's law); acids and bases; dissolved and precipitated chemicals; and

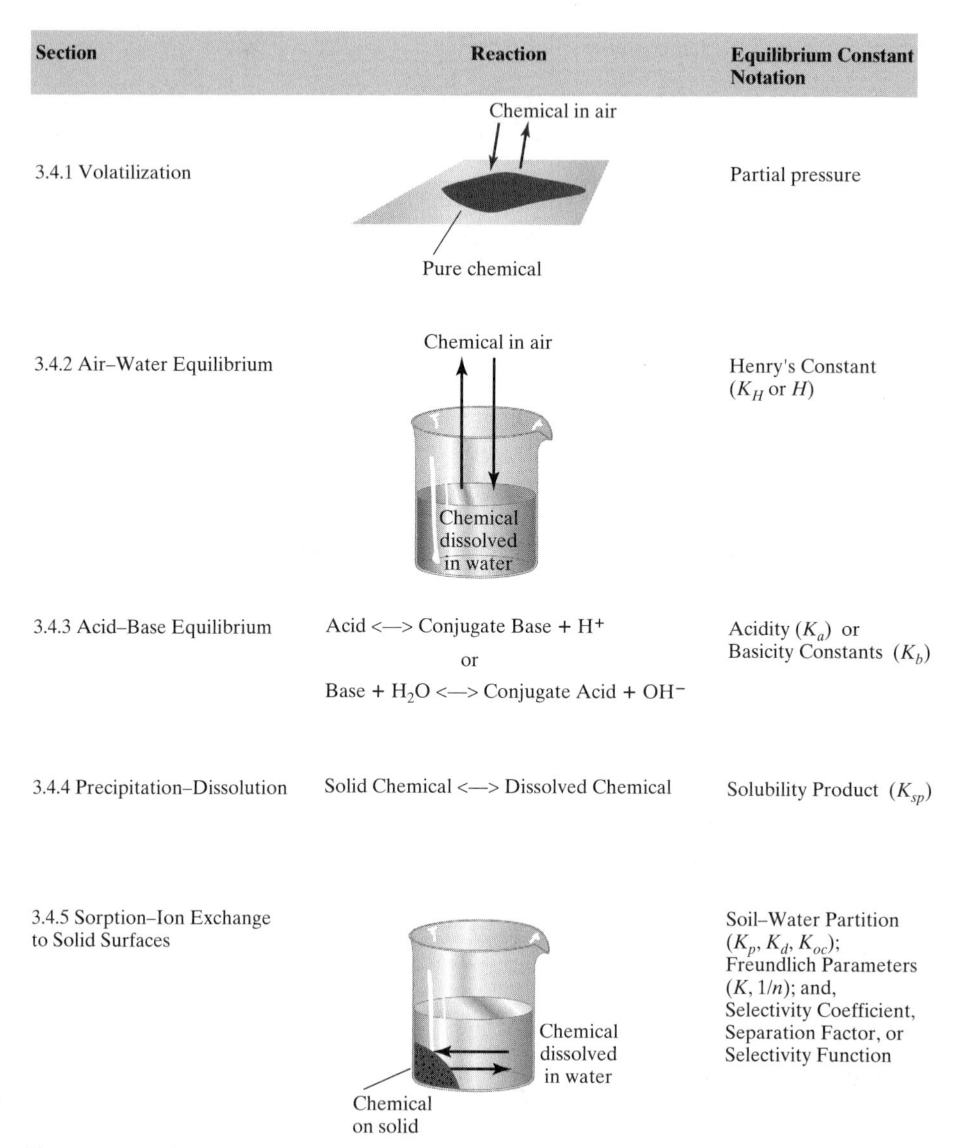

Figure 3-10. The equilibrium processes described in Section 3.4.

chemicals dissolved in the water and adsorbed on a solid (adsorption and ion exchange). Figure 3-10 outlines these equilibrium processes. This section reviews each of these processes in detail.

3.4.1 Volatilization

One feature of environmental problems is that they seldom are confined to just one medium. For example, a lot of the mercury discharged into the environment is first emitted as an air pollutant, but its most damaging effects occur in lakes after it moves through the atmosphere, is deposited into a lake, and then undergoes a biological transformation process called *methylation*. This process resulted in approximately 1,300 fishing advisories in lakes throughout the United States in 1995.

Similarly, as shown in Figure 3-11, pesticides such as DDT, which had its use banned in the United States, is still sprayed on fields in Central America where it then volatilizes. Airborne DDT can then be transported to the United States; as a result DDT is found in remote lakes and fish in the northernmost reaches of the lower 48 states, including Isle Royale National Park (MI), located in Lake Superior. Also, lead that was initially used in paint and gasoline is now found primarily in the soil where it still can pose a health hazard to children who ingest the contaminated soil. Finally, Examples 3.8 and 3.9 showed how acid rain originates with the burning of fossil fuels, with the subsequent emission of air pollutants. In all these examples, pollutants are emitted to the atmosphere and cause damage after they are deposited on land or surface waters.

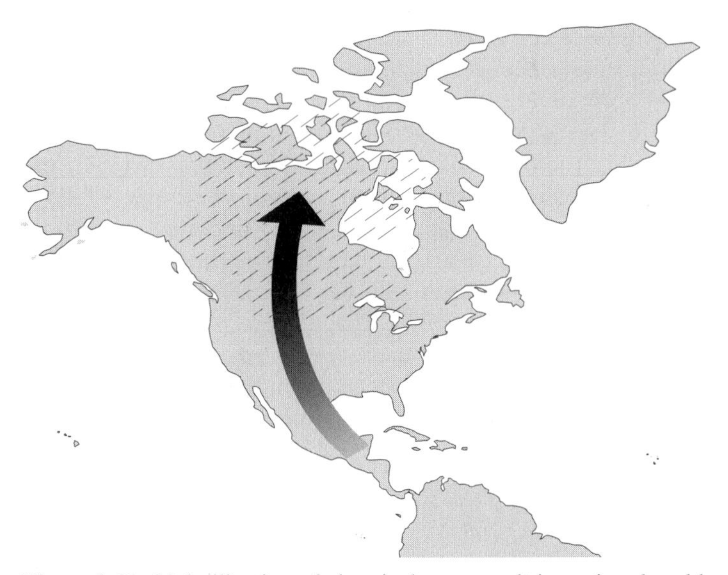

Figure 3-11. Volatilization of chemicals can result in regional and long-range transport of the chemicals to places far away where adverse environmental effects can be detected. Persistent organic pollutants (POPs) become concentrated in the food chain where they can cause toxic effects on animal reproduction, development, and immunological function. The U.S. State Department has termed POPs "one of the great environmental challenges the world faces." POPs include polychlorinated biphenyls (PCBs), polychlorinated dibenzo-*p*-dioxins and furans, and pesticides such as DDT, toxaphene, chlordane, and heptachlor. Though banned for use in many developed countries, they are still manufactured for export and remain widely used and unregulated in developing countries. These chemicals volatilize more easily in the warm surface temperatures found in the southern United States and subtropical and tropical regions, then condense and deposit in high latitudes where temperatures are cooler (Wania and MacKay, 1996). This process of exporting toxic chemicals to other countries that then return by atmospheric transport has been termed "the circle of poison."

A key step in the transfer of pollutants between different environmental media is volatilization. All liquids and solids exist in equilibrium with a gas or vapor phase. *Volatilization (synonymous with evaporation) is the transformation of a compound from its liquid state to its gaseous state.* Sublimation is the word used for transformation from the solid state. The reverse reaction is termed condensation. Everyone has firsthand experience with this phenomenon with water. The water vapor in the atmosphere (i.e., the humidity) is a function of temperature. Modern refrigerators prevent frost buildup by maintaining a low humidity inside the refrigerator; any ice that forms is sublimed, or is vaporized. Similarly, the amount of snow on the ground decreases in periods between snowfalls partially due to the sublimation or volatilization of the snow. Many organic pollutants volatilize much more readily than water. The fumes from gasoline, paint thinners, waxes, and glue attest to the volatility of organic chemicals that are contained in these commonly used products.

The equilibrium between a gas and liquid or solid phase is determined by the saturated vapor pressure of a compound. *Saturated vapor pressure is defined as that partial pressure of the gas phase of a substance that exists in equilibrium with the liquid phase of the substance at a given temperature.* It is important to realize that in this case the equilibrium of concern is between a liquid chemical or solid chemical and the atmosphere, and not the equilibrium between a chemical that was first dissolved in water (the latter process is discussed in Section 3.4.2). The more volatile a compound is, the higher is its saturated vapor pressure. For example, the saturated vapor pressure of the solvent tetrachloroethene (PCE) is 0.025 atm at 25°C, while the saturated vapor pressure of lindane, a pesticide, is 10^{-6} atm at the same temperature. Clearly, lindane is much less volatile than is PCE. For the sake of comparison, water at 25°C has a slightly higher saturated vapor pressure (0.031 atm) than PCE. In other words, there would be about as much PCE in a closed room's atmosphere as there is water if containers or spills of PCE were left exposed to the air in the presence of containers of water.

The equilibrium between gas and liquid phases can be expressed in the usual form of a chemical reaction with an equilibrium constant:

$$H_2O_{(l)} \longleftrightarrow H_2O_{(g)} \quad \textbf{(3-42)}$$

Equation 3-42 indicates that liquid water is in equilibrium with gaseous water (i.e., water vapor). The equilibrium constant (called the *saturated vapor partial pressure*) for this reaction is

$$K = \frac{[H_2O_{(g)}]}{[H_2O_{(l)}]} = P_{H_2O} \quad \textbf{(3-43)}$$

where P_{H_2O} is the partial pressure of water. Because the concentration of a pure liquid is defined as 1 (remember Rule 1 in Section 3.3.2), the equilibrium constant is simply equal to the concentration in the vapor phase (called the *saturated vapor pressure*). One way of expressing gas-phase concentrations is as partial pressures, and hence the equilibrium constant for volatilization is often expressed in units of atmospheres.

If a mixture of miscible (i.e., mutually soluble) liquids, rather than a pure liquid was present, the denominator in Equation 3-43 would be the concentration of the individual liquid (A) in mole fractions, X_A:

$$K = \frac{[A_{(g)}]}{[A_{(l)}]} = \frac{P_A}{X_A} \qquad \textbf{(3-44)}$$

Equation 3-44 is known as Raoult's law. The equation constant, K, equals the saturated vapor pressure. Raoult's law is useful whenever a mixture of chemicals (e.g., gasoline, diesel fuel, kerosene) is spilled.

The vapor pressure for all compounds increases with temperature, and at the boiling point of the compound the vapor pressure equals atmospheric pressure. There are many consequences of this statement. First, atmospheric concentrations of volatile substances can be expected to be higher in summer than in winter, in the day versus in the night, and in warmer (southern) locations compared to northern locations. Second, for any group of liquid chemicals exposed to the air, the equilibrium gas-phase concentrations will decrease in order of increasing boiling points.

EXAMPLE 3.12. USE OF VAPOR PRESSURE

The chemical 1,4-dichlorobenzene (1,4-DCB) is used in an enclosed area. At 20°C (68°F) the saturated vapor pressure of DCB is 5.3×10^{-4} atm. What would be the concentration in the air of the enclosed area (units of g/m^3) at 20°C? The molecular weight of 1,4-DCB is 147 g/mole.

SOLUTION

Rearrange the Ideal Gas Law ($PV = nRT$) to solve for the concentration of DCB in the air:

$$\frac{n}{V} = \frac{P}{RT} = \frac{5.3 \times 10^{-4}\ \text{atm}}{\dfrac{0.08205\ \text{L-atm}}{\text{mole} - \text{K}}(293\ \text{K})}$$

$$= 2.2 \times 10^{-5}\,\frac{\text{moles}}{\text{L}} \times \frac{1{,}000\ \text{L}}{\text{m}^3} \times \frac{147\ \text{g}}{\text{mole}} = \frac{3.2\ \text{g}}{\text{m}^3}$$

EXAMPLE 3.13. CALCULATION OF AIR CONCENTRATION IN A CONFINED AREA

On a Friday afternoon, a worker spills one liter of tetrachloroethene (PCE) on a laboratory floor. The worker immediately closes all the windows and doors and turns off the ventilation in order to not contaminate the rest of the building. The worker notifies the appropriate safety authority, but it is Monday morning before the safety official stops by with a crew to clean up the laboratory. Should the cleanup crew bring a mop or an air pump to clean up the room? The size of the laboratory is 30 × 40 × 10 ft (volume of 340 m^3) and the temperature in the room is 25°C. The vapor pressure of PCE is 0.025 atm, the liquid density of PCE at 25°C is 1.62 g/cm^3, and the molecular weight of PCE is 166 g/mole.

SOLUTION

PCE is a volatile chemical. The problem asks how much of the one liter of spilled PCE remained on the floor versus how much volatilized into the air. If any PCE remained on the floor, the partial pressure of PCE in the air would be 0.025 atm. The Ideal Gas Law can be used to solve for the number of moles present in the air:

$$n = \frac{PV}{RT} = \frac{(0.025\ \text{atm})(340\ \text{m}^3)\left(\dfrac{1{,}000\ \text{L}}{\text{m}^3}\right)}{\dfrac{0.08205\ \text{L} - \text{atm}}{\text{mol} - K}(298\ K)} = 348\ \text{moles}$$

The density of PCE can be used to determine that the one-liter spill weighs 1,620 g. Using the molecular weight of PCE, the one-liter spill would contain 9.8 moles of PCE. This is much less than the amount that could potentially volatilize into the air in the room, assuming equilibrium has been attained. Thus it can be concluded that no PCE would remain on the floor and it would be entirely in the air. The cleanup crew should arrive at work equipped with air pumps and filters.

3.4.2 Air–Water Equilibrium

Organic chemicals discharged into lakes and streams can be transferred from the water into the atmosphere. For example, paper mills discharged PCBs and dioxins for many years into the Fox River near Green Bay, Wisconsin. Much of the PCBs and dioxins sorbed onto the sediments and settled to the bottom of the river or were carried downstream. Today, the majority of discharges of these chemicals have ceased, but now the pollutants are desorbing from the sediments back into the water column, and then partitioning from the water into the air. Therefore, the river and bay are a potential source of air pollution because of this transfer of pollutants from sediment to water to the atmosphere.

The partitioning of compounds between air and water has other important applications. For example, the chemical composition of natural waters is influenced by CO_2 that dissolves from the atmosphere into water. Also, the discharge of organic wastes into surface waters can lead to consumption of oxygen in the water, which may adversely influence water quality. Natural reaeration processes, however, restore oxygen to the water because of chemical equilibration between oxygen in the air and water. In other words, the oxygen from the air dissolves

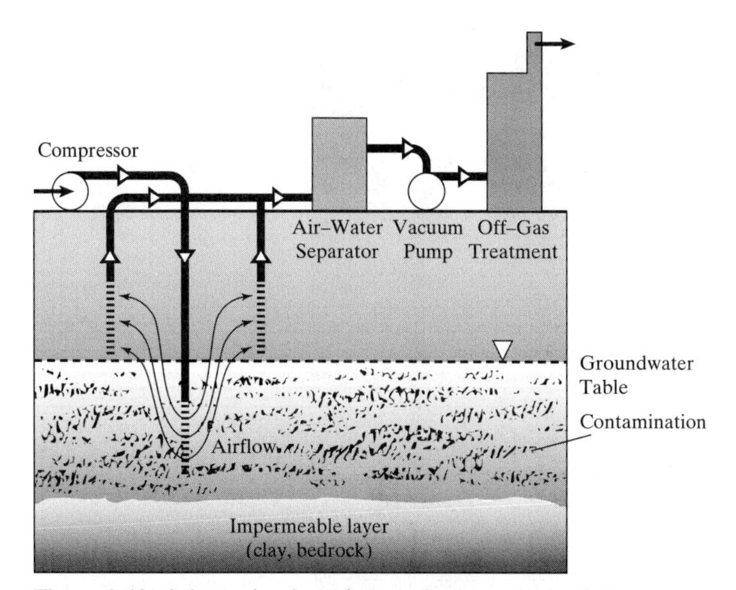

Figure 3-12. Schematic of an air-sparging process used to treat contaminated groundwater. Injected air strips chemicals from the saturated zone into the overlying unsaturated zone. Vapors stripped into the unsaturated zone must then be removed by a method such as vacuum extraction and treated above ground with activated carbon, biofiltration, or thermal oxidation.

into the water until it reaches its equilibrium (or saturation) concentration. Engineers use this equilibration between air and water to maintain the necessary concentration of oxygen in sewage-treatment plants and to calculate how much organic waste can safely be discharged to surface waters. Another important application of air–water partitioning is in the stripping of gases or volatile substances from water. Air sparging is used to strip volatile pollutants from groundwater to the unsaturated zone (vadose zone) where they can be treated and air-stripping towers can be designed to make drinking water safe by stripping hazardous chemicals out of water. Figure 3-12 is a schematic of an air-sparging process for treating contaminated groundwater, and Figure 3-13 depicts an air-stripping tower used to treat a drinking-water supply contaminated with volatile organic chemicals.

The Henry's constant, K_H (some authors use the term H), is used to describe a chemical's equilibrium between the air and water phases. Henry's law is just a special case of Raoult's law (Equation 3-44) applied to dilute systems. Remember, most environmental situations are dilute. Because the mole fraction of a dissolved substance in a dilute system is a very small number, concentrations such as moles/L are typically used rather than mole fractions. Note that Equation 3-44 may also be useful because it can be used to estimate Henry's constants in the absence of reliable experimental data. That is, a Henry's constant for a particular chemical can be determined by dividing the saturated vapor pressure of the chemical by its aqueous solubility.

Figure 3-13. Photo of air-stripping tower used at Wausau, Wisconsin, to remove some halogenated solvents (trichloroethene and perchloroethene) from a drinking-water supply. It treats about 5 million gallons per day, which is about one-half of the city's water needs. (Courtesy Dave Hand.)

The units of Henry's constant vary depending on whether a reaction is written as gas phase going to aqueous phase or aqueous phase going to gas phase. In addition, Henry's constants may also be unitless. Thus, it is important to use the proper units, understand why particular units are used, and be able to convert between different units.

3.4.2.1 Henry's Constant with Units for a Gas Dissolving in a Liquid

The following reaction describes the transfer of a gas (in this case, oxygen) from the atmosphere to water (depicted in Figure 3-10).

$$O_{2(g)} \longleftrightarrow O_{2(aq)} \quad \textbf{(3-45)}$$

The equilibrium expression for this reaction is

$$K_H = \frac{[O_{2(aq)}]}{[O_{2(g)}]} = \frac{[O_{2(aq)}]}{P_{O_2}} \quad \textbf{(3-46)}$$

The value of the Henry's constant, K_H, at 25°C for oxygen is 1.29×10^{-3} moles/L-atm. Note that in this case the units of K_H are moles/L-atm. *The reaction was written as gas going to aqueous because in this case we are concerned with how the composition of the gas affects the composition of the aqueous solution.* The preceding expression shows that the equilibrated dissolved oxygen saturation concentration in surface waters is a function of the partial pressure of oxygen in the atmosphere and the Henry's constant.

EXAMPLE 3.14. USING HENRY'S CONSTANT TO DETERMINE THE SOLUBILITY OF OXYGEN

Calculate the concentration of dissolved oxygen (units of moles/L and mg/L) in a water equilibrated with the atmosphere at 25°C. The Henry's law constant for oxygen at 25°C is 1.29×10^{-3} mole/L-atm.

SOLUTION

The partial pressure of oxygen in the atmosphere is 0.21 atm. Equation 3-46 can be rearranged to yield:

$$K_H \times P_{O_2} = [O_{2(aq)}] = \left(1.29 \times 10^{-3} \frac{\text{mole}}{\text{L-atm}}\right) \times 0.21 \text{ atm} = 2.7 \times 10^{-4} \frac{\text{mole}}{\text{L}}$$

Thus the solubility of oxygen at this temperature is 2.7×10^{-4} moles/L. If this is multiplied by the molecular weight of oxygen (32 g/mole), the solubility can be reported as 8.7 mg/L.

The value for the concentration of dissolved oxygen in water equilibrated with the atmosphere is 14.4 mg/L at 0°C and 9.2 mg/L at 20°C. This demonstrates that oxygen solubility in water is dependent on the water temperature (one reason trout like colder waters). The reason for this is that for the reaction described in Equation 3-45, the change in heat of formation (ΔH^0) at standard conditions is −3.9 kcal. Because ΔH^0 is negative, Equation 3-45 could be written as

$$O_{2(g)} \longleftrightarrow O_{2(aq)} + \text{heat} \qquad \textbf{(3-47)}$$

An increase in the temperature (or adding heat to the system) will, according to LeChatelier's principle, favor the reaction that tends to diminish the increase in temperature. The effect is to drive the reaction in Equation 3-47 to the left, which consumes heat, diminishing the temperature increase in the process. Therefore, more oxygen will be in the gas phase at an increased temperature, and thus the solubility of dissolved oxygen at the increased temperature will be lower.

3.4.2.2 Dimensionless Henry's Constant for a Species Transferring from a Liquid to a Gas

In the case where an individual is interested in the transfer of a chemical dissolved in the aqueous phase to the atmosphere, the chemical equilibrium between the gas and liquid phase chemical is described by a reaction written in reverse of Equation 3-45. Examples of this are in the engineered stripping of ammonia and volatile organic chemicals (e.g., benzene, trichloroethene) from water or the natural transfer of chemicals from a lake or river to the atmosphere. For example, for the chemical trichloroethene (TCE):

$$\text{TCE}_{(aq)} \longleftrightarrow \text{TCE}_{(g)} \tag{3-48}$$

In this case the equilibrium expression is written as

$$K_H = \frac{[\text{TCE}_{(g)}]}{[\text{TCE}_{(aq)}]} \tag{3-49}$$

Table 3-3. Unit Conversion of Henry's Constants (Hand et al.)[1]

$$H\left(\frac{\text{L}_{\text{H}_2\text{O}}}{\text{L}_{\text{Air}}}\right) = \frac{H\left(\frac{\text{L} \cdot \text{atm}}{\text{mol}}\right)}{RT}$$

$$H\left(\frac{\text{L} \cdot \text{atm}}{\text{mol}}\right) = H\left(\frac{\text{L}_{\text{H}_2\text{O}}}{L_{\text{Air}}}\right) \times RT$$

$$H\left(\frac{\text{L}_{\text{H}_2\text{O}}}{\text{L}_{\text{Air}}}\right) = \frac{H(\text{atm})}{RT \times 55.6\,\frac{\text{mol H}_2\text{O}}{\text{L}_{\text{H}_2\text{O}}}}$$

$$H\left(\frac{\text{L} \cdot \text{atm}}{\text{mol}}\right) = \frac{H(\text{atm})}{55.6\,\frac{\text{mol H}_2\text{O}}{\text{L}_{\text{H}_2\text{O}}}}$$

$$H(\text{atm}) = H\left(\frac{\text{L} \cdot \text{atm}}{\text{mol}}\right) \times 55.6\,\frac{\text{mol H}_2\text{O}}{\text{L}_{\text{H}_2\text{O}}}$$

$$H(\text{atm}) = H\left(\frac{\text{L}_{\text{H}_2\text{O}}}{\text{L}_{\text{Air}}}\right) \times RT \times 55.6\,\frac{\text{mol H}_2\text{O}}{\text{L}_{\text{H}_2\text{O}}}$$

$$R = 0.08205\,\frac{\text{atm} \cdot \text{L}}{\text{mol K}}$$

[1]Equations reprinted from Hand et al. in Water Quality and Treatment, by permission. Copyright 1998, American Water Works Association).

where the gas phase TCE is described by units of moles/liter of gas, not as partial pressure. Accordingly, the Henry's constant, K_H, has units of "moles/liter of gas" divided by "moles/liter of water" which cancel out. Therefore, the Henry's constant in this case is termed "dimensionless" by some users. In fact, it really has units of liters of water per liters of air!

Other units of Henry's constant include "atm" and "L-atm/mole." Dimensionless Henry's constants and those with units can be related using the ideal gas law. Several unit conversions of Henry's constant are provided in Table 3-3.

EXAMPLE 3.15. CONVERSION BETWEEN DIMENSIONLESS AND NONDIMENSIONLESS HENRY'S CONSTANTS

The Henry's constant for the reaction transferring oxygen from *air to water* is 1.29×10^{-3} moles/L-atm at 25°C. What is the dimensionless K_H for the transfer of oxygen from *water to air* at 25°C?

SOLUTION

The problem is requesting a Henry's constant for the reverse reaction, therefore, the Henry's constant provided equals $[1.29 \times 10^{-3} \text{ moles/L-atm}]^{-1} = 775$ L-atm/mole for the reaction of aqueous oxygen going to gaseous oxygen. Using the Ideal Gas Law:

$$K_H \text{ (dimensionless)} = \frac{\dfrac{775 \text{ L-atm}}{\text{mole}}}{\left(\dfrac{0.08205 \text{ L-atm}}{\text{mole K}}\right)(298 \text{ K})} = 32$$

Note that the units on the dimensionless Henry's constant of 32 are really mole/L of air divided by mole/L of water, which equals L of water per L of air.

EXAMPLE 3.16. USE OF HENRY'S CONSTANT IN PREDICTING THE FATE OF AN ORGANIC POLLUTANT BETWEEN WATER AND AIR

Hexachlorobenzene (C_6Cl_6) is a chemical produced mainly as an unwanted by-product during the production of pesticides and other organic chemicals. This compound is widely distributed in the environment and bioaccumulates strongly. The dimensionless Henry's law constant for this compound is 0.054. If the concentration in the air above Lake Michigan is measured to be 50 pmole/m^3 and the concentration in the water of Lake Michigan is measured to be 100 pmole/L, would hexachlorobenzene (HCB) tend to move from the water into the air or from the air into the water?

SOLUTION

At equilibrium, $K = [HCB_{(g)}]/[HCB_{(aq)}]$, which is the reaction of transferring aqueous HCB going to gaseous HCB. Therefore, the equilibrium water concentration can be predicted from the air concentration:

$$\frac{\frac{50 \text{ pmole}}{\text{m}^3}}{0.054} = \frac{930 \text{ pmoles}}{\text{m}^3} \times \frac{\text{m}^3}{1{,}000 \text{ L}} = \frac{0.93 \text{ pmoles}}{\text{L}}$$

Because the equilibrium concentration of HCB in the water predicted from the measured gas concentration is much less than the measured value of 100 pmole/L, the system is not yet at equilibrium. Therefore, HCB will continue to move from the water to the air until equilibrium is established.

3.4.3 Acid–Base Chemistry

Acid–base calculations are required in instances where neutralization of an acidic or basic solution is required, and for numerous treatment processes in which it is necessary to maintain pH within a certain range. For example, pH may need to be kept constant during chemical- and biological-treatment processes to provide an optimal environment for microorganisms during bioremediation of organic pollutants, eliminate odors (caused by volatile acids) at an industrial plant, or minimize the corrosion of pipes. In addition, acid/base chemistry can control the fate and treatment of many chemical species.

3.4.3.1 pH

By definition, the pH of a solution is

$$\text{pH} = -\log\{H^+\} \quad \textbf{(3-50)}$$

where $\{H^+\}$ is the activity (i.e., the apparent concentration) of the hydrogen ion and has units of moles/L. The pH scale in aqueous systems ranges from 0 to 14, with acidic solutions having a pH below 7, basic solutions having a pH above 7, and neutral solutions having a pH near 7. Most natural waters have a pH near neutral. In fact, 95% of all natural waters have a pH between 6 and 9, and most rain water, not impacted by anthropogenic acid-rain emissions, has a pH of approximately 5.6 due to the presence of dissolved carbon dioxide from the atmosphere.

The concentrations of OH^- and H^+ are related to one another through the equilibrium reaction for the dissociation of water:

$$H_2O \longleftrightarrow H^+ + OH^- \quad \textbf{(3-51)}$$

Table 3-4. Dissociation Constant for Water at Various Temperatures

Temperature, °C	K_w*	pH of Neutral Solution
0	0.12×10^{-14}	7.47
15	0.45×10^{-14}	7.18
20	0.68×10^{-14}	7.08
25	1.01×10^{-14}	7.00
30	1.47×10^{-14}	6.92

*Values of K_w obtained from Stumm and Morgan, 1996.

The equilibrium (also called dissociation) constant (K_w) for Equation 3-51 equals 10^{-14} at 25°C. Thus,

$$K_w = 10^{-14} = [H^+][OH^-] \tag{3-52}$$

This allows the determination of the concentration of H^+ or OH^- if the other is known. The van't Hoff relationship (see Section 3.3.4) can be used to determine the K_w at other temperatures. Table 3-4 shows the K_w at various temperatures of environmental significance. Note that at 25°C $[H^+]$ equals $[OH^-]$; thus $[H^+] = 10^{-7}$ and the pH of a neutral water would equal 7.00. However, at 15°C, $[H^+]$ equals $10^{-7.18}$ and thus the pH of a neutral solution at this temperature would equal 7.18.

3.4.3.2 Definition of Acids and Bases and Their Equilibrium Constants

Acids and bases are substances that react with hydrogen ions (H^+). *An acid is defined as a species that can release or donate a hydrogen ion (also called a proton). A base is defined as a chemical species that can accept or combine with a proton.* The reaction of a monoprotic (*monoprotic means the acid can donate one proton and the base can accept one proton*) acid, HA, with a base, B^-, is illustrated in Equation 3-53. If the reaction is reversible, then according to the preceding definition, A^- is a base and HB is an acid. It is always the case that an acid donates a proton to a base, and in the process itself becomes a base. For acid–base pairs like those shown in Equation 3-53, A^- is termed the conjugate base of acid HA, and HB is termed the conjugate acid of base B^-.

$$HA + B^- \longleftrightarrow A^- + HB \tag{3-53}$$

One of the remarkable features of water is that it can act as both an acid and a base. This feature makes acids and bases important to humans and to all living organisms. Note that in Equation 3-54 where a base is added to water, water acts as an acid and donates a proton to a base (B):

$$B^- + H_2O \longleftrightarrow HB + OH^- \tag{3-54}$$

However, in the case where an acid is added to water, water accepts a proton from the acid (HA) and acts as a base:

$$HA + H_2O \longleftrightarrow H_3O^+ + A^- \quad \textbf{(3-55)}$$

This substance H_3O^+ is nothing other than a water molecule with a proton attached and looks like:

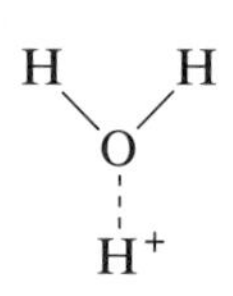

In point of fact, no free H^+ ions actually exist in water. They are always combined with one or more water molecules. However, for the sake of simplicity the water molecules attached to the H^+ are generally ignored and Equation 3-55 is generally written as

$$HA \longleftrightarrow H^+ + A^- \quad \textbf{(3-56)}$$

The forward reaction of Equation 3-56 is termed the dissociation or deprotonation of the acid, HA. The reverse reaction is termed the protonation of the base. All acids can undergo such reactions, and this becomes the basis for distinguishing their relative strengths. Acids that have a strong tendency to dissociate (Reaction 3-56 goes far to the right) are called strong acids, while acids that have less of a tendency to dissociate (Reaction 3-56 goes a little to the right) are called weak acids.

The strength of an acid is indicated by the magnitude of the equilibrium constant for the dissociation reaction. The equilibrium constant for the reaction depicted in Equation 3-56 is

$$K_a = \frac{[H^+][A^-]}{[HA]} \quad \textbf{(3-57)}$$

At equilibrium, a strong acid will dissociate and show high concentrations of H^+ and A^- and a smaller concentration of HA. This means that when a strong acid is added to water, the result is a much larger negative free-energy change than when adding a weaker acid. Thus for strong acids the equilibrium constant, K_a, will be large (and ΔG would be very negative). Similarly, the K_a for a weak acid would be small (and ΔG would be less negative).

Chemists have adopted the habit of taking the negative logarithm of equilibrium constants and concentrations and denoting them by the prefix "p." For example, pH equals $-\log\{H^+\}$. Hence pK_a is the negative logarithm of the acid dissociation constant (i.e., p$K_a = -\log(K_a)$). Table 3-5 shows values of equilibrium constants for some acids and bases of environmental importance. Table 3-5 shows that the pK_a of a weak acid is larger than the pK_a of a strong acid.

The pK_a of an acid is related to the pH at which the acid will dissociate. Strong acids are those that have a pK_a below 2. They can be assumed to dissociate

Table 3-5. Some Common Acids and Bases and Their Equilibrium Constants When Added to Water at 25°C

Acid	Name	$pK_a = -\log K_a$	Base	Name	$pK_b = -\log K_b$
HCl	Hydrochloric	−3	Cl^-	Chloride ion	17
H_2SO_4	Sulfuric	−3	HSO_4^-	Bisulfate ion	17
HNO_3	Nitric	−1	NO_3^-	Nitrate ion	15
HSO_4^-	Bisulfate	1.9	SO_4^{2-}	Sulfate ion	12.1
H_3PO_4	Phosphoric	2.1	$H_2PO_4^-$	Dihydrogen phosphate	11.9
CH_3COOH	Acetic	4.7	CH_3COO^-	Acetate ion	9.3
H_2CO_3*	Carbon dioxide and carbonic acid	6.3	HCO_3^-	Bicarbonate	7.7
H_2S	Hydrogen sulfide	7.1	HS^-	Bisulfide	6.9
$H_2PO_4^-$	Dihydrogen phosphate	7.2	HPO_4^{2-}	Monohydrogen phosphate	6.8
HCN	Hydrocyanic	9.2	CN^-	Cyanide ion	4.8
NH_4^+	Ammonium ion	9.3	NH_3	Ammonia	4.7
HCO_3^-	Bicarbonate	10.3	CO_3^{2-}	Carbonate	3.7
HPO_4^{2-}	Monohydrogen phosphate	12.3	PO_4^{3-}	Phosphate	1.7
NH_3	Ammonia	23	NH_2^-		−9

*Values obtained from Stumm and Morgan, 1996.

completely in water in the pH range 3.5–14. HCl, HNO_3, H_2SO_4, and $HClO_4$ are four very strong acids commonly encountered in environmental situations. Likewise, their conjugate bases (Cl^-, NO_3^-, SO_4^{2-} and ClO_4^-) are so weak that in the pH range of 0–14, they are assumed to never exist with attached protons.

Though most natural waters exhibit pH values in the range of 6–8, the pH range of water is restricted to 0–14. The pH is restricted to this range because of another special characteristic of water. The dissociation reaction for water acting as an acid can be written as

$$H_2O \longleftrightarrow H^+ + OH^- \tag{3-58}$$

The equilibrium constant for this reaction at 25°C is 10^{-14}, or the pK_a for water is 14. Hence water is a very weak acid because it has a very high pK_a. The expression for the equilibrium quotient is

$$K_a = \frac{[H^+][OH^-]}{[H_2O]} = [H^+][OH^-] = 10^{-14} \tag{3-59}$$

This is the same expression as Equation 3-52. When the concentration of H^+ is 1 M or 10^0, then the concentration of OH^- must be 10^{-14} M. Why is the pH of water never less than zero? The concentration of water is about 55.6 moles/L. Hence the maximum possible concentration of hydrogen ions that could occur is if all the water molecules were protonated. Accordingly, if the H^+ ion concentration was 55.6 M, the pH would be -1.74. However, pH is not actually defined as the negative logarithm of $[H^+]$, but rather the negative logarithm of $\{H^+\}$, the activity. (See Section 3.1 for a discussion of the difference between concentration and activity.) The "effective" concentration or activity of H^+ never reaches 55.6 M, only about 1 M. Hence the lowest possible pH is approximately 0, and for similar reasons the maximum pH (this occurs when all water exists as OH^- ions) is approximately 14.

EXAMPLE 3.17. ACID–BASE EQUILIBRIUM

What percentage of total ammonia (i.e., $NH_3 + NH_4^+$) is present as NH_3 at a pH of 7? The pK_a for NH_4^+ is 9.3; therefore,

$$K_a = 10^{-9.3} = \frac{[NH_3][H^+]}{[NH_4^+]}$$

SOLUTION

The problem is requesting:

$$\frac{[NH_3]}{([NH_4^+] + [NH_3])} \times 100\%$$

In order to solve this problem another independent equation is required because the one expression written above has two unknown values. The equilibrium expression for the NH_4^+/NH_3 system provides the second required equation:

$$10^{-9.3} = \frac{[NH_3][H^+]}{[NH_4^+]} = \frac{[NH_3][10^{-7}]}{[NH_4^+]}$$

Thus, at pH = 7, $[NH_4^+] = 200[NH_3]$. This expression can be substituted into the first expression, yielding

$$\frac{[NH_3]}{(200[NH_3] + [NH_3])} \times 100\% = 0.5\%$$

Note that at this neutral pH, almost all of the total ammonia of a system exists as ammonium ion (NH_4^+). In fact, only 0.5% exists as NH_3! The form of total ammonia most toxic to aquatic life is NH_3. It is toxic to several fish species at concentrations above 0.2 mg/L. Thus wastewater discharges with a pH < 9 have most of the total ammonia in the less toxic NH_4^+ form. This is one reason why some wastewater discharge permits for ammonia specify that the pH of the discharge must also be less than 9.

EXAMPLE 3.18. ACID–BASE EQUILIBRIUM CALCULATION

One nonbiological method to remove ammonia from water is to run the water through an air-stripping tower. $NH_{3(aq)}$ is in equilibrium with $NH_{3(g)}$. However, the other form of ammonia, NH_4^+, is not in equilibrium with the gaseous phase. Thus, in order to strip ammonia from water, to what pH must the water be adjusted to ensure efficient removal of total ammonia? The pK_a for NH_4^+ is 9.3.

SOLUTION

We must ensure that most of the total aqueous ammonia (NH_3 + NH_4^+) is in the form of NH_3 which is the only aqueous species in equilibrium with gaseous ammonia, $NH_{3(g)}$. The amount of $NH_{3(aq)}$ at different pHs can be determined, as was done in Example 3.17:

$$\text{at pH} = 6 \text{ the } \%NH_{3(aq)} \text{ is } 0.052\%$$

$$\text{at pH} = 8 \text{ the } \%NH_{3(aq)} \text{ is } 4.8\%$$

$$\text{at pH} = 10 \text{ the } \%NH_{3(aq)} \text{ is } 83.3\%$$

$$\text{at pH} = 11 \text{ the } \%NH_{3(aq)} \text{ is } 98\%$$

Because only $NH_{3(aq)}$ is in equilibrium with NH_3, an operator would typically calculate the lime dosage to raise the pH to approximately 11 before running the wastestream through an air-stripping tower.

EXAMPLE 3.19. STRONG ACID–BASE EQUILIBRIUM CALCULATION

What is the pH of a river immediately after a train wreck releases 400 liters of concentrated (14 M) hydrochloric acid over a period of 1 hour? The river flow rate (Q) is 100 m^3/s, and assume that the stream is pristine and dilute (i.e., no other acids or bases are present). The pK_a for HCl is −3.

SOLUTION

Assume the acid is mixed evenly into the volume of water that flows past the train in one hour. In one hour, a volume of river water that flows past the train wreck equals

$$V = Q \times t = \frac{100 \text{ m}^3}{\text{s}} \times \frac{3{,}600 \text{ s}}{\text{h}} \times 1 \text{ h} = 3.6 \times 10^5 \text{ m}^3$$

The total concentration of the acid in this volume is then

$$C_{\text{total,acid}} = \frac{Vol_{acid} \times Conc_{acid}}{Vol_{river}} = \frac{(400 \text{ L})(14 \text{ M})}{(3.6 \times 10^5 \text{ m}^3)} \times \frac{\text{m}^3}{10^3 \text{ L}} = 1.6 \times 10^{-5} \text{ M}$$

The total acid concentration equals the sum of dissociated and undissociated forms of the acid ($HCl + Cl^-$). The equation for the equilibrium constant will allow the acid concentration to be related to the concentration of H^+ and hence to the pH:

$$K_a = \frac{[H^+][Cl^-]}{[HCl]} = 10^{+3}$$

As seen from its equilibrium constant ($pK_a = -3$) HCl is a very strong acid and will completely dissociate. Therefore at equilibrium $[Cl^-] \gg [HCl]$. A mass balance on the added acid therefore results in

$$C_{\text{total,acid}} = [HCl] + [Cl^-] \cong [Cl^-] = 1.6 \times 10^{-5} \text{ M}$$

The problem states that there are no other acids present, and hence the only source of H^+ is the dissociation of the hydrochloric acid. Hence $[H^+]$ must equal $[Cl^-]$ because each molecule of HCl dissociates to yield one H^+ and one Cl^-. Thus the concentration of H^+ also is 1.6×10^{-5} M, which equals a pH of 4.8. This is the pH of the impacted river.

3.4.3.3 The Carbonate System, Alkalinity, and Buffering Capacity

The carbonate system is a very important example of how the chemical composition of the atmosphere influences the chemistry of lakes, rivers, and groundwater. The carbonate system is important when studying the effect of increased carbon dioxide emissions on global climate and the possible uptake of some released carbon dioxide by the world's oceans. For example, one reason the planet Venus is so much hotter than Earth is that Venus had no water to uptake atmospheric carbon dioxide, unlike the vast oceans of Earth that have served as a sink for a lot of Earth's gaseous carbon dioxide. In addition, carbon dioxide levels in subsurface environments are indicative of biological processes, some of which are critical to the fate and remediation of organic chemicals that have contaminated the subsurface. Figure 3-14 shows the important components of the carbonate system.

The concentration of dissolved carbon dioxide in water equilibrated with the atmosphere (partial pressure of CO_2 is $10^{-3.5}$ atm) is 10^{-5} moles/L. This is a significant amount of carbon dioxide dissolved in water. This reaction can be written as

$$CO_{2(g)} \longleftrightarrow CO_{2(aq)} \qquad \text{where } K_H = \frac{10^{-1.5} \text{ moles}}{\text{L-atm}} \tag{3-60}$$

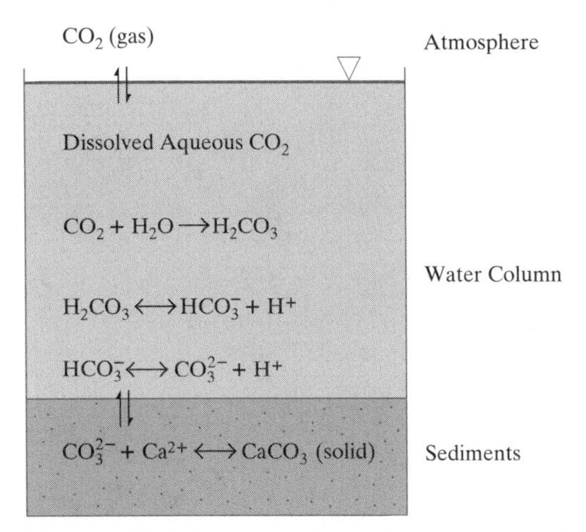

Figure 3-14. Diagram of the carbonate system showing its important components.

Upon dissolving in water, dissolved CO_2 undergoes a hydration reaction by reacting with water to form carbonic acid:

$$CO_{2(aq)} + H_2O \longleftrightarrow H_2CO_3 \qquad K = 10^{-2.8} \tag{3-61}$$

This reaction has important implications for the chemistry of water in contact with the atmosphere. First, the water in contact with the atmosphere (e.g., rain) has the relatively strong acid, carbonic acid, dissolved in it.

Thus, the pH of rain water, not impacted by anthropogenic emissions, will be below 7. In fact, the pH of "unpolluted" rain water is approximately 5.6. Thus, acid rain, which typically has measured pH values of 3.5 to 4.5, is approximately 10–100 times more acidic than natural rain water, but not 10,000 times more acidic if natural rain water had a pH of 7.0. In addition, because natural rain water is slightly acidic and the partial pressure of carbon dioxide in soil may also be high from biological activity, water that contacts rocks and minerals can dissolve ions into solution. Thus, inorganic constituents dissolved in fresh water, and the dissolved salts in the oceans, have their origin in minerals and the atmosphere. Carbon dioxide from the atmosphere provides an acid that can react with the bases of rocks, releasing the rock constituents into water where they can remain either dissolved (see Equation 2-12 for example) or precipitate into a solid phase.

Figure 3-15 shows how the dissolution of limestone by groundwater can remove large amounts of rock and form a surface topography known as karst topography. This topography consists of sinkholes, large solution cavities (caves and caverns), and small lakes and ponds. Some of these cavities serve as conduits for groundwater transport. Mammoth Cave National Park (Kentucky) (the world's longest cave with 330 miles mapped so far!), Carlsbad Caverns National Park (New Mexico), Wind Cave National Park (South Dakota), and Moravian Karst National Park (Czech Republic) were all created by the dissolution of limestone by slightly acidic water. When waters rich in dissolved carbon dioxide

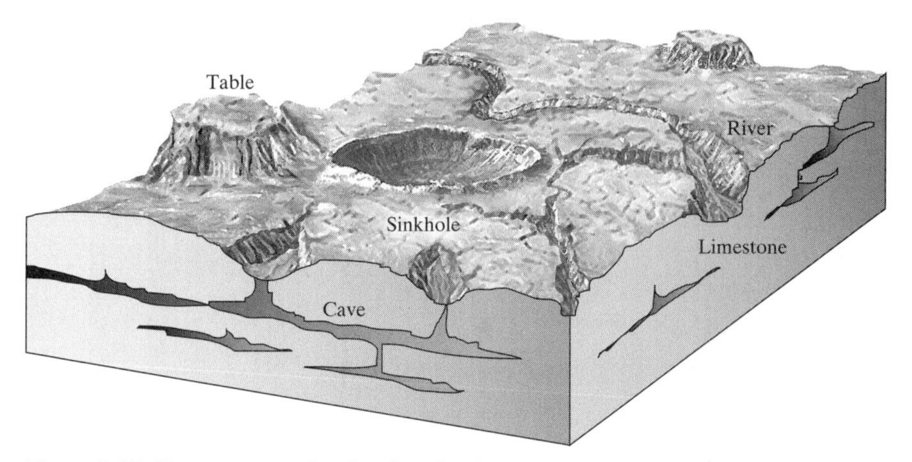

Figure 3-15. Karst topography showing the development of sinkholes, cavities, small lakes, and ponds. This topography is the result of dissolution of limestone from the presence of naturally occurring, acidic water. Redrawn from *Geology and America's National Park Areas*, by Ellwood, Brooks, B., © 1996, with permission of Prentice-Hall, Inc., Upper Saddle River, N.J.

and calcium ion drip from the ceiling of caves, some of the dissolved carbon dioxide degasses, which concentrates the calcium, which then crystallizes as calcite forming the familiar stalagmites (on the floor) and stalactites (hanging from ceilings). Figure 3-16 shows the result of this acid–base and precipitation–dissolution chemistry.

It is difficult to analytically distinguish the difference between $CO_{2(aq)}$ and true H_2CO_3. Therefore, the term $H_2CO_3^*$ has been defined to equal the concentration of $CO_{2(aq)}$ plus the concentration of true H_2CO_3. However, $H_2CO_3^*$ can be approximated by $CO_{2(aq)}$ because true H_2CO_3 makes up only about 0.16% of $H_2CO_3^*$. Thus, the concentration of $H_2CO_3^*$ in waters equilibrated with the atmosphere is approximately 10^{-5} M.

$H_2CO_3^*$ is in equilibrium with bicarbonate ion as follows:

$$H_2CO_3^* \longleftrightarrow HCO_3^- + H^+ \qquad K_{a1} = 10^{-6.3} \tag{3-62}$$

and bicarbonate is in equilibrium with carbonate ion as follows:

$$HCO_3^- \longleftrightarrow CO_3^{2-} + H^+ \qquad K_{a2} = 10^{-10.3} \tag{3-63}$$

Note that by our definition of an acid and base, bicarbonate can act as either an acid or a base. In addition, bicarbonate and carbonate are common bases in water.

The total inorganic carbon content of a water sample is defined as

$$\text{Total inorganic carbon} = [H_2CO_3^*] + [HCO_3^-] + [CO_3^{2-}] \tag{3-64}$$

In the pH range of most natural waters (pH 6–9), $H_2CO_3^*$ and CO_3^{2-} are small relative to HCO_3^-; therefore, HCO_3^- is the most dominant component in Equation 3-64. It can be seen from Equation 3-64 and the previous equilibrium reac-

Figure 3-16. Stalagmite and stalactite formation in Carlsbad Caverns National Park (New Mexico). These formations are formed from the dissolution and subsequent precipitation of minerals such as calcite ($CaCO_3$). (Raymond G. Barnes/Tony Stone Images/New York, Inc.)

tions (Equations 3-62 and 3-63) that the total inorganic carbon of a system can be measured by first acidifying a water sample so that almost all of the total inorganic carbon is converted H_2CO_3*. This compound is in equilibrium with gaseous CO_2, and therefore can be air-stripped and the gaseous CO_2 measured.

Alkalinity. *Alkalinity is a measure of a water's capacity to neutralize acids.* The alkalinity of natural waters is "usually" due to naturally occurring salts of weak acids. Bicarbonate is the dominant weak acid in most natural waters (pH of 6–8), with minor contributions from carbonate and hydroxide, OH^-. Accordingly, the alkalinity of these waters can be defined as

$$\text{Alkalinity (moles/L)} = [HCO_3^-] + 2[CO_3^{2-}] + [OH^-] - [H^+] \qquad \textbf{(3-65)}$$

Note that there is the integer of 2 in front of the carbonate term because one mole of carbonate can consume 2 moles of acidity, H^+. Also, in most natural waters of neutral pH (pH 6–8), the concentration of bicarbonate is significantly greater than that of carbonate or hydroxide; therefore, the total alkalinity can be approximated by the bicarbonate concentration.

In some natural waters and industrial waters, other salts of weak acids that may be important in determining a solution's alkalinity are borates, phosphates, ammonia, or organic acids. For example, anaerobic digester supernatant and municipal wastewaters contain large amounts of bases such as ammonia (NH_3), phosphates (HPO_4^{2-} and PO_4^{3-}), and bases of various organic acids. The bases of silica ($H_3SiO_4^-$) and boric acid ($B(OH)_4^-$) can contribute to alkalinity in the oceans. However, in the majority of natural waters alkalinity is caused primarily by HCO_3^-, CO_3^{2-}, and OH^-. This is because in most natural systems, these are the three most commonly available bases.

Buffering Capacity. In acid–base chemistry, the buffering capacity is defined as a water's ability to resist changes in pH when either acidic or alkaline material is added. In most natural waters the buffering capacity is due primarily to the bases (OH^-, CO_3^{2-}, HCO_3^-) and acids (H^+, H_2CO_3, HCO_3^-). Many lakes in the United States (e.g., in New England and the Upper Midwest) have a low buffering capacity and thus have been strongly influenced by acidic deposition (i.e., acid rain). This is because the geology of the basins that underlie these lakes is such that the dissolution of the underlying rocks and minerals does not result in the release of alkalinity.

EXAMPLE 3.20. DETERMINING THE pH OF WATER USING CARBONATE-SYSTEM EQUILIBRIUM EXPRESSIONS

Acid rain is an example of the effect of mixing a strong acid with a weak base. Examples 3.8 and 3.9 already showed that acid rain can result from the burning of fossil fuels and the subsequent oxidation of the SO_2 to sulfuric acid. When this strong acid enters a lake it reacts with the weak bases that are present in the lake water. The ability of a lake to neutralize inputs of strong acids is referred to as the *alkalinity*; also referred to as the *acid-neutralizing capacity* (ANC). The ANC of a lake depends primarily on the concentration of bicarbonate (HCO_3^-) in the lake. Bicarbonate can accept a proton and thus act as a base according to the following reaction:

$$H_2CO_3^* \longleftrightarrow HCO_3^- + H^+$$

The equilibrium quotient for this reaction is

$$K_{a1} = 10^{-6.3} = \frac{[HCO_3^-][H^+]}{[H_2CO_3^*]}$$

Many lakes in Michigan's Upper Peninsula have low concentrations of bicarbonate because the rocks underlying these lakes weather (i.e., dissolve) very slowly. A typical concentration of bicarbonate in the small ponds in this region is 50 μM. $H_2CO_3^*$ cannot be measured directly, but the total inorganic carbon dissolved in the water can be easily measured. The total inorganic carbon for a lake at a slightly acidic pH approximates the sum of the concentrations of HCO_3^- and $H_2CO_3^*$ because the concentration of CO_3^{2-} is very small at these pHs.

A typical value for the total inorganic carbon of a small pond in this region would be 200 μM. Estimate the pH of this typical pond.

SOLUTION

The pH of the lake can be calculated from the equilibrium expression and an understanding of the carbonate system:

$$[H^+] = \frac{10^{-6.3}[H_2CO_3^*]}{[HCO_3^-]} = \frac{10^{-6.3} \times [200 \times 10^{-6} - 50 \times 10^{-6}]}{50 \times 10^{-6}} = 10^{-5.82}$$

The pH of this pond is estimated to be 5.82. This pond is acidic (pH < 7), and it has been shown that certain organisms cannot grow at this pH. Many lakes in the north central United States are naturally slightly acidic because of the types of rocks that underlie the lakes. Thus, only a little acid rain is needed to acidify them further. This problem also occurs in many other parts of the world.

3.4.4 Precipitation–Dissolution

Precipitation–dissolution reactions involve the dissolution of a solid to form soluble species (or the reverse process whereby soluble species react to precipitate out as a solid). Common precipitates include hydroxide, carbonate, and sulfide precipitates. A reaction that sometimes occurs in homes is the precipitation of $CaCO_3$. This compound forms a scale in tea kettles, hot water heaters, and pipes. Much effort is devoted to preventing excessive precipitation of $CaCO_3$ in industrial settings, and the process of removing divalent cations from a water is referred to as water softening. Figure 3-17 shows scale precipitation in a water distribution pipe. Figure 3-16 showed how precipitation–dissolution reactions form natural features such as stalagmites and stalactites.

In municipal wastewater-treatment plants one solid ($FeCl_3$) is sometimes added at a point just prior to the final clarifier(s), where it first dissolves to form Fe^{3+} and Cl^-. The Fe^{3+} then reacts with dissolved phosphorus and precipitates out of solution as a ferric iron (+3)/phosphate solid. This reaction is performed to remove excess P from a treatment plant's effluent that could ultimately cause eutrophication in a receiving water body.

The reaction common to all of these situations is the conversion of a solid salt into dissolved components. In this example, the solid is calcium carbonate:

$$CaCO_{3(s)} \longleftrightarrow Ca^{2+} + CO_3^{2-} \qquad \textbf{(3-66)}$$

Here the subscript (s) denotes that the species is a solid. The equilibrium constant for such a reaction is referred to as the solubility product, K_{sp}. For the reaction in Equation 3-66, the K_{sp} would be given by

$$K_{sp} = \frac{[Ca^{2+}][CO_3^{2-}]}{[CaCO_{3(s)}]} = [Ca^{2+}][CO_3^{2-}] \qquad \textbf{(3-67)}$$

Figure 3-17. Scale formation in a water distribution pipe. (Sheila Terry/Science Photo Library/Photo Researchers.)

Because the activity (which we are assuming equals concentration) of a solid is defined to be equal to one (Rule 2 of Section 3.3.2), the equilibrium constant is equal to the product of the concentration of the two dissolved ions. Thus, if the equilibrium constant and the concentration of one of the species is known, the concentration of the other specie(s) can be determined. No precipitate will form if the product of the concentration of the ion pairs is less than the K_{sp} (in Equation 3-67 Ca^{2+} and CO_3^{2-} are the ion pairs). This solution is described as being undersaturated. Likewise, if the product of the concentrations of the ion pair exceeds the K_{sp}, the solution is described as being supersaturated and the solid species will precipitate out until the product of the ion pair equals the K_{sp}. Table 3-6 shows some important solubility products of use in environmental engineering and science.

EXAMPLE 3.21. INFLUENCE OF pH ON PRECIPITATION OF METALS

Metals can be removed from wastewater as hydroxide precipitates by raising the pH to 8–11. In what order would the divalent cations, Cd, Zn, Mg, and Cu, be removed as the pH of a solution is raised assuming that the molar concentration

Table 3-6. Examples of Some Common Precipitate–Dissolution Reactions, the Associated K_{sp}, and the Significance in Environmental Engineering

Equilibrium Equation	K_{sp} at 25°C	Significance
$CaCO_3$ (s) $\longleftrightarrow$ $Ca^{2+} + CO_3^{2-}$	3.3×10^{-9}	Hardness removal, scaling
$MgCO_3$ (s) $\longleftrightarrow$ $Mg^{2+} + CO_3^{2-}$	3.5×10^{-5}	Hardness removal, scaling
$Ca(OH)_2$ (s) $\longleftrightarrow$ $Ca^{2+} + 2OH^-$	6.3×10^{-6}	Hardness removal
$Mg(OH)_2$ (s) $\longleftrightarrow$ $Mg^{2+} + 2OH^-$	6.9×10^{-12}	Hardness removal
$Cu(OH)_2$ (s) $\longleftrightarrow$ $Cu^{2+} + 2OH^-$	7.8×10^{-20}	Heavy metal removal
$Zn(OH)_2$ (s) $\longleftrightarrow$ $Zn^{2+} + 2OH^-$	3.2×10^{-16}	Heavy metal removal
$Al(OH)_3$ (s) $\longleftrightarrow$ $Al^{3+} + 3OH^-$	6.3×10^{-32}	Coagulation
$Fe(OH)_3$ (s) $\longleftrightarrow$ $Fe^{3+} + 3OH^-$	6×10^{-38}	Coagulation, iron removal
$CaSO_4$ (s) $\longleftrightarrow$ $Ca^{2+} + SO_4^{2-}$	4.4×10^{-5}	Flue-gas desulfurization

Adapted with permission of the McGraw-Hill Companies from Sawyer et al., Chemistry for Environmental Engineering, 1998, McGraw-Hill Companies. K_{sp} obtained from Morel and Hering, 1993.

of all metals equals 9.6×10^{-6}? The K_{sp} for formation of metal hydroxide precipitates are 5.0×10^{-15}, 3.2×10^{-16}, 6.9×10^{-12}, and 7.8×10^{-20}, respectively.

SOLUTION

The metals all have the following generic reaction where Me is the divalent cation.

$$Me(OH)_{2(s)} = Me^{2+} + 2OH^-$$

Therefore, first set up the $K_{sp} = [Me^{2+}][OH^-]^2$ for each metal cation and solve for $[OH^-]$. The equilibrium concentration of each divalent cation was given as 9.6×10^{-6} moles/L and each individual K_{sp} was provided. Then use the dissociation constant for water (Equation 3-52) to solve for $[H^+]$, which can be converted to pH. This resulting pH is where precipitation of the cations will begin to occur:

Cation	pH at Which Precipitate Begins to Form
Cd	9.36
Zn	8.76
Mg	10.9
Cu	6.95

The metals will precipitate out in the order Cu > Zn > Cd > Mg.

EXAMPLE 3.22. USE OF REACTION STOICHIOMETRY TO CALCULATE SLUDGE PRODUCTION

How many kg of hydroxide sludge with a water content of 95% (5% solids) are generated daily for a wastewater plant treating 1 MGD of a copper waste with 10^{-3} M Cu^{2+}? Assume all the Cu is removed as $Cu(OH)_2$ (MW = 97.5).

SOLUTION

$$\frac{10^6 \text{ gal}}{\text{day}} \times \frac{3.78 \text{ L}}{\text{gal}} \times 10^{-3} \frac{\text{moles } Cu^{2+}}{\text{L}} \times \frac{\text{mole } Cu(OH)_2}{\text{mole } Cu^{2+}} \times \frac{97.5 \text{ g}}{\text{mole } Cu(OH)_2} \times \frac{\text{kg}}{1{,}000 \text{ g}}$$

$$= \frac{370 \text{ kg dry sludge}}{\text{day}} \times \frac{100 \text{ kg total sludge}}{5 \text{ kg dry solids}} = \frac{7{,}400 \text{ kg wet sludge}}{\text{day}}$$

EXAMPLE 3.23. PRECIPITATION/DISSOLUTION EQUILIBRIUM

What pH is required to reduce a high concentration of dissolved Mg^{2+} down to 43 mg/L? The K_{sp} for the following reaction is $10^{-11.16}$:

$$Mg(OH)_{2(s)} \longleftrightarrow Mg^{2+} + 2OH^-$$

SOLUTION

In this situation the dissolved magnesium is removed from solution by precipitating out as an hydroxide. First convert the concentration of Mg^{2+} from mg/L to moles/L

$$[Mg^{2+}] = \frac{43 \text{ mg}}{\text{L}} \times \frac{\text{g}}{1{,}000 \text{ mg}} \times \frac{1 \text{ mole}}{24 \text{ g}} = 0.0018 \text{ M}$$

Then, the equilibrium relationship is written as

$$10^{-11.16} = \frac{[Mg^{2+}][OH^-]^2}{[Mg(OH)_{2(s)}]}$$

Substituting values for all the known parameters

$$10^{-11.16} = \frac{[0.0018][OH^-]^2}{1}$$

Solve for $[OH^-] = 6.2 \times 10^{-5}$ M. This results in $[H^+] = 10^{-9.79}$ M; so pH = 9.79. At this pH any magnesium in excess of 0.0018 M will precipitate out as $Mg(OH)_{2(s)}$ because the solubility of Mg^{2+} would be exceeded.

EXAMPLE 3.24. DETERMINING THE SOLUBILITY OF SOLID ELECTROLYTES

Calculate the solubility of $CaSO_{4(s)}$ (MW = 136 g/mole) in a solution with no other sources of Ca^{2+} or SO_4^{2-}. $CaSO_{4(s)}$ is a component of gypsum, which is used as drywall. Landfills containing construction refuse often have leachate with high concentrations of Ca^{2+} and SO_4^{2-} as a result of the dissolution of this substance. The solubility product (K_{sp}) for $CaSO_{4(s)}$ is $10^{-4.59}$.

SOLUTION

The solubility is the amount (moles/L) of the calcium sulfate that can dissolve in water. From the stoichiometry, for $CaSO_{4(s)}$, each x moles that dissolve in water produces $1x$ moles of Ca^{2+} and $1x$ moles of SO_4^{2-}. Thus the solubility would be equal to the concentration of Ca^{2+} or the concentration of SO_4^{2-} in a solution in equilibrium with the solid $CaSO_{4(s)}$. Because the concentrations of both ions would be equal, the equilibrium relationship can be written as

$$K_{sp} = 10^{-4.59} = [Ca^{2+}][SO_4^{2-}] = [1x][1x]$$

Solve for x, which equals $10^{-2.3}$ M. This is the maximum concentration of Ca^{2+} and SO_4^{2-} that will result if one attempts to dissolve solid calcium sulfate into one liter of water. That is, only $10^{-2.3}$ mole (0.68 g) of calcium sulfate can be dissolved in the one liter of water. After that, no additional solid can be dissolved in the water. Any additional solid added to the water will not dissolve but will remain as a precipitate. This value is termed the chemical's *solubility*.

EXAMPLE 3.25. EFFECT OF TEMPERATURE ON THE SOLUBILITY OF A SOLID ELECTROLYTE

Typically, as temperature is increased, the solubility of a solid electrolyte will also increase (for solid electrolytes, K_{sp} gets higher). However, notable exceptions to this rule are $CaCO_{3(s)}$, $Ca_3(PO_4)_{2(s)}$, $CaSO_{4(s)}$, and $FePO_{4(s)}$ where the K_{sp} decreases with higher temperatures so the solubility decreases. This is shown in the following reaction for $FePO_{4(s)}$.

$$FePO_{4(s)} = Fe^{3+} + PO_4^{2-} + \text{heat}$$

The ΔH^0 for this reaction is -18.7 kcal so heat is released by this reaction as just shown.

What is the solubility of $Ca_3(PO_4)_{2(s)}$ at 25°C and 10°C? The K_{sp} is 1.3×10^{-32} at 25°C. The ΔH_f^0 for $Ca_3(PO_4)_{2(s)}$ is -986.2 kcal/mole, for Ca^{2+} is -129.77 kcal/mole, and for PO_4^{3-} is -306.9 kcal/mole.

SOLUTION

The equilibrium quotient (Equation 3-39) for this reaction at 25°C is

$$1.3 \times 10^{-32} = [Ca^{2+}]^3[PO_4^{3-}]^2$$

From the reaction stoichiometry, for every x moles of $Ca_3(PO_4)_{2(s)}$ that dissolves $3x$ moles of Ca^{2+} are released and $2x$ moles of PO_4^{3-} are released. Therefore, the solubility of $Ca_3(PO_4)_{2(s)}$ at 25°C is found as follows.

$$1.3 \times 10^{-32} = [3x]^3[2x]^2$$

x can be solved for as 1.9×10^{-7} M, the solubility of $Ca_3(PO_4)_{2(s)}$ at 25°C.

The solubility of $Ca_3(PO_4)_{2(s)}$ at 10°C can be found similarly, except the K_{sp} at 10°C must be determined. For this the van't Hoff relationship (Equation 3-41) is required along with the heat of formation (ΔH^0). At 10°C the ΔH^0 for the reaction is found as

$$\Delta H^0 = [3(-129.77) + 2(-306.9)] - [(1)(-986.2)] = -16.91 \text{ kcal}$$

Remember that this assumes that the values of ΔH_f^0 (measured at 25°C) did not change significantly at 10°C. Equation 3-41 can be written as

$$\ln \frac{K_{sp} \text{ at } T_2}{K_{sp} \text{ at } T_{25}} = \frac{-\Delta H^0}{R}\left[\frac{1}{T_{25}} - \frac{1}{T_2}\right]$$

Substituting values in results in a K_{sp} at 10°C equal to 5.89×10^{-32}. The solubility at 10°C can then be found as was done before for 25°C.

$$5.89 \times 10^{-32} = [3x]^3[2x]^2$$

Solve for x, which equals 2.8×10^{-7} M. In this case the solubility increased as the temperature decreased. Again, this is an exception to the rule that solubility increases as temperature increases. Note that the solubility of organic chemicals (e.g., chloroform, benzene, PCBs) also increases with increasing temperature.

3.4.5 Sorption and Ion Exchange to Solid Surfaces

3.4.5.1 Introduction

The processes of sorption, adsorption, and ion exchange are important in the atmosphere, in surface waters, in soils, in groundwaters, and in engineered treatment systems. If chemicals become associated with particles in the atmosphere they can be deposited to land by gravitational settling or rain and snow. In lakes, association of chemicals with particles can influence the toxicity of a chemical. For example, sorption of aluminum on particles prevents the aluminum from reacting with fish gill tissues (which reduces toxicity of the aluminum), but for particle-feeding organisms such as the water flea this could be catastrophic (because the water flea is exposed to additional aluminum through feeding). Also, sorption onto particles that settle to the bottom of lakes helps to remove pollutants from the lake water but could concentrate chemicals in the sediments. In subsurface systems like the unsaturated zone and groundwater there is obviously a tremendous surface area, derived from solids, onto which chemicals can sorb. Substances that sorb to subsurface solids are not as likely to be transported quickly in the subsurface, or at least their movement will be slower relative to the movement of water and substances that do not sorb appreciably. In addition, some substances sorb so strongly that bacteria cannot remove them from the solids and break them down (Mihelcic et al., 1993). Hence sorption can adversely influence bioremediation of contaminated soil and groundwater. In addition, sorption and ion-exchange processes are used to treat water and gases that are contaminated with a wide assortment of organic and inorganic pollutants.

3.4.5.2 Sorption and Adsorption

Adsorption is the physical and/or chemical process in which a substance accumulates at a solid–liquid interface. The solid can be a natural material (e.g., surface soil, harbor or river sediment, aquifer material) or it can be of anthropogenic origin (e.g., activated carbon). The word "sorption" is used to describe the interaction of organic solutes with soils and sediments. *Sorption is the combined process of adsorption of a solute at a surface and partitioning of the solute into the organic carbon that has coated the surface of a particle.* The sorbate (or adsorbate) is the substance being transferred at the interface from the liquid to the solid phase. The sorbent (or adsorbent) is the solid phase onto which the adsorbate accumulates. Though sorption and adsorption can also take place from the vapor phase, the following discussion will deal only with sorption and adsorption from the liquid phase. Figure 3-18 shows a schematic of the sorption process.

Why does adsorption occur? From a thermodynamic viewpoint molecules always prefer to be in a lower energy state. A molecule adsorbed onto a surface has a lower energy state on a surface as compared to being in the aqueous phase. Therefore, during the process of equilibration, the molecule is driven to the sur-

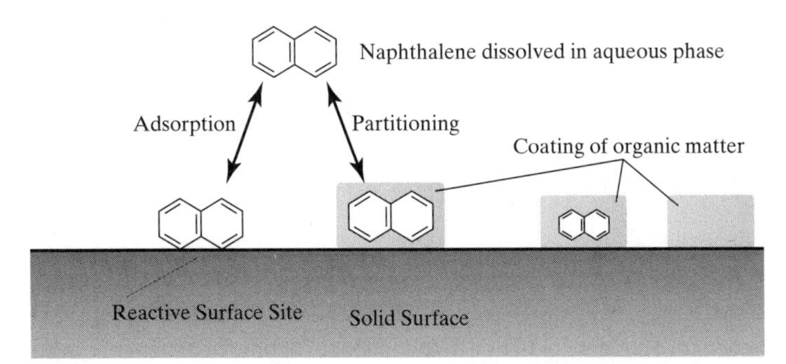

Figure 3-18. Sorption of an organic chemical typically takes place when the organic solute (the sorbate) adsorbs either onto reactive surface sites (adsorption) or partitions into organic matter that has coated the particle (the sorbent). This process influences the mobility, natural degradation, and engineered remediation of organic pollutants.

face and a lower energy state. Attraction of a molecule to a surface can be caused by physical and/or chemical forces. Electrostatic forces are the basic physical principle governing the interactions between most adsorbates and adsorbents. These forces include dipole–dipole interactions, dispersion interactions or London–van der Waals force, and hydrogen bonding. During sorption to soils and sediments, a phenomenon termed *hydrophobic partitioning* can also account for the interaction with a surface.

Simply put, if equilibrium is achieved between the organic chemical (the sorbate) and the solid surface (the sorbent), an equilibrium constant, K, can be used to relate the aqueous- and solid-phase concentrations of a chemical (i.e., $[C_{\text{solid}}] = K[C_{\text{aqueous}}]$). Understanding this will allow you to do Examples 3-27 to 3-29. The remainder of this section discusses sorption and adsorption in more depth.

The Freundlich Isotherm Equation. Although some say the Freundlich isotherm equation is empirical, it can be derived from thermodynamic relationships, assuming a heterogeneous surface. It is written as

$$q = KC^{1/n} \tag{3-68}$$

where q is the equilibrium solid-phase concentration (mg sorbate/g sorbent), C is the equilibrium liquid-phase concentration (mg/L), K is the Freundlich capacity parameter (mg sorbate/g sorbent)(L water/mg sorbate)$^{1/n}$, and $1/n$ is the Freundlich intensity parameter (unitless). In gas-phase systems, C would represent the gas-phase concentration of the chemical.

Figure 3-19 shows the Freundlich isotherm equation plotted for different values of $1/n$. Note that for values of $1/n$ less than one, the isotherm is considered favorable for sorption because low values of the sorbate liquid-phase concentration yield large values of the solid-phase concentration. This means that the sorbate prefers to be sorbed onto the surface. However, at higher aqueous concen-

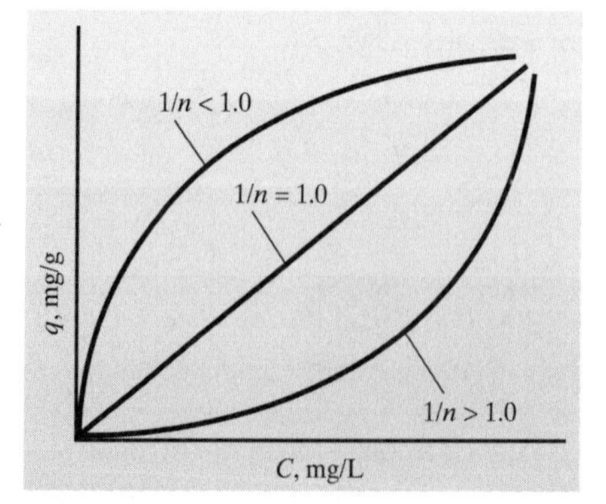

Figure 3-19. Freundlich isotherm plotted for different values of $1/n$.

trations the sorptive capacity of the solid decreases as the active sorption sites become saturated with sorbate molecules. For $1/n$ values greater than one the isotherm is considered to be unfavorable for sorption because high values of the liquid-phase sorbate concentration are required to get sorption to occur on the sorbent. However, it shows that as sorption occurs, the surface is modified by the sorbing chemical and made more favorable for additional sorption. If the $1/n$ value equals one, the isotherm is termed a linear isotherm. A linear isotherm is commonly observed with dilute aqueous phases and natural sorbents, as is discussed later in this section.

In order to determine the Freundlich parameters (K, $1/n$) for a particular chemical solute and sorbent, the Freundlich isotherm is linearized by taking the log of both sides of the Equation 3-68, which results in

$$\log q = \frac{1}{n} \log C + \log K \quad \textbf{(3-69)}$$

One reason for using Equation 3-69 is that it can be fitted to experimentally determined q and C data in order to determine the specific Freundlich parameters (K and $1/n$) specific to a particular chemical solute and sorbent. Values of K and $1/n$ for a wide range of synthetic adsorbents and chemicals are provided by Sontheimer et al. (1988). The Freundlich isotherm parameters are used in Example 3.31.

Linear Isotherm Equation. In many soil and sediment systems (aquatic or subsurface) the concentration of the chemical of concern is relatively low (i.e., a dilute system). In these situations it has been observed on numerous occasions that $1/n$ is very close to one. For the case where $1/n$ equals 1, the Freundlich isotherm equation (Equation 3-68) can be written as

$$q = K \times C \quad \textbf{(3-70)}$$

where K is the equilibrium constant between an aqueous- and solid-phase solute and has units such as cm^3/gm or L/kg. In this case, K is referred to as a *soil-* or *sediment-water partition coefficient.* The soil–water partition coefficient is sometimes referred to as a *partition coefficient* (K_p) or *distribution coefficient* (K_d). Equation 3-70 is useful because if equilibrium is obtained between a chemical in the aqueous (or gaseous) phase and solid phase, and K and either the solid or aqueous (or vapor) phase concentration is known, the remaining unknown concentration can be solved for. The problem with this K is that it is chemical and sorbent specific. Thus, although K can be measured for every relevant system, this would be time-consuming and costly. Fortunately, when the solute is a neutral, nonpolar organic chemical, the soil–water partition coefficient can be normalized for organic carbon, where it then becomes chemical-specific, but no longer sorbent-specific.

Soil–Water Partition Coefficients Normalized to Organic Carbon (K_{oc}) and Octanol–Water Partition Coefficients (K_{ow}). It has been shown that for soils and sediments with a fraction organic carbon (f_{oc}) greater than 0.001 (0.1%) and low equilibrium solute concentrations ($< 10^{-5}$ molar or $\frac{1}{2}$ the aqueous solubility) the soil–water partition coefficient (K) can be normalized to the soil's organic carbon content. K can be normalized for a soil or sediment's fraction of organic carbon as follows:

$$\frac{K}{f_{oc}} = K_{oc} \tag{3-71}$$

Here K_{oc} is called the *soil–water partition coefficient normalized to organic carbon.* K_{oc} has units of cm^3/g organic carbon (or L/kg organic carbon) and f_{oc} is the fraction organic carbon for a specific soil or sediment (1% organic carbon equals an f_{oc} of 0.01). A sand–gravel aquifer might have 0.1% or less organic carbon (f_{oc} less than or equal to 0.001), a surface soil might contain 1–5% organic carbon, and lake sediments may be even higher, especially if they originate from a eutrophic water body where the sediments are high in biological solids (e.g., algae). It has been found by many that for systems with a relatively high amount of organic carbon (greater than 0.1%), K_{oc} can be directly correlated to a parameter called the *octanol–water partition coefficient, K_{ow}, of a chemical.*

The octanol–water partition coefficient of a specific chemical can be determined by a relatively easy-to-understand experiment. If an organic chemical, A, is mixed in a sealed container that contains equal volumes of octanol ($C_8H_{17}OH$) and water, after equilibration and separation of the water and the octanol phase the concentration of A is measured in these two phases. The equilibrium constant for this system is the octanol–water partition coefficient, or K_{ow}, and equals

$$K_{ow} = \frac{[A]_{\text{octanol}}}{[A]_{\text{water}}} \tag{3-72}$$

Values of K_{ow} range over many orders of magnitude; therefore, K_{ow} is usually reported as log K_{ow}. Table 3-7 lists some typical values of log K_{ow} for a wide variety of chemicals. Values of K_{ow} for environmentally significant chemicals range from approximately 10^1 to 10^7 (log K_{ow} range of 1–7). The higher the value, the greater the tendency of the compound to partition from the water into an organic phase. Chemicals with high values of K_{ow} are hydrophobic or "water-fearing." Such compounds will have a very high tendency to sorb to particles in the environment.

The magnitude of an organic chemical's K_{ow} can tell a lot about the chemical's ultimate fate in the environment. For example, the values in Table 3-7 indicate that very hydrophobic chemicals such as 2,3,7,8-TCDD are likely to bioaccumulate in the lipid portions of humans and animals. Conversely, chemicals such as benzene, trichloroethene (TCE), tetrachloroethene (PCE), and toluene are

Table 3-7. Examples of log K_{ow} for Some Environmentally Significant Chemicals

Chemical	Log K_{ow}
Phthalic acid	0.73
Benzene	2.13
Trichloroethene	2.42
Tetrachloroethene	2.53
Toluene	2.73
2,4-D	2.81
Naphthalene	3.36
1,2,4,5-Tetrachlorobenzene	4.05
Phenanthrene	4.46
Pyrene	5.09
2,3,7,8-TCDD	6.64

Values obtained from Baker et al, 1997.

frequently identified as groundwater contaminants because they are relatively soluble in groundwater recharge that is infiltrating vertically toward an underlying aquifer. This is in comparison to pyrene or 2,3,7,8-TCDD, which are most likely to be confined near the soil's surface in the location of the spill.

The fact that K_{oc} correlates well with K_{ow} should make sense because hydrophobic chemicals that have a high K_{ow} will be attracted to a "like" phase; in this case, the hydrophobic natural organic material that coats particles in the environment. Figure 3-20 shows how K_{oc} and K_{ow} are linearly correlated for a 72 chemical data set that spans many ranges of hydrophobicity. K_{ow} has also been

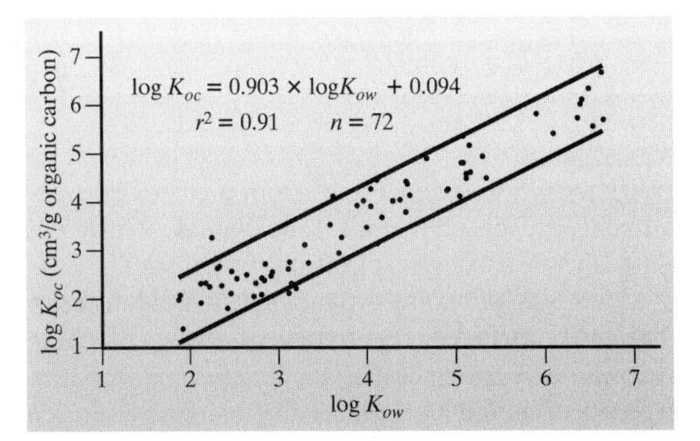

Figure 3-20. Scatter plot of log K_{oc} (cm^3/g organic carbon) versus log K_{ow} for 72 chemical data set. Note that the correlation shown in Equation 3-73 was fit to these data. The heavy lines represent the ~90% confidence interval on the correlation. From Baker et al. (1997). Copyright WEF, reprinted with permission.

correlated to other environmental properties such as bioconcentration factors (see Chapter 5, for example) and aquatic toxicity.

Estimating K_{oc}. K's are sometimes normalized to organic matter (therefore, K_{om}). When obtaining organic normalized soil–water partition coefficients from the literature a user must be careful of whether the soil–water partition coefficient (K) was normalized to organic carbon (K_{oc}) or organic matter (K_{om}). This is because organic carbon is not the same as organic matter (om). Organic matter consists of organic carbon, organic nitrogen, organic sulfur, organic phosphorus, and so on. Thus, the amount of organic matter in a sample would be expected to be greater than organic carbon. In fact, it has been found that approximately 50 to 59% of organic matter is organic carbon; thus, the ratio of organic matter to organic carbon is approximately 1.7 to 2.0 (Nelson and Summer, 1982). The advantage of using K_{oc} rather than K_p or K_d is that "chemical-specific" values can be obtained from the literature or predicted. Thus, after a soil or sediment's specific f_{oc} is determined (see Nelson and Summer for how this is done), the "site- and chemical-specific" K can be obtained from Equation 3-71.

Over 70 correlations are available in the literature to predict K_{oc} (see Baker et al., 1977, for a discussion of them) from knowledge of other properties of the chemical or its chemical structure. The data in Figure 3-20 were correlated and the following relationship for estimating K_{oc} from K_{ow} was obtained (Baker et al., 1997):

$$\log K_{oc} \frac{\text{cm}^3}{\text{g}} = 0.903 \log K_{ow} + 0.094 \qquad (n = 72,\ r^2 = 0.91) \tag{3-73}$$

The 95% confidence interval for a value of log K_{oc} obtained from this correlation is

$$\pm 0.66 \left\{ \frac{138 + (\log K_{ow} - 3.92)^2}{136} \right\}^{1/2} \tag{3-74}$$

The correlation provided in Equation 3-73 was developed from a critically evaluated data set of 72 chemicals that represented 11 different classes of organic chemicals. These 11 chemical classes were aromatic amines; aromatic acids; pesticides; phthalates; chlorinated alkanes; chlorinated alkenes; alkyl-substituted aromatics; polychlorinated biphenyls; chlorinated aromatics; phenols and chlorinated phenols; and polycyclic aromatic hydrocarbons. The correlation accounts for 91% of the variability in the data for chemicals with log K_{ow} ranging from 1.7 to 7.0. The data set is available in Baker et al. (1997).

Individuals seeking values of K_{oc} should first consult a data set that has undergone a quality check or use an appropriate, statistically validated correlation (like Equation 3-73) to estimate the value of K_{oc}; K_{oc} can then be related to the site-specific K by knowledge of the system's organic carbon content using Equation 3-71.

EXAMPLE 3.26. DETERMINATION OF K_{oc} FROM K_{ow}

The log K_{ow} for anthracene is 4.45. What is anthracene's soil–water partition coefficient normalized to organic carbon along with the 95% confidence interval?

SOLUTION

Use an appropriate correlation such as is provided in Equation 3-73. Note that this correlation requests the log K_{ow}, not K_{ow}:

$$\log K_{oc} = 0.903(4.45) + 0.094 = 4.11 \text{ cm}^3/\text{g organic carbon}$$

Therefore, $K_{oc} = 10^{4.11}$ cm^3/g organic carbon.

For this estimated value, the 95% confidence interval on log K_{oc} is given by Equation 3-74:

$$\pm 0.66\left\{\frac{138 + (4.45 - 3.92)^2}{136}\right\}^{1/2} = \pm 0.66 \text{ cm}^3/\text{g organic carbon}$$

Thus log $K_{oc} = 4.11 \pm 0.66$ cm^3/g or $K_{oc} = 1.29 \times 10^4$ cm^3/g and the 95% confidence interval is [2.82×10^3, 5.89×10^4). That is, a user can be 95% confident that the true value of K_{oc} lies somewhere between this range. Note that a user of this correlation has 95% confidence that the estimated value of log K_{oc} is within only approximately ±2/3 base-10 log units. This is actually pretty good for a K_{oc} correlation.

EXAMPLE 3.27. USE OF K_{oc} TO PREDICT WATER CONCENTRATION

Anthracene has contaminated harbor sediments, and the solid portion of sediments is in equilibrium with the pore water. If the organic carbon content of sediments is 5% and the solid sediment anthracene concentration is 50 μg/kg sediment, what is the pore water concentration of anthracene at equilibrium?

SOLUTION

From Example 3.26, the log K_{oc} for anthracene was estimated to be 4.11 cm^3/g organic carbon. An organic carbon (OC) content of 5% means that $f_{oc} = 0.05$.

Therefore, using Equation 3-71, the sediment specific partition coefficient, K, is equal to

$$K \frac{\text{cm}^3}{\text{g sediment}} = \frac{10^{4.11}\ \text{cm}^3}{\text{g OC}} \times \frac{0.05\ \text{g OC}}{\text{g sediment}} = 664\ \text{cm}^3/\text{g sediment}$$

The equilibrium aqueous-phase concentration, C, is then derived from the equilibrium expression given in Equation 3-70:

$$C = q/K = \frac{\dfrac{50\ \mu\text{g}}{\text{kg sediment}} \times \dfrac{\text{kg}}{1000\ \text{g}}}{\dfrac{644\ \text{cm}^3}{\text{g sediment}}} \times \frac{\text{cm}^3}{\text{mL}} \times \frac{1{,}000\ \text{mL}}{\text{L}} = \frac{0.078\ \mu\text{g}}{\text{L}}$$

Note that the aqueous-phase concentration of anthracene is relatively low compared to the sediment-phase concentration (50 ppb in the sediments and 0.078 ppb in the pore water). This is because anthracene is hydrophobic. Its water solubility is low (and K_{ow} is high) so it wants to partition to the solid phase. Also, the solid phase is high in organic carbon content. A sand–gravel aquifer would be much lower in organic carbon (f_{oc} very low); therefore, less of the anthracene would partition from the aqueous to the solid phase.

EXAMPLE 3.28. USE OF K TO PREDICT SOIL CONCENTRATION OF METALS

Equilibrium constants for inorganic species with solid surfaces are also used. For example, assume that lead leaches from mine tailings into a stream that passes through a small wetland. The lead sorbs to the organic soil of the wetland with a K value of 10^4 L/kg. Predict the concentration of Pb in the soil (in μg/g) after 25 years of exposure to leachate with a Pb concentration of 25 μg/L.

SOLUTION

Initially, the concentration of Pb in the soil may have been close to zero, and much of the inflowing Pb would have been removed from the water by sorption onto the soil. Eventually, however, enough Pb will have sorbed onto the soil such that the soil is in equilibrium with the Pb in the inflowing water. It is this equilibrium concentration that the problem requests. From the definition of K in Equation 3-70:

$$K = \frac{[\text{Pb}_{\text{sorbed}}]}{[\text{Pb}_{\text{dissolved}}]}$$

Hence the concentration of sorbed Pb is simply equal to the partition coefficient (K) times the dissolved concentration (25 μg/L). The sorbed concentration would be 10^4 L/kg $\times$ 25 μg/L = 25 $\times$ 10^4 μg/kg or 250 μg/g.

EXAMPLE 3.29. APPLICATION OF SOIL–WATER PARTITION COEFFICIENTS TO LAKE PARTICULATE PROBLEM

A typical value of f_{oc} for particles suspended in the upper waters (termed the *epilimnion*) of Lake Superior is 0.2. What is the ratio of sorbed to dissolved PCB in these upper waters? Assume the log K_{oc} (cm^3/g organic carbon) for PCB equals 6.

SOLUTION

Use an equilibrium expression to determine the ratio of sorbed to dissolved PCB (i.e., q/C). Therefore

$$\frac{[\mathrm{PCB}_{\mathrm{sorbed}}]}{[\mathrm{PCB}_{\mathrm{dissolved}}]} = 10^6 \times 0.2 = 2 \times 10^5 \text{ cm}^3/\text{g total}$$

One must examine the units in order to understand this ratio. The units of this ratio (K) are cm^3/g of particles. Thus, the units of sorbed PCB are mg/kg, and the units of dissolved PCB are mg/L. Thus, on every kilogram of suspended particles there are 2 $\times$ 10^5 times more PCB than there is in every liter of water. In Lake Superior there are typically only 0.5 mg of suspended solids per liter of water; therefore, in each liter of water only one-tenth of the PCB gets sorbed onto particles. This result can be obtained by multiplying the ratio of 2 $\times$ 10^5 L/kg by the concentration of suspended solids (0.5 mg/L $\times$ 10^{-6} kg/mg) to yield the dimensionless ratio of 0.1. However, sedimentation of these suspended particles is one method by which PCBs are removed from the water column.

The retardation factor can be defined as the ratio of the average linear velocity of groundwater over the velocity of a chemical whose movement is retarded because of sorption to the solid material in the aquifer. Thus, if the retardation factor equals 10, the average velocity of the chemical would be about 10 times slower than the average velocity of the groundwater. The retardation factor is defined as

$$1 + \frac{\rho_b}{\eta} K \tag{3-75}$$

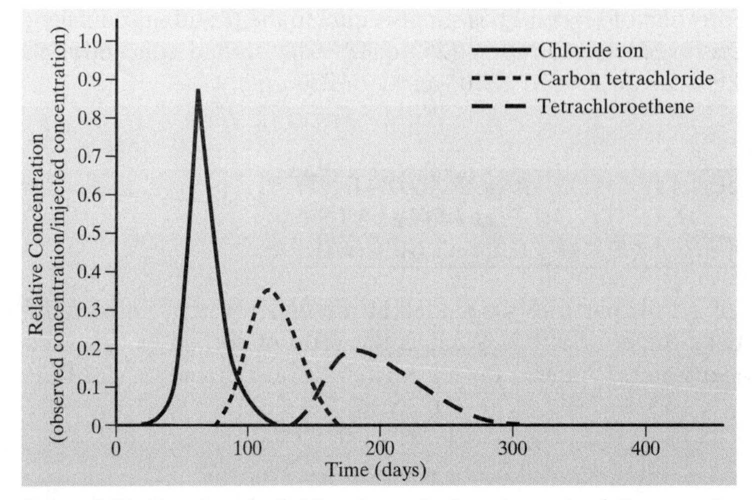

Figure 3-21. Results of a field scale study that shows the influence of retardation on chemical solute transport in the subsurface. The study was conducted in Borden, Ontario. The average groundwater velocity at this site is approximately 30 m/yr. Twelve cubic meters of water containing chloride ion, carbon tetrachloride, and tetrachloroethene were injected into a sand aquifer. A monitoring well located 5 meters downgradient was then sampled on a regular basis for the 3 chemical solutes (field data from Roberts et al., 1986).

Chloride ion is an inert tracer; therefore, its movement is not retarded by sorption with the solid phase and it moves at the same velocity as the groundwater. The next chemical to reach the monitoring well is carbon tetrachloride, followed by tetrachloroethene (also referred to as tetrachloroethylene, perchloroethene, and PCE). Thus, one might expect that carbon tetrachloride is less hydrophobic than tetrachloroethene. However, the log octanol-water partition coefficient (K_{ow}) of carbon tetrachloride is reported as 2.73 and the log K_{ow} of tetrachloroethene is reported as 2.53 (data provided by the American Institute of Chemical Engineers, Design Institute for Physical Property Data (AIChE/DIPPR) Project 912). What is going on here? While this figure clearly demonstrates how retardation influences chemical transport in a subsurface environment, it also shows some of the problems in estimating chemical transport in the subsurface using simplified expressions such as the retardation factor (Equation 3-75) and K_{ow} to predict soil-water partitioning (as is done with Equation 3-73).

The subsurface is a very complex environment. In this case, the organic carbon content of the aquifer is very low (0.02%) so sorption to organic matter may not be the dominant mechanism controlling retardation. In addition, there is also some uncertainty in the reported values of the octanol-water partition coefficient (K_{ow}) so perhaps tetrachloroethene is slightly more hydrophobic. For example, AIChE/DIPPR reports that the water solubility of carbon tetrachloride is 757 mg/L and the water solubility of tetrachloroethene is 484 mg/L. These values contradict the reported values for K_{ow}, which lead us to believe that carbon tetrachloride was more hydrophobic. In fact some correlations used to estimate K_{oc} use values of water solubility instead of K_{ow}. In either case, a more detailed understanding of the subsurface hydrogeology and chemical properties is required before we can completely understand the results of the field scale study.

where η is the volume of voids/total volume and ρ_b is the soil bulk mass density (both values are discussed further in Chapter 4). The retardation factor is useful for predicting the relative movement of chemicals as shown in Example 3.30. Figure 3-21 shows how the groundwater contamination plume from two chemicals with different sorptive properties (therefore, different retardation factors) will spread at different rates because of sorption. Thus, hydrophobic chemicals, which sorb more strongly, will typically move through the subsurface at a slower rate than a less hydrophobic chemical.

EXAMPLE 3.30. HOW K_{ow} and f_{oc} INFLUENCE THE RETARDATION FACTOR

A typical value of subsurface porosity is 0.3 and soil bulk mass density is 2.1 g/cm^3. If the subsurface material's fraction of organic carbon is 0.01 (1% organic carbon), determine the retardation factors for toluene ($\log K_{ow} = 2.73$) and 1,3,5-trichlorobenzene ($\log K_{ow} = 4.49$).

SOLUTION

First determine the K_{oc} of each solute. Use an appropriate correlation for these chemicals such as Equation 3-73.

For toluene, $\log K_{oc} = 0.903(2.73) + 0.094 = 2.56$

$$K_{oc} = 10^{2.56} = \frac{360\ \text{cm}^3}{\text{g organic carbon}}$$

$$K = K_{oc} \times f_{oc} = 360(0.01) = \frac{3.6\ \text{cm}^3}{\text{g aquifer material}}$$

For 1,3,5-TCB, $\log K_{oc} = 0.903(4.49) + 0.094 = 4.15$

$$K_{oc} = 10^{4.15} = \frac{14{,}100\ \text{cm}^3}{\text{g organic carbon}}$$

$$K = K_{oc} \times f_{oc} = 14{,}100(0.01) = \frac{141\ \text{cm}^3}{\text{g aquifer material}}$$

Use Equation 3-75 to determine the retardation factor. For toluene,

$$\log K_{oc} = 1 + \left[\frac{\dfrac{2.1\ \text{g}}{\text{cm}^3}}{0.3} \times \frac{3.6\ \text{cm}^3}{\text{g}}\right] = 26$$

Similarly, the retardation factor for 1,3,5-TCB equals 990.

This shows that the toluene moves about 26 times slower than water and 1,3,5-TCB 990 times slower than water. Note that how much a chemical's movement is retarded (or slowed) relative to the velocity of the groundwater is primarily influenced by the hydrophobicity of the chemical and the amount of organic carbon contained in the soil or aquifer. This problem also suggests that toluene will move more quickly vertically through the unsaturated zone with rain recharge compared to 1,3,5-TCB. Also both chemicals should move slower than the groundwater when moving horizontally toward a downgradient well.

Activated Carbon. Activated carbon particles are charcoal granules that have been carefully prepared from the pyrolysis of organic materials such as bituminous coal, peat, lignite, wood, and coconut shells. In this process, the carbonaceous materials are converted to mixtures of gas, tars, and ash. The tar is burned off and the gases are allowed to escape, creating a highly porous charcoal material that contains a very high internal surface area (500–1,000 m^2/g). The special pyrolysis process provides active or adsorption sites on the internal surfaces of the charcoal where adsorption of chemicals can occur. This provides an ideal adsorbent; high surface area and active sites. Activated carbon adsorbents are used for water treatment, industrial solvent recovery, and treatment of contaminated air streams. Figure 3-22 depicts one of these applications.

Activated carbon can be broken into two types: powdered activated carbon (PAC) and granular activated carbon (GAC). PAC consists of very small particles that are typically less than 0.05 mm in diameter. The typical size range

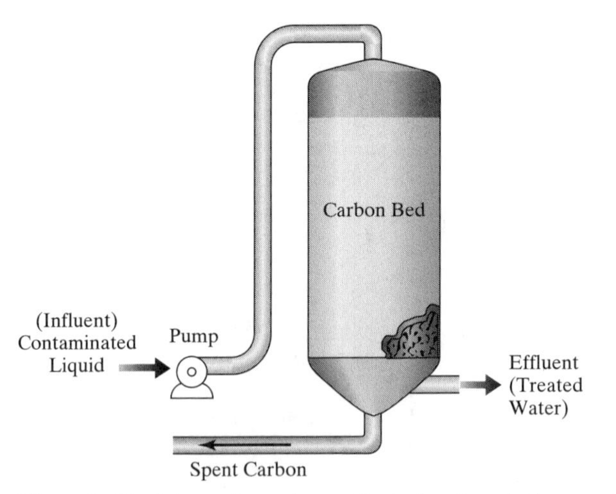

Figure 3-22. Adsorption using activated carbon is the most widely used treatment technology for the purification of air and water. It is also widely used by industry for purifying gases and liquids, and recovering solvents from by-product streams that are used in various production processes. This figure shows a schematic diagram of a fixed-bed activated carbon system. Contaminated liquid is pumped through a bed of activated carbon. Items such as the bed's volume and type of activated carbon then determine the effluent quality.

Table 3-8. Classes of Organic Chemicals That Readily Adsorb on Activated Carbon

Aromatic solvents (e.g., benzene, toluene, nitrobenzenes)
Chlorinated aromatics (e.g., PCBs, chlorophenols)
Polycyclic aromatic hydrocarbons (e.g., acenaphthene, benzo(*a*)pyrene)
Pesticides and herbicides (e.g., DDT, aldrin, chlordane, atrazine)
Halogenated aliphatics (e.g., trichloroethene, carbon tetrachloride, bromoform, chloroalkyl ethers)
High-molecular-weight hydrocarbons (e.g., dyes, gasoline and jet fuel components, amines, some humics)

for GAC is from 0.3 to 3 mm (Sontheimer et al., 1988). Synthetic polymeric adsorbents are primarily used for industrial solvent recovery and separation/purification processes though future applications may include water treatment. Silica-based adsorbents are mostly used for vapor-phase applications. Table 3-8 lists some readily adsorbed classes of organic chemicals while Table 3-9 lists some classes of poorly adsorbed organic chemicals.

The usual process scheme for applying adsorption to the treatment of drinking water involves the addition of PAC before or with the flocculation chemicals and the subsequent separation of the carbon with the floc particles. Another approach is to add the PAC before a sand filter and to separate it in the filter bed. GAC is typically packed into a column, and a contaminated wastestream is then passed through the column. At contaminated subsurface sites GAC may be provided in easily replaced 55-gallon drums that are used to treat contaminated groundwater and vapors from the unsaturated zone that have been brought aboveground.

EXAMPLE 3.31. USE OF FREUNDLICH ISOTHERM

Atrazine contained in agricultural runoff has contaminated a reservoir that is used as a source of drinking water. The atrazine concentration in the reservoir was measured to be 0.012 mg/L (12 ppb). In order to treat the reservoir water so that atrazine is removed below the drinking water standard of 0.003 mg/L (3 ppb), powdered activated carbon is added to a contact basin (a mixing tank) to adsorb the atrazine. The PAC is then removed in a settling tank located down-

Table 3-9. Classes of Organic Chemicals That Poorly Adsorb on Activated Carbon

Low-molecular-weight ketones, acids, and aldehydes
Sugars and starches
Very-high-molecular-weight or colloidal organics
Low-molecular-weight aliphatics

gradient. Assume that the city treats 10^6 gallons of drinking water per day, and that the Freundlich isotherm parameters for atrazine and this particular type of PAC are $K = 287$ mg/g $(\text{L/mg})^{1/n}$ and $1/n = 0.335$. What concentration is found on the PAC (in mg atrazine/g PAC) given that the aqueous concentration is lowered to the drinking-water standard?

SOLUTION

Use the Freundlich isotherm to relate the solid- and aqueous-phase concentrations:

$$q = \frac{287 \text{ mg}}{\text{g}} (0.003)^{0.335}$$

Solve for q, which equals 41 mg atrazine/g PAC. This problem is further developed in Example 3-32, where a mass balance is used to determine the amount of PAC required on a daily basis.

EXAMPLE 3.32. USE OF FREUNDLICH ISOTHERM AND MASS BALANCE IN DETERMINING POWDERED ACTIVATED CARBON DOSAGE DURING DRINKING-WATER TREATMENT

(This problem requires some understanding of mass balances. They are covered in Chapter 4.) Atrazine contained in agricultural runoff has contaminated a reservoir that is used as a source of drinking water. The atrazine concentration in the reservoir was measured to be 0.012 mg/L (12 ppb). In order to treat the reservoir water so that atrazine is removed below the drinking water standard of 0.003 mg/L (3 ppb), powdered activated carbon is added to a contact basin (a mixing tank) to adsorb the atrazine. The PAC is then removed in a settling tank located just downgradient of the contact basin. Assume that the city treats 10^6 gallons of drinking water per day, and that the Freundlich isotherm parameters for atrazine and this particular type of PAC are $K = 287$ mg/g $(\text{L/mg})^{1/n}$ and $1/n = 0.335$ (the same as were used in Example 3.31). What mass of PAC must be placed in the contact basin daily to ensure that atrazine is removed to concentrations that satisfy the drinking-water standard?

SOLUTION

Assume that the contact basin and separator can be described by a completely mixed flow reactor (CMFR) and that the contact basin is defined as the control volume (see Chapter 4 for a discussion of CMFR and control volume). A mass balance can be performed around the control volume where C_0 is the influent aqueous atrazine concentration, C is the effluent atrazine aqueous concentration, Q is the plant flow rate, S_0 is the atrazine concentration on the PAC being added to the control volume, and S is the concentration of atrazine of the PAC leaving the control volume. However, the incoming PAC is clean, so S_0 equals zero. The rate at which PAC is added to the contact basin and also removed from the separator is M. This term is what the problem is requesting. Therefore, the addition and removal rates of the PAC are assumed to equal one another. 10^6 gallons per day equals 3,780 m^3/day.

The mass balance in words is given as atrazine mass in untreated influent plus atrazine mass on added PAC = atrazine mass in treated effluent plus atrazine mass adsorbed to PAC after treatment (remember, the PAC added to the system is free of atrazine; therefore, S_0 = zero). And the mass balance in mathematical form is:

$$Q \times C_0 + S_0 \times M = Q \times C + M \times S$$

$$\frac{3{,}780 \text{ m}^3}{\text{day}} \times \frac{0.012 \text{ mg}}{\text{L}} \times \frac{1{,}000 \text{ L}}{\text{m}^3} = \left(\frac{3{,}780 \text{ m}^3}{\text{day}} \times \frac{0.003 \text{ mg}}{\text{L}} \times \frac{1{,}000 \text{ L}}{\text{m}^3}\right) + \text{M} \times \text{S}$$

$$\frac{4.536 \times 10^4 \text{ mg}}{\text{day}} = \frac{1.134 \times 10^4 \text{ mg}}{\text{day}} + \text{M} \times \text{S}$$

Therefore, the difference, or 3.402×10^4 mg of atrazine, must be removed every day by the added PAC. If equilibrium is assumed between the aqueous and adsorbed phase atrazine, the effluent concentration of atrazine on the PAC can be related to the effluent aqueous atrazine concentration using the Freundlich isotherm (Equation 3-68), as was done in Example 3-31.

$$\frac{3.402 \times 10^4 \text{ mg}}{\text{day}} = M \times \frac{287 \text{ mg}}{\text{g}} \times (0.003)^{0.335}$$

M equals 830 g PAC/day (which is the solution to this problem). This equals a dosage of 0.22 mg PAC/L of treated water.

Note that very little PAC is required (less than 2 lb/day) to treat a million gallons of contaminated water. This is because activated carbon has a huge surface area on a weight basis (500–1,000 m^2/g for both GAC and PAC!). Also, this calculation provides an answer that is really the lowest possible usage rate. This is because we have assumed that instantaneous equilibrium takes places between the aqueous- and adsorbed-phase atrazine. In reality, equilibration may take longer to attain because the atrazine molecules dissolved in the aqueous phase must diffuse through the liquid and then into the highly porous structure of the PAC. Also, there could be other chemicals dissolved in the water that might interfere with the adsorption of atrazine.

EXAMPLE 3.33. PARTITIONING OF CHEMICAL BETWEEN AIR, WATER, AND SOIL PHASES

(This problem requires some understanding of mass balances. They are covered in Chapter 4.) A sealed 1-liter reactor contains 500 mL water, 200 mL soil (1% organic carbon and density of 2.1 g/cm^3), and 300 mL of air. The temperature of the reactor is 25°C. One hundred μg of TCE are added to the reactor. The reactor is then incubated until equilibrium is achieved between all three phases. The Henry's constant for TCE is 10.7 L-atm/mole at 25°C, and TCE has a log K_{ow} of 2.42. Assuming that no chemical or biological degradation of the TCE occurs during the incubation, what is the aqueous-phase concentration of TCE at equilibrium? What is the mass of TCE in the aqueous, air, and sorbed phases after equilibrium is attained?

SOLUTION

First, set up a simplified mass balance that equates the total mass of TCE added to the mass of TCE in each phase *at equilibrium*.

$$\begin{aligned}\text{Total mass of TCE added} &= [\text{mass of aqueous TCE}] \\ &+ [\text{mass of gaseous TCE}] \\ &+ [\text{mass of sorbed TCE}] \\ 100\ \mu\text{g} &= [V_{aq} \times C_{aq}] + [V_{\text{air}} \times C_{\text{air}}] + [M_{\text{soil}} \times C_{\text{sorbed}}]\end{aligned}$$

The problem is requesting C_{aq}. The three known parameters are $V_{aq} = 500$ mL; $V_{\text{air}} = 300$ mL; and mass of soil $= V_{\text{soil}} \times$ density of soil (200 mL $\times$ cm^3/mL $\times$ 2.1 g/cm^3) = 420 gm. The three unknowns are C_{aq}, C_{air}, C_{sorbed}; however, C_{air} can be related to C_{aq} by a Henry's constant, and C_{sorbed} can be related to C_{aq} by a soil–water partition coefficient. Accordingly:

Convert the Henry's constant to dimensionless form. $K_H = 10.7$ L-atm/mole (by units one can tell this Henry's constant is for the reaction written in the following direction, $C_{aq} \longleftrightarrow C_{\text{air}}$). Convert to dimensionless form using the ideal gas law (see Table 3-3)

$$\frac{\dfrac{10.7\ \text{L-atm}}{\text{mole}}}{\dfrac{0.08205\ \text{L-atm}}{\text{mole K}}(298\ \text{K})} = 0.44$$

Therefore, the Henry's constant, $0.44 = C_{\text{air}}/C_{aq}$, so $C_{\text{air}} = 0.44\ C_{aq}$

Determine the soil–water partition coefficient. Remember $K = K_{oc} \times f_{oc}$ and 1% organic carbon means $f_{oc} = 0.01$. Because K_{oc} and K are not provided, estimate K_{oc} from Equation 3-73 or similar correlation ($\log K_{oc} = 0.903 \times 2.42 + 0.094 = 2.28$). Therefore,

$$K_{oc} = 10^{2.28} \text{ and } K = 10^{2.28} \times 0.01 = \frac{1.9\ \text{cm}^3}{\text{g}} \text{ and } C_{\text{sorbed}} = \frac{1.9\ \text{cm}^3}{\text{g}} \times C_{aq}$$

Accordingly, substitute into mass balance so all concentrations are in terms of C_{aq}:

$$100\ \mu g = [500\ mL \times C_{aq}] + [300\ mL \times 0.44C_{aq}] + \left[420\ g \times \frac{1.9\ cm^3}{g} \times \frac{mL}{cm^3} \times C_{aq}\right]$$

$$100\ \mu g = C_{aq}\left\{[50^\circ\ mL] + [300\ mL \times 0.44] + \left[420\ g \times \frac{1.9\ cm^3}{g} \times \frac{mL}{cm^3}\right]\right\}$$

$$100\ \mu g = C_{aq}[500\ mL + 132\ mL + 798\ mL]$$

Solve for C_{aq}, which equals 0.070 μg/mL = 0.070 mg/L = 70 ppb.

The total mass of TCE in the aqueous phase is 35 μg; in the air phase it is 9.2 μg; sorbed to soil it is 55.8 μg. Note that the mass of chemical found in each of the three phases is a function of the combined effects of partitioning between each phase. The amount of chemical that partitions to each phase is based upon the physical/chemical properties of the chemical (e.g., Henry's constant, log K_{ow}) and soil/sediment properties (f_{oc}). This is very important when determining where a chemical migrates in the environment or an engineered system, as well as determining what method of treatment should be selected.

3.4.5.3 Ion Exchange

Ion exchange is a physical–chemical process by which ions are transferred from the liquid to a solid phase or vice versa. Ions held by electrostatic forces to charged functional groups on the surface of a solid are exchanged for ions of similar charge in a solution that is in contact with the solid. The exchange phase could be natural (e.g., clay particles in the subsurface or an aquatic system) or engineered to treat contaminated waters. The exchange of an ion A with a solid B can be described by the following equation:

$$aA + bB_{(ads)} = aA_{(ads)} + bB \qquad \textbf{(3-76)}$$

where a and b are the number of moles of the chemicals A and B, respectively. The superscript "ads" represents the concentration of the chemical in the adsorbed phase. The equilibrium expression can be written as:

$$K = \frac{[A_{(ads)}]^a \times [B]^b}{[A]^a \times [B_{(ads)}]^b} \qquad \textbf{(3-77)}$$

For ion exchange, K is referred to as either a selectivity coefficient or selectivity function for natural systems and a separation factor when used in designing ion-exchange systems for water treatment.

The selectivity of an ion for a given ion-exchange resin will depend upon the ionic *charge* and *size* of the ion. Two general rules-of-thumb are:

Rule 1. At low aqueous phase concentrations ($<$1,000 mg/L TDS) and at typical surface and groundwater temperatures, the extent of exchange increases with increasing valence of the exchanging ion as follows:

$$Th^{4+} > Al^{3+} > Ca^{2+} > Na^{+} \quad \text{and} \quad PO_4^{3-} > SO_4^{2-} > Cl^{-}$$

Rule 2. At low aqueous concentrations ($<$1,000 mg/L TDS) and typical surface and groundwater temperatures, the extent of exchange increases with increasing atomic number (decreasing hydrated radius) of the exchanging ion as follows:

$$Cs^{+} > Rb^{+} > K^{+} > Na^{+} > Li^{+} \quad \text{and} \quad Ba^{2+} > Sr^{2+} > Ca^{2+} > Mg^{2+} > Be^{2+}$$

At high aqueous-phase ionic concentrations, the differences in exchange "potentials" of ions of different valence (Na^+ versus Ca^{2+} or NO_3^- versus SO_4^{2-}) diminish; in some cases the ion of lower valence has the higher exchange potential. In general the hydrated radius is inversely proportional to the nonhydrated radius. Synthetic resins prefer the ion with the smallest hydrated radius; the smaller-radii ions are being held more tightly by the resin.

Ion-exchange media are composed of insoluble matrices that have fixed-charged sites. These sites will exchange ions associated with them for aqueous-phase ions. Several types of exchange media include natural exchangers (greensand, clay, peat, aluminosilicates (zeolites) and synthetic organic resins (strong acid cation exchange resins (SAC), weak acid cation exchange resins (WAC), strong base anion exchange resins (SBA), weak base anion exchange resins (WBA)). Ion-exchange resins can either be strong or weak, based upon the functional pH ranges of the resins. The total number of exchange sites per unit of resin is the total exchange capacity and is independent of the experimental conditions. However, the apparent capacity depends on the experimental conditions such as pH and solution concentrations, and is usually lower than the total capacity. The capacity of a resin also depends upon the presaturated ion (i.e., H^+ and Na^+), because the density is different for each form of the resin.

3.5 OXIDATION/REDUCTION

Some chemical reactions occur because electrons are transferred between different chemical species. The species-losing electron(s) (the electron donor) is oxidized, while the species-accepting electron(s) (the electron acceptor) is reduced. These reactions are called *oxidation/reduction reactions* or *redox reactions*. One easy-to-learn acronym states "LEO goes GER" or loss of electrons equals oxidation, gain of electrons equals reduction. Oxidation/reduction reactions control the fate and speciation of many metals and organic pollutants in natural systems, and many treatment processes employ redox chemistry. The most commonly used treatment processes involving redox reactions are probably aerobic biological processes where organic compounds are oxidized to CO_2 and oxygen is reduced to water. These processes are discussed in depth in Chapter 5.

Another reason for understanding redox chemistry is that the oxidation state of a chemical may control its mobility in the environment or its toxicity. For example, hexavalent chromium (Cr^{6+}) is bright orange in color. One use of it was a reagent in the leather-making industry. It is considered very hazardous to humans. Trivalent chromium (Cr^{3+}) is green in color and is much less hazardous. There are many sites around the country where soil and groundwater have been contaminated by chromium waste products from tanneries. Accordingly, many tanneries have switched from the use of hexavalent to trivalent chromium. This is because of consideration of worker safety as well as potential impact on chromium discharged to the environment. The substitution of a less hazardous form of chromium in the leather tannery industry is an example of "pollution prevention."

Ammonia nitrogen (oxidation state of −3) can be converted through nitrification and denitrification to N_2 gas (oxidation state of 0). In addition important atmospheric pollutants include NO (oxidation state of +2) and NO_2 (oxidation state of +4). This conversion of nitrogen to different compounds occurs through many redox reactions. In addition, acid rain is caused by emissions of SO_2 (sulfur oxidation state of +4), which is oxidized in the atmosphere to sulfate ion, SO_4^{2-} (sulfur oxidation state of +6). It is sulfate ion that returns to the Earth's surface in dry or wet deposition as sulfuric acid. In the remainder of this section we discuss two examples of common redox reactions, the COD test and oxidation of organic matter.

The chemical oxygen demand (COD) test is used to characterize the strength of a wastewater. In this test, a sample containing an unknown amount of organic matter is added to a 250-mL flask. Also added to the flask are $AgSO_4$ (a catalyst to ensure complete oxidation of the organic matter); a strong acid (H_2SO_4), dichromate ($Cr_2O_7^{2-}$, a strong oxidizing agent); and, $HgCl_2$ (to provide Hg^{2+} ion that complexes chloride ion, Cl^-). Chloride ion interferes with the test and is present in high amounts in many wastewater samples. This is because it can be oxidized to Cl^0 by dichromate as well as by organic matter. However, the complexed form of Cl^- is not oxidized. Thus, if uncomplexed Cl^- is allowed to be oxidized to Cl^0, it would result in a false-positive COD value if not accounted for.

The sample and all the reagents are combined and the sample is refluxed for 3 h. Then the sample is cooled to room temperature and the dichromate that remains in the system is determined by titration with ferrous ammonium sulfate. The following reactions explain the COD test where glucose has been assumed to represent the organic matter present in the sample:

$$C_6H_{12}O_6 + 6H_2O \longleftrightarrow 6CO_2 + 24H^+ + 24e^- \quad \textbf{(3-78)}$$

$$6e^- + 14H^+ + Cr_2O_7^{2-} \longleftrightarrow 2Cr^{3+} + 7H_2O \quad \textbf{(3-79)}$$

Equations 3-78 and 3-79 are termed half-reactions. Equation 3-78 is the half-reaction for the oxidation of glucose, and Equation 3-79 is the half-reaction for the reduction of hexavalent chromium to trivalent chromium. Electrons are written as e^-. The overall reaction is found by first multiplying Equation 3-79 by 4 to balance the number of electrons in both reactions, followed by adding both half-reactions to obtain:

$$C_6H_{12}O_6 + 32H^+ + 4Cr_2O_7{}^{2-} \rightarrow 6CO_2 + 8Cr^{3+} + 22H_2O \quad \textbf{(3-80)}$$

Note that the glucose is oxidized (donates electrons) and the chromium is reduced (accepts electrons) from the hexavalent form (Cr^{6+}) to the trivalent form (Cr^{3+}). Thus, the COD test determines how much of the hexavalent chromium is reduced during the COD test. The amount of hexavalent chromium is then related to the amount of organic matter that was oxidized.

Engineered biological remediation of natural and synthetic organic chemicals typically consists of adding oxygen to convert the compounds to end products such as carbon dioxide and water. In addition, discharge of wastes to surface waters may contain organic matter that will react in the receiving lake or river. Chapter 5 covers the oxidation of organic matter and the associated consumption of oxygen. If the organic matter is assumed to be represented by glucose, the glucose can be oxidized to carbon dioxide and oxygen is reduced to water as follows:

$$C_6H_{12}O_6 + 6H_2O \longleftrightarrow 6CO_2 + 24H^+ + 24e^- \quad \textbf{(3-81)}$$

$$O_{2(aq)} + 4H^+ + 4e^- \longleftrightarrow 2H_2O \quad \textbf{(3-82)}$$

with an overall reaction of (after multiplying Equation 3-82 by 6 to balance the electrons)

$$C_6H_{12}O_6 + 6O_{2(aq)} \longleftrightarrow 6CO_2 + 6H_2O \quad \textbf{(3-83)}$$

Note that in this reaction the mole of glucose (the electron donor) donates 24 total electrons, which are accepted by the 6 moles of oxygen (the electron acceptor).

3.6 PHOTOCHEMISTRY AND THE ATMOSPHERE (URBAN SMOG AND THE OZONE HOLE)

3.6.1 Introduction

The stratosphere is the region of the atmosphere 15 to 50 km above the Earth's surface. The troposphere is the region from ground level to 15 km. Ozone concentrations in the stratosphere are in the ppm_V range. This ozone filters out some of the harmful ultraviolet (UV) rays that are emitted by the Sun. Tropospheric ozone concentrations reach several hundred ppb$_V$ in polluted urban areas. At these concentrations, ozone (called smog) can be harmful to humans and plants. Figure 3-23 shows the effect on visibility from smog in Los Angeles. In this section we discuss the general chemistry of tropospheric ozone (i.e., urban smog) and depletion of stratospheric ozone (i.e., the ozone hole). A more detailed chemical description is available elsewhere (Baird, 1995). This chemistry is interesting in that it uses photochemical reactions that are catalyzed by chemicals occurring naturally in the environment or are emitted by human activities. Photochemistry is extremely important in the study of atmospheric chemistry as well as some photochemical reactions that occur in the upper waters of lakes and rivers. An example of photochemistry in our daily lives is the fading of fabric dyes that have been exposed to sunlight. Another example is photosynthesis, which is perhaps the most important photochemical reaction to the world (photosynthesis is discussed in Chapter 5).

The different types of light can be differentiated according to the size of their wavelengths. Table 3-10 shows the entire electromagnetic spectrum. Photochemical reactions from UV light help control the chemistry of the ozone hole and urban smog. Infrared light is important in understanding the greenhouse effect, as will be shown in Example 4.11.

Light can be envisioned to consist of small bundles of energy called *photons*. These bundles of energy can be absorbed or emitted by matter. The energy of a photon, E (units of joules), is equal to

$$E = \frac{hc}{\lambda} \tag{3-84}$$

where h equals Planck's constant (6.626×10^{-34} J-s), c is the speed of light (3×10^8 m/s), and λ is the light's specific wavelength. Equation 3-84 shows that greater energy is contained in photons with a shorter wavelength.

This "light energy" can be absorbed by a molecule. A molecule that absorbs light energy has its energy increased by rotational, vibrational, and electronic excitation. The molecule then typically has a very short time (small fraction of a second) to either use the energy in a photochemical reaction or lose it (most likely as heat). *All atoms and molecules have a favored wavelength at which they absorb light.* That is, an atom or molecule will absorb light with a specific range of wavelength. Greenhouse gases such as water vapor, CO_2, N_2O, and CH_4 absorb energy emitted by the Earth as infrared light, while the major components of the atmosphere (N_2, O_2, Ar) are incapable of absorbing infrared light. It is this "capture" of energy released by the Earth that partially contributes to the warming of the Earth's surface. Anthropogenic emissions of greenhouse gases such as CO_2, CH_4, and CFCs have increased the amount of this energy that is "captured."

Another example of molecules absorbing light energy is in the filtering of the UV light that enters the Earth's atmosphere. O_2 molecules located above the

Figure 3-23. Photo showing visibility of mountains in Los Angeles on a day of good air quality (*a*) and a day where air quality is poor due to high levels of ozone (*b*). (Courtesy Tom Eichhorn/South Coast Air Quality Management District.)

Table 3-10. The Electromagnetic Spectrum

Wavelength (nm)	Range
<50	X-rays
50–400	Ultraviolet (UV)
400–750	Visible (400–450 = violet, and 620–750 = red)
>750	Infrared

stratosphere filter out (or absorb) most of the incoming UV light in the 120–220-nm range, and other gases such as N_2 filter out the UV light with wavelengths smaller than 120 nm. This means that no UV light with a wavelength below 220 nm reaches the surface of the Earth. All of the UV light in the 220–290-nm range is filtered out by ozone (O_3) molecules in the stratosphere with a little help from O_2 molecules. However, O_3 alone filters a fraction of UV light in the 290–320-nm range and the remainder makes it to our planet's surface. Overexposure to this portion of the light spectrum can result in malignant and nonmalignant skin cancer and adversely affect the human immune system and plant and animal growth. Most of the UV light in the 320–400-nm range reaches the Earth's surface, but fortunately this is the least harmful type of UV light to the Earth's biological systems.

3.6.2 Stratospheric Ozone Chemistry

The "ozone layer" is actually a misnomer because there is no actual layer in the stratosphere. Instead the middle part of the stratosphere (15–35 km) contains most of the protective ozone. Ozone concentrations never exceed 10 ppm_V in the stratosphere, so there is never really a lot around. In fact, if all of the ozone in the stratosphere were moved to the Earth's surface at 1-atm pressure, it would form a layer only approximately 0.3 mm thick. Thus, when people talk about the "ozone layer" they mean that most of the stratospheric ozone is located in the lower half of the stratosphere, about 15–35 km above the Earth's surface. Sometimes the thickness of the ozone layer is measured by Dobson units (DU). One DU equals 0.001-mm thickness of pure ozone at the density O_3 would have at 1-atm pressure. Ozone concentrations near the equator are approximately 250 DU, above North America about 350 DU, and in subpolar regions approximately 450 DU. This difference in the amount of O_3 between the equator and polar regions is caused by stratospheric winds that transport O_3 away from the equator toward both polar regions.

The "ozone hole" that appears each spring over the Antarctic lasts for several months and has resulted in measured O_3 levels as low as 90 DU. Their relative closeness to the southern polar region is one reason why countries such as Australia and New Zealand are world leaders in treating the "ozone layer" as an important public health issue.

Ozone Cycling in the Stratosphere. In the stratosphere, ozone is created, destroyed, and then recreated by several natural chemical reactions. However, the introduction of some anthropogenic chemicals has disrupted ozone's natural cycle. This has resulted in enhanced destruction rates of O_3, which has led to reduced ozone levels.

Stratospheric O_3 is formed by the reaction of atomic oxygen (O) with dimolecular oxygen (O_2) according to the following reaction:

$$O + O_2 \rightarrow O_3 \quad \textbf{(3-85)}$$

The O required in Equation 3-85 is derived from the reaction of O_2 with UV photons (λ = 241 nm) according to the following reaction:

$$O_2 + \text{UV photon} \rightarrow 2O \quad \textbf{(3-86)}$$

However, in the stratosphere the majority of oxygen exists as O_2, so only a little O is available. Therefore, even though there is little O in the stratosphere relative to O_2, small amounts of O created here will react with the abundant O_2 to form ozone (O_3) according to Equation 3-85.

Equations 3-85 and 3-86 explain the natural formation of ozone in the stratosphere. They also provide insight as to why the concentration of ozone is much, much higher in the stratosphere (pp**m**$_V$ levels) versus the troposphere (pp**b**$_V$ levels). This is because the stratosphere contains much more O than the troposphere, where O is not produced by natural mechanisms in large amounts except under human-induced conditions of smog formation (discussed later in this section).

Ozone is destroyed naturally in the stratosphere by the reaction of ozone with UV photons with λ = 320 nm:

$$O_3 + \text{UV photon} \rightarrow O_2 + O^* \quad \textbf{(3-87)}$$

Here, the molecule with the * superscript is defined as being in an "*excited state*" because of a change in its electron configuration. The molecule in this "excited state" has more energy compared to its "*ground state*," which is defined as the lowest energy state of a molecule. These "excited" oxygen molecules can either (1) react with O_2 to form more O_3, or (2) destroy O_3 to create O_2. Fortunately, the activation energy required for this destruction of ozone is quite high (18 kJ/mole), so this natural destruction reaction occurs at a very slow rate.

Catalytic Destruction of Ozone. The destruction of ozone in the atmosphere occurs in the presence of catalysts (referred to as "X" in Equations 3-88 and 3-89 below). These catalysts are either nonchlorine-containing chemicals that are produced by natural processes over the world's oceans, or are "halogen-containing" (usually chlorine) catalysts emitted by human activities. It is the accelerated emissions of these chlorine-containing catalysts by human activities that has resulted in decreases in stratospheric ozone, reported in the news as the "ozone hole."

The catalytic destruction of ozone occurs in a two-step process:

$$X + O_3 \rightarrow XO + O_2 \quad \textbf{(3-88)}$$

$$XO + O \rightarrow X + O_2 \quad \textbf{(3-89)}$$

The overall reaction for this two-step process is

$$O_3 + O \rightarrow 2\ O_2 \quad \textbf{(3-90)}$$

The catalyst, X, refers to an atom or molecule that has one unpaired electron. These atoms or molecules are called "free radicals" and will be designated by placing a "•" after the appropriate species (e.g., hydroxyl free radical is designated as OH•). Note that in Equations 3-88 to 3-89 the catalyst (X) can be recycled again and again, destroying more and more ozone. In fact, some anthropogenic catalysts can assist in the destruction of close to 10^4 ozone molecules before they are removed from the stratosphere!

Nonchlorine Catalysts. One of the most common natural catalysts is the free radical of nitric oxide (NO•), which is produced when an N_2O molecule rises into the stratosphere and happens to react with one of the excited atomic oxygen molecules (O*), whose formation was described by Equation 3-87. Although most of the N_2O and O* react to form N_2 and O_2, some reactions do produce NO•. NO• can be substituted for X into Equations 3-88 and 3-89 as follows:

$$NO\bullet + O_3 \rightarrow NO_2\bullet + O_2 \quad \textbf{(3-91)}$$

$$NO_2\bullet + O \rightarrow NO\bullet + O_2 \quad \textbf{(3-92)}$$

The overall reaction of Equations 3-91 and 3-92 results in the destruction of O_3 as provided by Equation 3-90.

Chlorine Catalysts. Free-radical chlorine atoms are very efficient catalysts in the destruction of ozone. Thus, the greatest threat to stratospheric O_3 is from chlorine-containing chemicals. Bromine-containing chemicals are also a problem and may account for up to 5% of the O_3 depletion, but their chemistry is not discussed here. If chlorine radicals (Cl•) are present in the stratosphere, O_3 destruction will occur according to the two-step reaction (Equations 3-88 and 3-89) followed by the O_3 destruction reaction (Equation 3-90) as follows:

$$Cl\bullet + O_3 \rightarrow ClO\bullet + O_2 \quad \textbf{(3-93)}$$

$$ClO\bullet + O \rightarrow Cl\bullet + O_2 \quad \textbf{(3-94)}$$

Remember that the overall reaction is written as

$$O_3 + O \rightarrow 2O_2 \quad \textbf{(3-95)}$$

Fortunately, 99% of stratospheric Cl is stored in nonreactive forms such as HCl and chlorine nitrate ($ClONO_2$). The amount of stratospheric chlorine has increased in recent decades, however, and during the Antarctic spring a lot of this stored chlorine is released into the active catalytic forms, Cl• and ClO•. A naturally occurring chlorine-containing chemical is chloromethane (CH_3Cl), which is formed over the world's oceans and may be transported up into the stratosphere. CH_3Cl molecules can react with UV photons (λ of 200–280 nm) to produce chlorine-free radicals, Cl•. The major anthropogenic source of chlorine is from the movement of chlorofluorocarbons (CFCs) into the stratosphere and subsequent release of chlorine free radicals. CFCs (known commercially as *freons*) were widely used in the northern hemisphere beginning in the 1930s. The three most commonly used CFCs were CFC-12 (CF_2Cl_2, used extensively as a coolant, refrigerant, and imbedded in rigid plastic foam); CFC-11 ($CFCl_3$, used to blow holes in soft plastic such as cushions, carpet padding, and car seats); and CFC-13 ($CF_2Cl\text{-}CFCl_2$, used to clean circuit boards). CFCs are relatively stable in the troposphere, but after being transported up into the stratosphere they can undergo photochemical reactions that release the catalytic chlorine free radical, Cl•. For example, the breakdown of CFC-12 occurs as follows:

$$CF_2Cl_2 + \text{UV photon (200–280-nm range)} \rightarrow CF_2Cl\bullet + Cl\bullet \quad \textbf{(3-96)}$$

Another Cl• can subsequently be released from the $CF_2Cl\bullet$.

3.6.3 Tropospheric Ozone Chemistry

Tropospheric ozone is also known as *urban ozone* or *smog*. It is not emitted by any smokestack or tailpipe, but is instead produced by a complex chemical process that involves hundreds of chemical reactions. In the troposphere high ozone levels can adversely affect the human respiratory system, agricultural crops and forests, and degrade materials such as rubber. Figure 3-23 showed how tropospheric ozone affects visibility in cities throughout the world.

Ozone production in the troposphere requires a source of nitrogen oxides, sunlight, and reactive hydrocarbons. The process is facilitated by natural features such as mountains, which can trap the emissions and limit their dilution by clean incoming air, or by a feature that causes the formation of air inversions, which trap pollutants close to ground level (e.g., being located near a large body of water). This local geography along with dense human populations contribute to the urban ozone problems of cities such as Los Angeles, Denver, Mexico City, and Rome. In addition, long-range transport of pollutants can carry an ozone-producing chemical soup to unfortunate regions that are located downwind. A simplified three-step reaction for tropospheric ozone production can be written as follows:

$$2\,NO + O_2 \longrightarrow 2\,NO_2 \quad \textbf{(3-97)}$$

$$NO_2 + h \longrightarrow NO\bullet + O \quad \textbf{(3-98)}$$

$$O + O_2 + M \longrightarrow O_3 + M \quad \textbf{(3-99)}$$

In these three chemical reactions, *h* refers to short wavelength light and M to a catalyst. Note that as with the catalytic destruction of stratospheric ozone, the catalyst used for tropospheric ozone production is released where it can assist in the production of additional ozone. The catalysts in this case are "reactive" hydrocarbons, which are produced by human activities such as in incomplete combustion and the evaporation of solvents and fuels. These "reactive" hydrocarbons can also be emitted by natural processes. For example, deciduous trees release hydrocarbons such as isoprene and conifer trees emit hydrocarbons such as pinene and limonene.

Equations 3-97 to 3-99 indicate that reduction of smog levels require efforts in reducing both the level of nitrogen oxides and the levels of reactive hydrocarbons. Nitrogen oxides are commonly referred to as NO_x, which equals the concentration of NO_2 plus the concentration of NO. NO_x is produced during any combustion process because oxygen and nitrogen make up a large percentage of the air (remember Table 2-2) used during combustion. For example, the reaction for NO production during any combustion process is given by

$$N_2 + O_2 = 2NO \tag{3-100}$$

This reaction is favored at high temperatures, so engineers have attempted to cool combustion temperatures in automobiles and power plants in order to reduce NO_x emissions. In addition, catalytic converters work by reducing the NO_x and hydrocarbon emissions from an automobile. In fact, they can reduce NO_x, CO, and hydrocarbon emissions up to 90%. However, using leaded fuel can "foul" the catalyst and destroy its effectiveness. Traffic planners can develop and encourage walking, cycling, mass transportation, or carpooling venues to decrease the amount of automobile emissions as well. In addition, traffic planners may substitute roundabouts for traffic signals to reduce air emissions. Finally, reactive hydrocarbons can be trapped at their source, as is done at many gasoline pumps. Also, control of hydrocarbons can occur by methods such as substitution of water-based paints for hydrocarbon-based paints or through destruction of hydrocarbon emissions at their source through catalytic oxidation or biofiltration.

CHAPTER PROBLEMS

3-1. How many grams of NaCl would you need to add to a one-liter water sample (pH = 7) so the ionic strength equaled 0.1M?

3-2. A first-order reaction that results in the destruction of a pollutant has a rate constant of 0.1/day. (a) How many days will it take for 90% of the chemical to be destroyed? (b) How long will it take for 99% of the chemical to be destroyed? (c) How long will it take for 99.9% of the chemical to be destroyed?

3-3. A strain of bacteria has been isolated that can cometabolize tetrachloroethane (TCA). This strain of bacteria can be used for the bioremediation of hazardous-waste sites contaminated with TCA. Assume that the biodegradation rate is independent of TCA concentration (i.e., the reaction is zero order). In a bioreactor it is observed that the rate for TCA removal was 1 μg/L-min. What water retention time would be required to reduce the concentration from 1 mg/L in the influent to 1 μg/L in the effluent of a reactor? Assume the reactor is completely mixed.

3-4. Assume that PO_4^{3-} is removed from municipal wastewater through precipitation with Fe^{3+} according to the following reaction: $PO_4^{3-} + Fe^{3+} \longrightarrow FePO_{4(s)}$.

The rate law for this reaction is

$$\frac{d[PO_4^{3-}]}{dt} = -k[Fe^{3+}][PO_4^{3-}]$$

(a) What is the reaction order with respect to PO_4^{3-}? (b) What order is this reaction overall?

3-5. Phosphate ion reacts in water to form monohydrogen phosphate according to the following equation:

$$PO_4^{3-} + H_2O \longleftrightarrow HPO_4^{2-} + OH^-$$

(a) Given that this is an ideal system, temperature is 298 K, and the total combined phosphate and monohydrogen phosphate is 10^{-4} M, what percentage of the total concentration is in the phosphate ion form at pH = 11? (b) Will the reaction proceed as written at pH = 9 when $[PO_4^{3-}] = 10^{-6.8}$ M, and $[HPO_4^{2-}] = 10^{-4}$ M, and if not, in which direction will it proceed? (c) Will the reaction proceed as written if 10^{-1} M NaCl solution is added to the system in part (b), and if not, in which direction will it proceed?

3-6. Ammonia, NH_3, is a common constituent of many natural waters and wastewaters. In treating water containing ammonia at a water-treatment plant, the ammonia reacts with the disinfectant hypochlorous acid, HOCl, in solution to form monochloroamine, NH_2Cl as follows:

$$NH_3 + HOCl \longrightarrow NH_2Cl + H_2O$$

The rate law for this reaction is

$$\frac{d[NH_3]}{dt} = -k[HOCl][NH_3]$$

(a) What is the reaction order with respect to NH_3? (b) What order is this reaction overall? (c) If the HOCl concentration is held constant and equals 10^{-4} M, and the rate constant equals 5.1×10^6 L/mole-s, calculate the time required to reduce the concentration of NH_3 to one-half its original value.

3-7. A treatability study is run to investigate the rate at which a hazardous chemical can be treated. The following data are collected at 25°C.

Concentration of Chemical (μg/L)	Time (days)
1.0	0
0.81	1
0.56	3
0.36	5
0.22	7
0.14	10

(a) Is this a zero- or first-order reaction? (b) What is the rate constant for this reaction? (c) What is the half-life for this reaction? (d) If Θ for this reaction is determined to be 1.1, what is the rate constant at 30°C?

3-8. A water-treatment plant is designed to remove dissolved manganese (Mn^{2+}) and iron (Fe^{2+}) from the water. The water is aerated to oxidize these reduced species and cause formation of a solid precipitate that can then be filtered out of the water. The reaction for Mn is

$$4Mn^{2+} + O_{2(aq)} + 8OH^- \rightarrow 4MnOOH_{(s)} + 2H_2O$$

(a) If the [Mn^{2+}] equals 25×10^{-6} M, pH is maintained at 7, temperature is 25°C, and the oxygen concentration is maintained at a constant value by blowing air equilibrated with the atmosphere into the solution, will the reaction proceed forward as written? To determine the order of the reaction and the rate constant, a laboratory experiment is conducted that yields the following results.

Time (min)	[Mn^{2+}](μM)
0	25
3	17.7
6	11.1
12	6.25

(b) Is this reaction first or zero order? (c) What is the rate constant?

3-9. Nitrogen dioxide (NO_2) concentrations are measured in an air-quality study and decrease from 5 ppm to 2 ppm in 4 min with a particular light intensity. (a) What is the first-order rate constant for this reaction? (b) What is the half-life of NO_2 during this study? (c) What would the rate constant need to be changed to in order to decrease the time required to lower the NO_2 concentration from 5 ppm to 2 ppm in 1.5 min?

3-10. You are provided the following rate constants for the hydrolysis of a variety of organic chemicals. Chloroform (7.13×10^{-10}/min) and trichloroethane (1.24×10^{-6}/min). (a) Calculate the half-lives in years for the hydrolysis of these chemicals. (b) Which chemical will persist longest in the environment given that hydrolysis is the only destruction reaction?

3-11. If the rate constant for the degradation of biochemical oxygen demand (BOD) at 20°C is 0.23/day, what is the value at 5°C and 25°C? Assume that Θ equals 1.1.

3-12. In environments where dissolved oxygen concentration is low, nitrate sometimes serves as an alternate electron acceptor in the oxidation of reduced species. Is it possible to oxidize sulfide with nitrate in natural waters under the following conditions at 25°C?

$$H^+ + NO_3^- + HS^- + H_2O \longleftrightarrow SO_4^{2-} + NH_4^+$$

The concentrations (moles/L) are $[H^+] = 10^{-8}$, $[NO_3^-] = [HS^-] = [SO_4^{2-}] = [NH_4^+] = 10^{-4}$. ΔG^0 for the preceding reaction is -2031.19 kJ/mole.

3-13. The reddish-colored mercury oxide (HgO) can decompose according to the following reaction:

$$HgO_{(s)} \rightarrow Hg_{(g)} + 1/2O_{2(g)}$$

Is this decomposition reaction a potential source of atmospheric mercury pollution at room temperatures of 25°C?

3-14. Will the following reaction proceed spontaneously under standard conditions?

$$Fe^{2+} + 4H^+ + O_{2(g)} \rightarrow 4Fe^{3+} + 2H_2O$$

3-15. Is the following reaction thermodynamically feasible? $-\Delta G_f^0$ for $NO_{(g)} =$ 20.69 kcal/mole; $NO_{2(g)} = 12.26$ kcal/mole; $O_{2(g)} = 0$ kcal/mole.

$$2NO_{(g)} + O_{2(g)} \rightarrow 2NO_{2(g)}$$

3-16. An engineer plans to treat wastes from a manufacturing process that utilizes chromium in a concrete tank. However, chromium may cause deterioration of concrete according to the following reactions:

$$2Fe^{3+} + (Fe, Mg)2SiO_{4(s)} \rightarrow Fe_2MgSiO_{4(s)} + 2Fe^{2+} + Mg^{2+}$$

$$3Fe^{2+} + CrO_4^{2-} + 8H^+ \rightarrow 3Fe^{3+} + Cr^{3+} + 4H_2O$$

$$Cr^{3+} + 3OH^- \rightarrow Cr(OH)_{3(s)}$$

The second reaction regenerates the Fe^{3+} required for the first reaction, and the process can continue until no more concrete or CrO_4^{2-} are available. Loss of Mg and Fe from the concrete in the first reaction can produce a more porous concrete that crumbles readily.

(a) Is the reaction shown in step 2 thermodynamically feasible? (b) Given the following concentrations in the reactor, will the integrity of the reactor be in danger? $[Fe^{3+}] = 10^{-10}$ M, $[Fe^{2+}] = 0.1$ μM, $[CrO_4^{2-}] = 10^{-6}$ M, $[Cr^{3+}] = 10$ nM, pH = 5.3.

3-17. (a) What is the solubility (in moles/L) of CaF_2 in pure water at 25°C? (b) What is the solubility of CaF_2 if the temperature is raised 10°C? (c) Does the solubility of CaF_2 increase, decrease, or remain the same if the ionic strength is raised? (Explain your answer.)

3-18. The chemical 1,4-dichlorobenzene (1,4-DCB) is sometimes used as a disinfectant in public lavatories. At 20°C (68°F) the vapor pressure is 5.3 ×

10^{-4} atm. (a) What would be the concentration in the air in units of g/m^3? The molecular weight of 1,4-DCB is 147 g/mole. (b) An alternative disinfectant is 1-bromo-4-chlorobenzene (1,4-CB). The boiling point of 1,4-CB is 196°C, whereas the boiling point of 1,4-DCB is 180°C. Which compound would cause the highest concentrations in the air in lavatories? (Explain your answer.)

3-19. The boiling temperatures of chloroform (an anesthetic), carbon tetrachloride (commonly used in the past for dry cleaning), and tetrachloroethene (previously used as a degreasing agent) are 61.7°C, 76.5°C, and 121°C. *The vapor pressure of a chemical is directly proportional to the inverse of the chemical's boiling point.* If a large quantity of these compounds were spilled in the environment, which compound would you predict to have higher concentrations in the air above the site? (Explain your answer.)

3-20. What would be the saturation concentration (mole/L) of oxygen (O_2) in a river in winter when the air temperature is 0°C if the Henry's law constant at this temperature is 2.28×10^{-3} mole L^{-1} atm^{-1}? What would the answer be in units of mg/L?

3-21. Polychlorinated biphenyls (PCBs) are a mixture of over 200 individual compounds. These pollutants have been spread widely throughout the environment. If the concentration of PCB-105 (one of the individual compounds) was 300 $pmole/m^3$ in the air above a large lake and the concentration in the surface water of the lake was 100 pmole/L, would this compound tend to move from the water into the air or from the air into the water? (Assume the air temperature is 25°C.) The Henry's law constant for this compound is 10 mole/L-atm.

3-22. On a cloudy, windy day in August the temperature above Lake Superior reached 25°C. The barometer read 760 mmHg, and a scientist aboard an EPA research vessel measured an oxygen concentration (partial pressure) of 0.18. If there is no water vapor in the air, the oxygen partial pressure would have been 0.2. Assuming that nitrogen, oxygen, and water vapor are the only important contributors to the gas pressure, calculate the equilibrium constant at 25°C for the reaction:

$$H_2O_{(liquid)} \longleftrightarrow H_2O_{(g)}$$

3-23. If the partial pressure of carbon dioxide at 25°C is $10^{-3.5}$ atm and its Henry's constant is $10^{-1.5}$ moles/L-atm, what is the molar concentration of dissolved carbon dioxide in rainwater equilibrated with the atmosphere? What is the Henry's constant for carbon dioxide in dimensionless form?

3-24. The log Henry constant (units of L-atm/mole and measured at 25°C) for trichloroethene is 1.03; for tetrachloroethene it is 1.44; for 1,2-dimethylbenzene it is 0.71; and for parathion it is −3.42. (a) What are the dimensionless Henry's constant for each of these chemicals? (b) Rank the chemicals in order of ease of stripping from water to air.

3-25. The dimensionless Henry's constant for trichloroethene (TCE) at 25°C is 0.4. A sealed glass vial is prepared that has an air volume of 4 mL overlying an aqueous volume of 36 mL. TCE is added to the aqueous phase so that initially it has an aqueous phase concentration of 100 ppb. After the system equilibrates, what will be the concentration (in units of μg/L) of TCE in the aqueous phase?

3-26. Determine the equilibrium pH of aqueous solutions of the following strong acids or bases:
(a) 15 mg/L of HSO_4^- (b) 10 mM NaOH (c) 2,500 μg/L of HNO_3

3-27. What would be the pH if 10^{-2} moles of hydrofluoric acid (HF) were added to one liter pure water? The pK_a of HF is 3.2.

3-28. When Cl_2 gas is added to water during the disinfection of drinking water, it hydrolyzes with the water to form HOCl. The disinfection power of the acid HOCl is 88 times better than its conjugate base, OCl^-. The pK_a for HOCl is 7.5. (a) What % of the total disinfection power (i.e., HOCl + OCl^-) exists in the acid form at a pH = 6? (b) At pH = 7?

3-29. Ammonia is a gas whose smell we detect in manure and in some cleaning solutions, among other things. The concentrations of this noxious gas can be controlled by regulating the pH of the water because only the nonprotonated form is volatile.

$$NH_4^+ \longleftrightarrow NH_3 + H^+$$

The pK_a for the preceding reaction is 9.3. What fraction of the total nitrogen ($NH_4^+ + NH_3$) is in the volatile form at pH 9.0?

3-30. The Henry's constant for H_2S is 0.1 moles/L-atm and

$$H_2S_{(aq)} \longleftrightarrow HS^- + H^+, \qquad K_a = 10^{-7}$$

If you bubble pure H_2S gas into a beaker of water, what is the concentration of HS^- at a pH of 5 in (a) moles/L; (b) mg/L; (c) ppm?

3-31. At a wastewater-treatment plant $FeCl_{3(s)}$ is added to remove excess phosphate from the effluent. Assume that the reactions that occur are

$$FeCl_{3(s)} \longleftrightarrow Fe^{3+} + 3Cl^-$$

$$Fe^{3+} + PO_4^{3-} \longleftrightarrow FePO_{4(s)}$$

The equilibrium constant for the second reaction is $10^{26.4}$. What concentration of Fe^{3+} would be needed to maintain the phosphate concentration below the limit of 1 mg P/L?

3-32. In a city's water distribution system for supplying drinking water, the concentration of Ca^{2+} is 1.5 mM and the concentration of CO_3^{2-} is 0.01 mM. Will calcite ($CaCO_{3(s)}$) precipitate or dissolve in the water pipes? Why? The reaction and equilibrium constant for the dissolution of calcite are

$$CaCO_{3(s)} \longleftrightarrow Ca^{2+} + CO_3^{2-} \qquad K = 10^{-8.34}$$

3-33. One method to remove metals from water is to raise the pH and cause them to precipitate as their metal hydroxides. (a) For the following reaction, compute the standard free energy of reaction:

$$Cd^{2+} + 2OH^- \longleftrightarrow Cd(OH)_{2(s)}$$

(b) The pH of water was initially 6.8 and was then raised to 8.0. Is the dissolved cadmium concentration reduced to below 100 mg/L at the final pH? Assume the temperature of the water is 25°C.

3-34. If the pH equals 7 and you add 10^{-6} moles of Fe^{3+} to one liter of water, how much precipitate will form (in mg Fe/L)?

3-35. Naphthalene has a log K_{ow} of 3.36. Estimate its soil–water partition coefficient normalized to organic carbon and also the 95% confidence interval of your estimate.

3-36. Atrazine is a herbicide widely used for corn and is a common groundwater pollutant in the corn-producing regions of the United States. The log K_{ow} for atrazine is 2.69. Calculate the fraction of total atrazine that will be adsorbed to the soil given that the soil has an organic carbon content of 2.5%. The bulk density of the soil is 1.25 g/cm^3; this means that each cubic centimeter of soil (soil plus water) contains 1.25 g of soil particles. The porosity of the soil is 0.4.

3-37. Alachlor is a widely used herbicide that has a log K_{ow} of 1.4. Calculate the fraction of alachlor that will be adsorbed to a flooded soil and the fraction that will be dissolved in the soil water given that the soil has an organic carbon content of 1%. The bulk density of the soil is 1.25 g/cm^3; this means that each cubic centimeter of soil (soil plus water) contains 1.25 g of soil particles.

3-38. Spills of organic chemicals that contact the ground sometimes reach the groundwater table, where they are then carried downgradient with the groundwater flow. The rate at which they are transported with the groundwater is decreased by sorption to the solids in the groundwater aquifer. The contaminated groundwater can reach wells, which is dangerous if the water is used for drinking. (a) For a soil having ρ_b of 2.3 g/cm^3, a porosity of 0.3, and the percent organic carbon equals 2%, determine the retardation factors of the following three chemicals: trichloroethene (log K_{ow} = 2.42); hexachlorobenzene (log K_{ow} = 5.50); and, dichloromethane (log K_{ow} = 1.15). (b) Which compound would be transported furthest, second furthest, and least furthest with the groundwater if these chemicals entered the same aquifer?

3-39. Chromium contamination is a problem at a number of hazardous-waste sites. At one site local concentrations of chromium reached 60,000 mg chromium/kg soil and 19,000 mg/L chromium in the water. (a) Assuming that the aqueous chromium reached equilibrium with the adsorbed chromium, determine a sorption coefficient (in L/kg) for the soil at the site. (b) At a site 200 ft downgradient, the aqueous chromium concentration was measured to be 1,000 mg/L. Compute the sorbed concentration in equilibrium with this aqueous concentration using the equilibrium constant determined in part (a). (c) In order to treat the contaminated groundwater at this site, a three-step process is used. First, the water is pumped out of the ground and the pH is then reduced from 3.5 to 2.3. Next the chromium is chemically reduced from hexavalent Cr(VI) to trivalent Cr(III). Finally, the chromium is removed from the water by raising the pH, causing $Cr(OH)_{3(s)}$ to precipitate. The equilibrium expression for this final reaction is

$$Cr(OH)_{3(s)} \longleftrightarrow Cr^{3+} + 3OH^- \qquad K_{sp} = 10^{-30}$$

If the pH of the water is raised to 8.5, what is the equilibrium concentration of the chromium(III) in mole/L? (d) What weight (in pounds) of precipitate is formed if groundwater containing 19,000 mg/L chromium is treated as described in part (c) and the volume of the precipitation tank is 6,000 gal?

3-40. Hexachlorobenzene (HCB) is one of ten common pollutants being monitored by the International Joint Commission for Great Lakes Research. Hexachlorobenzene has a log K_{ow} of 5.5, while benzene has a log K_{ow} of 2.1. Which compound would you predict to be more soluble in water? Why?

3-41. Your company's hydrogeologist reported that the velocity of groundwater in a contaminated aquifer was 1 m/day. The groundwater is contaminated with mercury from a municipal landfill. Given that the sorption constant of the mercury to the aquifer material is 5×10^3 L kg^{-1}, the porosity of the aquifer material is 0.15, and the bulk density of the aquifer material is 2.6 g cm^{-3}, determine the time (in years) before the mercury reaches a private well located 500 m downgradient.

3-42. Toluene has contaminated a city's drinking-water supply. An activated carbon system has been constructed so toluene concentrations are treated to below specified regulatory levels. If the toluene concentration in the water supply is 1,000 ppb, the treatment objective is 5 ppb, and the flow of water is 378 L/min (100 gpm), what mass of PAC must be placed in a contact basin on a daily basis to satisfy the treatment objective? The Freundlich parameters for toluene and the particular carbon are $K = 26.1$ mg/gm $(L/mg)^{1/n}$ and $1/n = 0.44$.

REFERENCES

Baird, C. 1995. Environmental Chemistry. W. H. Freeman. New York.

Baker, J. R., J. R. Mihelcic, D. C. Luehrs, and J. P. Hickey. 1997. Evaluation of estimation methods for organic carbon normalized sorption coefficients. Water Environment Research. 69:136–145.

CRC. 1991. Handbook of Chemistry and Physics. D. R. Lide, Ed. CRC Press, Boca Raton, Fla.

Finlayson-Pitts, B. J., and J. N. Pitts, Jr. 1986. Atmospheric Chemistry. John Wiley & Sons, New York.

Gossett, J. M. 1987. Measurement of Henry's law constants for C1 and C2 chlorinated hydrocarbons. Environmental Science & Technology. 21:202–208.

Hand, D. W., D. R. Hokanson, and J. C. Crittenden. 1998. Air stripping and aeration. *In* Water Quality and Treatment, Chap. 5, American Water Works Association, Denver, Colo.

Mihelcic, J. R., D. R. Lueking, R. J. Mitzell, and J. M. Stapleton. 1993. Bioavailability of sorbed and separate-phase chemicals. Biodegradation. 4:141–153.

Morel, F. M. M., and J. G. Hering. 1993. Principles and Applications of Aquatic Chemistry. John Wiley & Sons, New York.

Nelson, D. W., and L. E. Sommers. 1982. Total carbon, organic carbon, and organic matter. *In* Methods of Soil Analysis, A. L. Page, Ed. ASA and SSSA, Madison, Wis.

Perlinger, J. P., W. Angst, and R. P. Schwarzenbach. 1996. Kinetics of reduction of hexachloroethane by juglone in solutions containing hydrogen sulfide. Environmental Science & Technology. 30:3408–3417.

Roberts, P. V., M. N. Goltz, and D. A. Mackay. 1986. A natural gradient experiment on solute transport in a sand aquifer, 3. Retardation estimates and mass balances for organic solutes. Water Resources Research. 22(13):2047–2059.

Sawyer, C. N., P. L. McCarty, and G. F. Parkin. 1994. Chemistry for Environmental Engineering. McGraw-Hill, New York.

Schwarzenbach, R. P., P. M. Gschwend, and D. M. Imboden. 1993. Environmental Organic Chemistry. John Wiley & Sons, New York.

Snoeyink, V. L., and D. Jenkins. 1980. Water Chemistry. John Wiley & Sons, New York.

Sontheimer, H., J. C. Crittenden, and R. C. Summers. 1988. Activated Carbon for Water Treatment. DVGW-Forschungsstelle, Karlsruhe, Germany.

Subba-Rao, R. V., H. E. Rubin, and M. Alexander. 1982. Applied and Environmental Microbiology, 43:1139–1150.

Stumm, W., and J. J. Morgan. 1996. Aquatic Chemistry. John Wiley & Sons, New York.

Wania, F., and D. MacKay. 1996. Tracking the distribution of persistent organic pollutants. Environmental Science & Technology. 30(9):390A–396A.

Chapter 4

Physical Processes

Richard E. Honrath, Jr. and James R. Mihelcic

In this chapter, the reader will learn about the physical processes that are important in the movement of pollutants through the environment and in processes used to control and treat pollutant emissions. The chapter begins with a study of the use of material and energy balances. These tools are fundamental to environmental engineering and form the basis for reactor design. An understanding of mass and energy balances is essential to the solution of a greater number of environmental engineering and science problems.

The chapter continues with a discussion of the processes of advection and dispersion, which are responsible for the transport of pollutants through the environment. Transport of pollutants through the movement of wind and water currents (i.e., advection) is the mechanism by which pollutants can move great distances through the environment. On a smaller scale, however, molecular diffusion (i.e., random motion) is often more important than advection. On larger scales, turbulent dispersion can also be significant. For example, the movement of air pollutants on global scales can sometimes be best described as a process of turbulent dispersion.

The final section of this chapter extends this description of transport processes with a look at the movement of particles in fluids, and, in a reversal of that problem, the movement of a fluid through soil. The speed of particle settling and the velocity of groundwater flow is governed by the interplay of forces acting on the particle or the groundwater.

4.1 MASS BALANCES

The law of conservation of mass states that mass can neither be produced nor destroyed. Conservation of mass and conservation of energy (covered in Section 4.2) provide the basis for two tools that are used routinely in environmental engineering and science: the mass balance and the energy balance.

This principle of conservation of mass means that if the amount of a chemical somewhere (for example, in a lake) increases, then that increase cannot be the result of some "magical" formation. The chemical must have been either carried into the lake from elsewhere or produced via chemical or biological reaction from other compounds that were already in the lake. Similarly, if reactions produced the mass increase of this chemical, they must also have caused a corresponding decrease in the mass of some other compound(s).

Thus, conservation of mass provides a basis for compiling a budget of the mass of the chemical. In the case of a lake, this budget keeps track of the amounts of chemical entering the lake, the amount leaving the lake, and the amount formed or destroyed by chemical reaction. This budget can be balanced over a given time period, similarly to the way a checkbook is balanced. Equation 4-1 describes the mass balance:

$$\text{(Mass at time } t + \Delta t) = \text{(mass at time } t) + \begin{pmatrix} \text{mass that entered} \\ \text{from } t \text{ to } t + \Delta t \end{pmatrix} - \begin{pmatrix} \text{mass that exited} \\ \text{from } t \text{ to } t + \Delta t \end{pmatrix} + \begin{pmatrix} \text{net mass of chemical produced} \\ \text{from other compounds by} \\ \text{reactions between } t \text{ and } t + \Delta t \end{pmatrix} \tag{4-1}$$

Each term of Equation 4-1 has units of mass. This form of balance is most useful when there is a clear beginning and end to the balance period (Δt), so that the change in mass over the balance period can be determined. For example, when balancing a checkbook, a balance period of one month is often used.

In environmental problems, however, it is usually more convenient to work with values of *mass flux*—the *rate* at which mass enters or leaves a system. To develop an equation in terms of mass flux, the mass-balance equation is divided by Δt to produce an equation with units of mass per unit time. Dividing Equation 4-1 by Δt and moving the first term on the right (mass at time t) to the left-hand side yields the following equation.

$$\frac{\text{(Mass at time } t + \Delta t) - \text{(mass at time } t)}{\Delta t} = \frac{\begin{pmatrix} \text{mass entering from} \\ t \text{ to } t + \Delta t \end{pmatrix}}{\Delta t} - \frac{\begin{pmatrix} \text{mass exiting from } t \\ \text{to } t + \Delta t \end{pmatrix}}{\Delta t} + \frac{\begin{pmatrix} \text{net chemical} \\ \text{production} \\ \text{between } t \text{ and} \\ t + \Delta t \end{pmatrix}}{\Delta t} \tag{4-2}$$

Note that each term in Equation 4-2 has units of mass time^{-1}. The left-hand side of Equation 4-2 is equal to $\Delta m/\Delta t$. In the limit as $\Delta t \to 0$, the left-hand side becomes dm/dt, the rate of change of chemical mass in the lake. As $\Delta t \to 0$, the first two terms on the right side of Equation 4-2 become the rate at which mass enters the lake (the mass flux into the lake) and the rate at which mass exits the lake (the mass flux out of the lake). The last term of Equation 4-2 is the *net rate* of chemical production or loss.

The symbol $\dot{m}$ is used to refer to a mass flux with units of (mass time^{-1}). Making this substitution, the equation for mass balances is then

$$\begin{pmatrix} \text{mass} \\ \text{accumulation} \\ \text{rate} \end{pmatrix} = (\text{mass flux in}) - (\text{mass flux out}) + \begin{pmatrix} \text{net rate of} \\ \text{chemical} \\ \text{production} \end{pmatrix}$$

$$\text{or} \quad \frac{dm}{dt} = \dot{m}_{\text{in}} - \dot{m}_{\text{out}} + \dot{m}_{\text{reaction}} \tag{4-3}$$

Equation 4-3 is the governing equation for mass balances used throughout environmental engineering and science.

4.1.1 The Control Volume

A mass balance is only meaningful in terms of a specific region of space, which has boundaries across which the terms $\dot{m}_{in}$ and $\dot{m}_{out}$ are determined. This region is called the *control volume*. In the previous section, we used a lake as our control volume, and included mass fluxes into and out of the lake. Theoretically, any volume of any shape and location can be used as a control volume. Realistically, however, certain control volumes are more useful than others. *The most important attribute of a control volume is that it have boundaries over which $\dot{m}_{in}$ and $\dot{m}_{out}$ can be calculated.*

4.1.2 Terms of the Mass-Balance Equation for a CMFR

A well-mixed tank is an analog for many control volumes used in environmental situations. For example, in the lake example it might be reasonable to assume that the chemicals discharged into the lake are mixed throughout the entire lake. The term *completely mixed flow reactor* (CMFR) is used for such a system. (Other terms, most commonly continuously stirred tank reactor (CSTR), are also used for such systems.) A schematic diagram of a CMFR is shown in Figure 4-1, and two examples of CMFRs are depicted in Figure 4-2. The next sections describe each term in a mass balance of a hypothetical compound within the CMFR in Figure 4-1.

4.1.2.1 Mass Accumulation Rate, dm/dt

The rate of change of mass within the control volume, *dm/dt*, is referred to as the mass accumulation rate. To directly measure the mass accumulation rate, one would have to determine the total mass within the control volume of the compound for which the mass balance is being conducted. This would usually not be easy, and it is not usually necessary. If the control volume is well mixed, then the

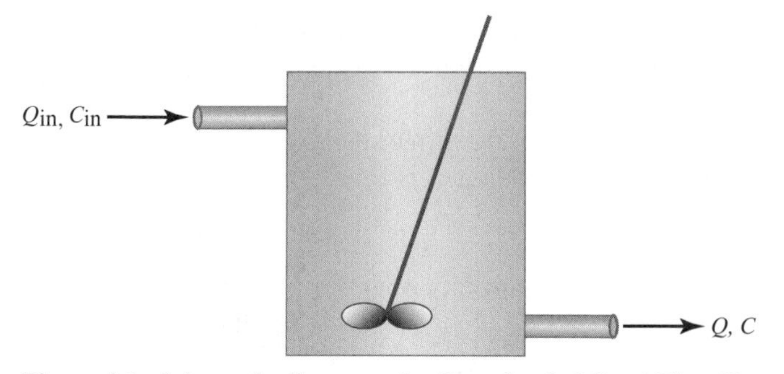

Figure 4-1. Schematic diagram of a Completely Mixed Flow Reactor (CMFR). The stir bar is used as a symbol to indicate that the CMFR is well mixed.

concentration of the compound is the same throughout the control volume, and the mass in the control volume is equal to the product of that concentration, C, and the volume, V. (To ensure that $C \times V$ has units of [mass time^{-1}], C must be expressed in [mass volume^{-1}] units.) Expressing mass as $C \times V$, the mass accumulation rate is equal to

$$\frac{dm}{dt} = \frac{d(VC)}{dt} \tag{4-4}$$

In most cases (and in all cases in this text), the volume is constant and can be moved outside the derivative, resulting in

$$\frac{dm}{dt} = V\frac{dC}{dt} \tag{4-5}$$

In any mass-balance situation, once a sufficient amount of time has passed conditions will approach *steady state*, meaning that conditions no longer change with time. In steady-state conditions, the concentration, and hence the mass, within the control volume remains constant. In this case, $dm/dt = 0$. If, however, insufficient time has passed since a flow, inlet concentration, reaction term, or other problem condition has changed, the mass in the control volume will vary with time, and the mass balance will be *non-steady state*.

How much time must pass before steady state is reached depends on the conditions of the problem. To see why, consider the approach to steady state of the amount of water in two large, initially empty sinks. In the first sink, the faucet is opened halfway, and the drain is opened slightly. Initially, the mass of water in the sink increases over time, since the faucet flow exceeds the flow rate out of the drain. Conditions are changing, so this is a non-steady-state situation. However, as the water level in the sink rises, the flow rate out of the drain will increase, and eventually the drain flow will equal the faucet flow. At this point, the water level will cease rising, and the situation will have reached steady state. If this experiment is then repeated with a second sink, but this time with the drain opened fully, the drain flow will increase more rapidly, and will equal the faucet

(*a*)

(*b*)

Figure 4-2. Examples of Completely Mixed Flow Reactors (CMFRs) in an engineered and a natural system. (*a*) Aerated wastewater lagoons use biological degradation to treat wastewater. The lagoons are generally earthen basins with a large surface area (up to several acres) and depth of 5 to 15 feet. With these conditions and a round or square layout, the lagoons usually attain completely mixed conditions. (Tom Hollyman/Photo Researchers.) (*b*) Onondaga Lake, located in Syracuse, New York, is an alkaline lake with a surface area of 11.7 km^2 and a mean depth of 12 m. Two tributaries account for 75% of the annual flow into the lake, and the local sewage treatment plant accounts for approximately 60% of the remainder. For many compounds, Onondaga Lake can be modeled as a CMFR. (Courtesy Martin Auer.)

flow while the water level in the sink is still low. In this case, then, steady state will be reached more rapidly. In general, the speed at which steady state is approached depends upon the magnitude of the mass flux terms, relative to the total mass in the control volume.

Determining whether a mass-balance problem is steady state or not is something of an art. However, if conditions of the problem have changed recently, then the problem is probably a non-steady state. On the other hand, if conditions have remained constant for a very long time, it is probably a steady-state problem. Treating a steady-state problem as non-steady state will always result in the correct answer, while treating a non-steady-state problem as steady state will not. (This does not mean that one should treat all problems as non-steady state, however. Non-steady-state solutions are generally more difficult, so it is advantageous to identify steady state whenever it is present.)

EXAMPLE 4.1. DETERMINING WHETHER A PROBLEM IS STEADY STATE

For each of the following mass-balance problems, determine whether a steady-state or non-steady-state mass balance would be appropriate.

(a) A mass balance on chloride (Cl^-) dissolved in a lake. Two rivers bring chloride into the lake, and one river removes chloride. No significant chemical reactions occur, as chloride is soluble and nonreactive. What is the annual average concentration of chloride in the lake?

(b) A degradation reaction within a well-mixed tank is used to destroy a pollutant. Inlet concentration and flow are held constant, and the system has been operating for several days. What is the pollutant concentration in the effluent, given the inlet flow and concentration and the first-order decay rate constant?

(c) The source of pollutant in problem (b) is removed, resulting in an instantaneous decline of the inlet concentration to zero. How long would it take until the outlet concentration reaches 10% of its initial value?

SOLUTION

(a) Over an annual period, river flows and concentrations can be assumed to be relatively constant. Since conditions are not changing, and since a single value independent of time is requested for chloride concentration, the problem is steady state.

(b) Again, conditions in the problem are constant and have remained so for a long time, so the problem is steady state. Note that the presence or absence of a chemical reaction does not provide any information on whether the problem is steady state or not.

(c) There are two clues that this problem is non-steady state. First, conditions have changed recently—the inlet concentration dropped to zero. Second, the solution requires calculation of a time period, which means that conditions must be varying with time.

4.1.2.2 *Mass Flux in, $\dot{m}_{in}$*

Often, the volumetric flow rate, Q, of each input stream entering the control volume is known. In Figure 4-1, the pipe has a flow rate of Q_{in}, with corresponding chemical concentration of C_{in}. The mass flux into the CMFR is then given by

$$\dot{m}_{in} = Q_{in} \times C_{in} \tag{4-6}$$

If it is not immediately clear how $Q \times C$ results in a mass flux, consider the units of each term:

$$\begin{array}{ccccc} \dot{m} & = & Q & \times & C \\ \left[\dfrac{\text{mass}}{\text{time}}\right] & = & \left[\dfrac{\text{volume}}{\text{time}}\right] & \times & \left[\dfrac{\text{mass}}{\text{volume}}\right] \end{array}$$

(Note that the concentration must be expressed in [mass volume^{-1}] units.)

If the volumetric flow rate is not known, it may be calculated from other parameters. For example, if the fluid velocity v and the cross-sectional area A of the pipe are known, then $Q = v \times A$.

In some situations, mass may enter the control volume through direct *emission* into the volume. In this case, the emissions are frequently specified in mass flux units (mass time^{-1}), which can be used in a mass balance directly. For example, if a mass balance is performed on the air pollutant carbon monoxide (CO) in the atmosphere over a city, estimates of the total CO emissions (in units of tons day^{-1}) from cars and power plants in the city would be used.

Another way to describe the flux is in terms of a *flux density*, J, times the area through which the flux occurs. J has units of *mass area*$^{-1}$ *time*$^{-1}$ and is discussed further when diffusion is covered in Section 4.3. This type of flux notation is most useful at interfaces where there is no fluid flow, such as the interface between the air and water at the surface of a lake.

Often, the mass flux in is composed of several terms. For example, a tank may have more than one inlet, or the air over a city may receive CO blowing from an upwind urban area in addition to its own emissions. In such cases, $\dot{m}_{in}$ is the sum of all individual contributions to mass input fluxes.

4.1.2.3 *Mass Flux out, $\dot{m}_{out}$*

In most cases, there is only one effluent flow from a CMFR. In such cases, the mass flux out may be calculated just as $\dot{m}_{in}$ was calculated in Equation 4-6:

$$\dot{m}_{out} = Q_{out} \times C_{out} \tag{4-7}$$

In the case of a well-mixed control volume, the concentration is constant throughout. Therefore, the concentration in flow exiting the control volume is referred to simply as C, the concentration in the control volume, and

$$\dot{m}_{\text{out}} = Q_{\text{out}} \times C \tag{4-8}$$

4.1.2.4 *Net Rate of Chemical Reaction, $\dot{m}_{\text{reaction}}$*

The term $\dot{m}_{\text{reaction}}$ ($\dot{m}_{rxn}$) refers to the net rate of production of a compound from chemical or biological reactions. It has units of mass time^{-1}. Thus, if other compounds react to form the compound, $\dot{m}_{rxn}$ will be greater than zero; if the compound reacts to form some other compound(s), resulting in a loss, $\dot{m}_{rxn}$ will be negative.

Although the chemical reaction term in a mass balance has units of mass time^{-1}, chemical reaction rates are usually expressed in terms of concentration, not mass. To calculate $\dot{m}_{rxn}$, then, we multiply the rate of change of concentration by the CMFR volume to obtain the rate of change of mass within the control volume:

$$\dot{m}_{rxn} = V \times \left(\frac{dC}{dt}\right)_{\text{reaction only}} \tag{4-9}$$

where the term $(dC/dt)_{\text{reaction only}}$ is obtained from the rate law for the reaction and is equal to the rate of change in concentration that would occur if the reaction took place in isolation, with no influent or effluent flows.

There are a number of possible forms for mass flux due to reaction. The most common include:

1. ***Conservative Compound.*** Compounds with no chemical formation or loss within the control volume are termed *conservative*. Conservative compounds are not affected by chemical or biological reactions, and thus $(dC/dt)_{\text{reaction only}} = \dot{m}_{\text{reaction}} = 0$. The term *conservative* is used for these compounds because their mass is truly conserved: what goes in is equal to what goes out.
2. ***Zero-order Decay.*** The rate of loss of the compound is constant. For a compound with zeroth-order decay, $(dC/dt)_{\text{reaction only}}$ equals $-k$ and $\dot{m}_{rxn}$ equals $-Vk$. Zero-order reactions are discussed in Chapter 3 (Section 3.2).
3. ***First-order Decay.*** The rate of loss of the compound is directly proportional to its concentration: $(dC/dt)_{\text{reaction only}}$ equals $-kC$. For such a compound, $\dot{m}_{rxn}$ equals $-VkC$. First-order reactions are discussed in Chapter 3 (Section 3.2).
4. Production at a rate dependent on the concentrations of other compounds in the CMFR. In this situation, the chemical is produced by reactions involving other compounds in the CMFR, and $(dC/dt)_{\text{reaction only}}$ is greater than zero.

4.1.2.5 Steps in Mass-Balance Problems

Solution of mass-balance problems involving CMFRs will generally be straightforward if the problem is done carefully. Most difficulties in solving mass-balance problems arise from uncertainty regarding the location of control volume boundaries or values of the individual terms in the mass balance. Therefore, it is suggested that the following steps be followed for each mass-balance problem.

1. Draw a schematic diagram of the situation, and identify the control volume and all influent and effluent flows. All mass flows that are known or are to be calculated must cross the control-volume boundaries, and it should be reasonable to assume that the control volume is well mixed.
2. Write the mass-balance equation in general form:

$$\frac{dm}{dt} = \dot{m}_{\text{in}} - \dot{m}_{\text{out}} + \dot{m}_{rxn}.$$

3. Determine whether the problem is steady state ($dm/dt = 0$) or non-steady state ($dm/dt = V\,dC/dt$).
4. Determine whether the compound being balanced is conservative ($\dot{m}_{rxn} = 0$) or nonconservative ($\dot{m}_{rxn}$ must be determined based on the reaction kinetics and Equation 4-9).
5. Replace $\dot{m}_{\text{in}}$ and $\dot{m}_{\text{out}}$ with known or required values, as just described.
6. Finally, solve the problem. This will require solution of a differential equation in non-steady-state problems and solution of an algebraic equation in steady-state problems.

4.1.3 Reactor Analysis: The CMFR

Reactor analysis refers to the use of mass balances to analyze pollutant concentrations in a control volume that is either a chemical reactor or a natural system modeled as a chemical reactor. Ideal reactors can be divided into two types: CMFRs (Completely Mixed Flow Reactors) and PFRs (Plug Flow Reactors). CMFRs are used to model well-mixed environmental reservoirs. Plug Flow Reactors (PFRs) (described in Section 4.1.5 below) behave essentially like pipes, and are used to model situations such as downstream transport in a river, in which fluid is not mixed in the upstream–downstream direction.

This section presents several examples involving CMFRs in different combinations of steady-state or non-steady-state conditions and conservative or nonconservative compounds, as summarized in Table 4-1. Example 4.2 demonstrates the use of CMFR analysis to determine the concentration of a substance resulting from the mixing of two or more influent flows. Examples 4.3 through 4.5 refer to the tank depicted in Figure 4-1 and demonstrate steady-state and non-steady-state situations with and without first-order chemical decay. Calculations analogous to those in Examples 4.3, 4.4, and 4.5 can be used to determine the concentration of pollutants exiting a treatment reactor, the rate of increase of pollutant

Table 4-1. Summary of CMFR Examples

Example Number	Form of *dm*/*dt*	Form of $\dot{m}_{\text{reaction}}$
Example 4.2	Steady state	Conservative
Example 4.3	Steady state	First-order decay
Example 4.4	Non-steady state	First-order decay
Example 4.5	Non-steady state	Conservative

concentrations within a lake resulting from a new pollutant source, or the period required for pollutant levels to decay from a lake or reactor once a source is removed.

EXAMPLE 4.2. STEADY-STATE CMFR WITH CONSERVATIVE CHEMICAL: THE MIXING PROBLEM

A pipe from a municipal wastewater-treatment plant discharges 1.0 $m^3\ s^{-1}$ of poorly treated effluent containing 5.0 mg L^{-1} of phosphorus compounds (reported as mg P L^{-1}) into a river with an upstream flow rate of 25 $m^3\ s^{-1}$ and a background phosphorus concentration of 0.010 mg P L^{-1} (see Figure 4-3). What is the resulting concentration of phosphorus (in mg L^{-1}) in the river just downstream of the plant outflow?

SOLUTION

To solve this problem, two mass balances will be applied to first determine the downstream volumetric flow rate (Q_d) and, second, determine the downstream phosphorus concentration (C_d). First a control volume must be selected. To en-

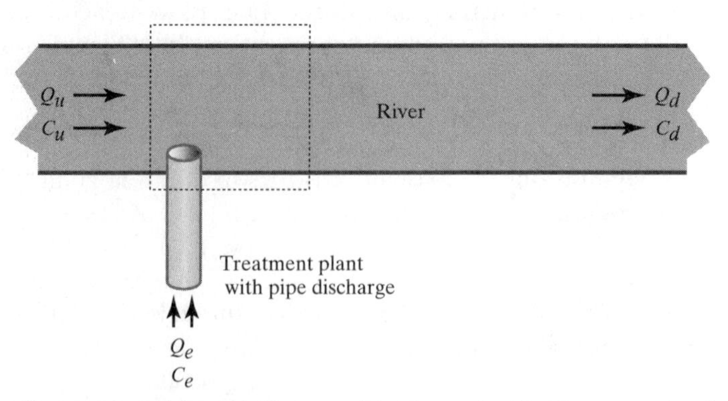

Figure 4-3. Mixing problem used in Example 4.2. The control volume is indicated by the area inside the dashed lines.

sure that the input and output fluxes cross the control volume boundaries, the control volume must cross the river upstream and downstream of the plant's outlet and must also cross the discharge pipe. The selected control volume is shown in Figure 4-3 as a dotted line. It is assumed to extend down river far enough that the discharged wastewater and the river water become well mixed before leaving the control volume. As long as that assumption is met, it makes no difference to the analysis how far downstream the control volume extends.

Before beginning the analysis, it should be determined whether this is a steady-state or non-steady-state problem, and whether the chemical reaction term will be nonzero. Because the problem statement does not refer to time at all, and it seems reasonable to assume that both the river and wastestream discharge have been flowing for some time and will continue to flow, this is a steady-state problem. In addition, this problem concerns the concentration resulting from rapid mixing of the river and effluent flows. Therefore, we can define our control volume to be small, and can safely assume that chemical or biological degradation is insignificant during the time spent in the control volume and thus treat this as a steady-state problem.

Step 1. Determine the downstream flow rate, Q_d. Q_d can be found by conducting a mass balance on the total river-water mass. In this case, the "concentration" of river water in (mass volume^{-1}) units is simply the density of the water, ρ:

$$\frac{dm}{dt} = \dot{m}_{\text{in}} - \dot{m}_{\text{out}} + \dot{m}_{rxn}$$

$$= \rho Q_{\text{in}} - \rho Q_{\text{out}} + 0$$

where the term $\dot{m}_{\text{reaction}}$ has been set to zero because the mass of water is conserved. Since this is a steady-state problem, $dm/dt = 0$. Therefore, as long as the density ρ is constant, $Q_{\text{in}} = Q_{\text{out}}$, or $(Q_u + Q_e) = 26 \text{ m}^3 \text{ s}^{-1} = Q_d$.

Step 2. Determine the phosphorus concentration downstream of the discharge pipe, C_d. C_d can be found by using the standard mass-balance equation with steady-state conditions and with no chemical formation or decay:

$$\frac{dm}{dt} = \dot{m}_{\text{in}} - \dot{m}_{\text{out}} + \dot{m}_{rxn}$$

$$0 = (C_u Q_u + C_e Q_e) - C_d Q_d + 0$$

and

$$C_d = \frac{C_u Q_u + C_e Q_e}{Q_d}$$

$$= \frac{(0.010 \text{ mg L}^{-1})(25 \text{ m}^3 \text{ s}^{-1}) + (5.0 \text{ mg L}^{-1})(1.0 \text{ m}^3 \text{ s}^{-1})}{26 \text{ m}^3 \text{ s}^{-1}}$$

$$= 0.20 \text{ mg L}^{-1}$$

EXAMPLE 4.3. STEADY-STATE CMFR WITH FIRST-ORDER DECAY

The CMFR shown in Figure 4-1 is used to treat an industrial waste product, using a reaction that destroys the pollutant according to first-order kinetics, with $k = 0.216\ \text{day}^{-1}$. The reactor volume is 500 m^3, the volumetric flow rate of the single inlet and exit is 50 $m^3\ \text{day}^{-1}$, and the inlet pollutant concentration is 100 mg L^{-1}. What is the outlet concentration after treatment?

SOLUTION

An obvious control volume is the tank itself. The problem requests a single, constant outlet concentration, and all problem conditions are constant. Therefore, this is a steady-state problem ($dm/dt = 0$). The mass-balance equation with a first-order decay term ($[dC/dt]_{\text{reaction only}} = -kC$ and $\dot{m}_{rxn} = -VkC$) is

$$\frac{dm}{dt} = \dot{m}_{\text{in}} - \dot{m}_{\text{out}} + \dot{m}_{rxn}$$

$$0 = QC_{\text{in}} - QC - VkC$$

Solving for C, we find that

$$C = C_{\text{in}} \times \frac{Q}{Q + kV}$$

or

$$C = C_{\text{in}} \times \frac{1}{1 + kV/Q}$$

The numerical solution is

$$C = 100\ \text{mg L}^{-1} \times \frac{50\ \text{m}^3\ \text{day}^{-1}}{50\ \text{m}^3\ \text{day}^{-1} + (0.216\ \text{day}^{-1})(500\ \text{m}^3)}$$

$$= 32\ \text{mg L}^{-1}$$

EXAMPLE 4.4. NON-STEADY-STATE CMFR WITH FIRST-ORDER DECAY

The manufacturing process that generates the waste in Example 4.3 has to be shut down, and, starting at $t = 0$, the concentration C_{in} entering the CMFR is set to 0. What is the outlet concentration as a function of time after the concentration is set to 0? How long does it take the tank concentration to reach 10% of its initial, steady-state value?

SOLUTION

The tank is again the control volume. In this case, the problem is clearly non-steady-state, because conditions change as a function of time. The mass-balance equation is

$$\frac{dm}{dt} = \dot{m}_{\text{in}} - \dot{m}_{\text{out}} + \dot{m}_{rxn}$$

$$V\frac{dC}{dt} = 0 - QC - kCV$$

and

$$\frac{dC}{dt} = -\left(\frac{Q}{V} + k\right)C$$

(a) To determine C as a function of time, the preceding differential equation must be solved. Rearranging and integrating

$$\int_{C_0}^{C_t} \frac{dC}{C} = \int_0^t -\left(\frac{Q}{V} + k\right) dt$$

Integration yields

$$\ln(C) - \ln(C_0) = -\left(\frac{Q}{V} + k\right)t$$

Because $(\ln(x) - \ln(y))$ is equal to $\ln(x/y)$,

$$\ln\left(\frac{C}{C_0}\right) = -\left(\frac{Q}{V} + k\right)t$$

which yields

$$\frac{C_t}{C_0} = e^{-(Q/V+k)t}$$

We can verify that this solution is reasonable by considering what happens at $t = 0$ and $t = \infty$. At $t = 0$, the exponential term is equal to 1 and $C = C_0$, as expected. As $t \rightarrow \infty$, the exponential term approaches zero, and concentration declines to zero, again as expected, since C_{in} is equal to zero.

We can now plug in values to determine the dependence of C on time. Example 4.3 provides Q and V. The initial concentration is equal to the concentration before C_{in} was set to zero, which was found to be 32 mg/L in Example 4.3. Plugging in these values yields the outlet concentration as a function of time

$$C_t = 32 \text{ mg L}^{-1} \times \exp\left[-\left(\frac{50 \text{ m}^3 \text{ day}^{-1}}{500 \text{ m}^3} + \frac{0.216}{\text{day}}\right)t\right]$$

$$= 32 \text{ mg L}^{-1} \times \exp\left(-\frac{0.316}{\text{day}}t\right)$$

This solution is plotted in Figure 4-4*a*.

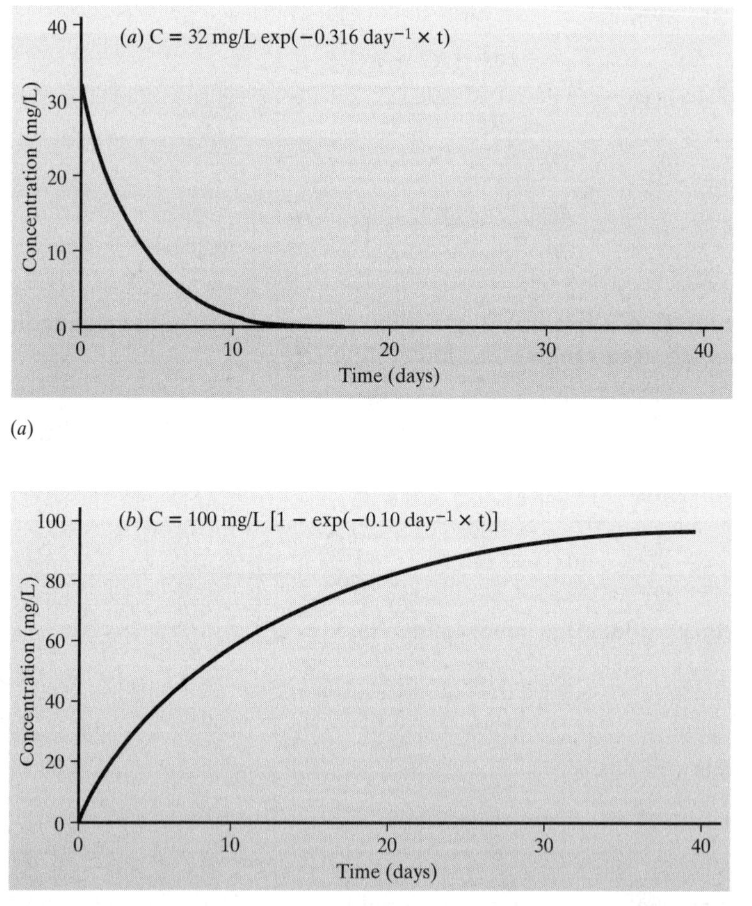

Figure 4-4. Concentration versus time profiles for the solutions to Examples 4.4 and 4.5. (*a*) First-order decay in concentration resulting from the removal of $\dot{m}_{\text{in}}$ at time zero. The decay in concentration results from the sum of chemical reaction loss ($\dot{m}_{rxn}$) and the mass flux out term ($\dot{m}_{\text{out}}$). (*b*) Exponential approach to steady-state conditions when a reactor is started with initial concentration equal to zero. In the absence of a chemical reaction loss term, concentration in the reactor exponentially approaches the inlet concentration.

(b) How long will it take the concentration to reach 10% of its initial, steady-state value? That is, at what value of t is $C_t/C_0 = 0.10$? At the time when $C_t/C_0 = 0.10$,

$$\frac{C}{C_0} = 0.10 = \exp\left(-\frac{0.316}{\text{day}}t\right)$$

Taking the natural logarithm of both sides,

$$\ln(0.10) = -2.303 = -\frac{0.316}{\text{day}}t$$

or

$$t = 7.3 \text{ days}$$

EXAMPLE 4.5. NON-STEADY-STATE CMFR, CONSERVATIVE SUBSTANCE

The CMFR reactor depicted in Figure 4-1 is filled with clean water prior to being started. After start-up, a waste stream containing 100 mg L^{-1} of a conservative pollutant is added to the reactor at a flow rate of 50 m^3 day^{-1}. The volume of the reactor is 500 m^3. What is the concentration exiting the reactor as a function of time after it is started?

SOLUTION

Again, the tank will serve as a control volume. We are told that the pollutant is conservative, so $\dot{m}_{rxn} = 0$. The problem asks for concentration as a function of time, so the mass balance must be non-steady state. The mass balance equation is

$$\frac{dm}{dt} = \dot{m}_{\text{in}} - \dot{m}_{\text{out}} + \dot{m}_{rxn}$$

$$V\frac{dC}{dt} = QC_{\text{in}} - QC + 0$$

and

$$\frac{dC}{dt} = -\left(\frac{Q}{V}\right)(C - C_{\text{in}})$$

Because of the extra term on the right (C_{in}), this equation cannot be immediately solved. However, with a change of variables we can transform the mass-balance equation into a simpler form that can be integrated directly, using the same method used in Example 4.4. Let $y = (C - C_{\text{in}})$. Then, $dy/dt =$

$dC/dt - d(C_{in}/dt)$. Since C_{in} is constant, $dC_{in}/dt = 0$, and thus $dy/dt = dC/dt$. Therefore, the last of the preceding equations is equivalent to

$$\frac{dy}{dt} = -\frac{Q}{V} y.$$

Rearranging and integrating,

$$\int_{y(0)}^{y(t)} \frac{dy}{y} = \int_0^t -\frac{Q}{V} dt$$

which yields

$$\ln\left(\frac{y(t)}{y(0)}\right) = -\frac{Q}{V} t$$

or

$$\frac{y(t)}{y(0)} = e^{-(Q/V)t}$$

Replacing y with $(C - C_{in})$, the following equation is obtained

$$\frac{C - C_{in}}{C_0 - C_{in}} = e^{-(Q/V)t}$$

Since clean water is present in the tank at start-up, $C_0 = 0$, and

$$\frac{C - C_{in}}{-C_{in}} = e^{-(Q/V)t}$$

Rearranging results in

$$C - C_{in} = -C_{in}e^{-(Q/V)t}$$

or

$$C = C_{in} \times (1 - e^{-(Q/V)t})$$

This is the solution to the question posed in the problem statement. Note what happens as $t \to \infty$: $e^{-(Q/V)t} \to 0$ and $C \to C_{in}$. This is not surprising, since this is a conservative substance. If the reactor is run for a long enough period, the concentration in the reactor will eventually reach the inlet concentration. This final equation (plotted in Figure 4-4*b*) provides C as a function of time, and can be used to determine how long it would take for the concentrations to reach, say, 90% of the inlet value.

4.1.4 The Batch Reactor

A reactor that has no inlet or outlet flows is termed a *batch reactor*. A batch reactor is essentially just a tank in which a reaction is allowed to occur. After one batch is treated, the reactor is emptied and a second batch can be treated. Because there are no flows, $\dot{m}_{\text{in}} = 0$ and $\dot{m}_{\text{out}} = 0$. Therefore, the mass-balance equation reduces to

$$\frac{dm}{dt} = \dot{m}_{rxn} \tag{4-10}$$

or

$$V\frac{dC}{dt} = V\left(\frac{dC}{dt}\right)_{\text{reaction only}} \tag{4-11}$$

or

$$\frac{dC}{dt} = \left(\frac{dC}{dt}\right)_{\text{reaction only}} \tag{4-12}$$

Thus, in a batch reactor, the change in concentration with time is simply that resulting from chemical reaction alone. For example, for a first-order decay reaction, $r = -kC$ and thus

$$\frac{dC}{dt} = -kC \tag{4-13}$$

or

$$\frac{C_t}{C_0} = e^{-kt} \tag{4-14}$$

as discussed in Section 3.2.2.

4.1.5 The Plug-Flow Reactor

The *plug-flow reactor* (PFR) is used to model the chemical transformation of compounds as they are transported in systems resembling "pipes." The "pipe" may represent a river, a region between two mountain ranges through which air flows, or a variety of other engineered or natural conduits through which liquids

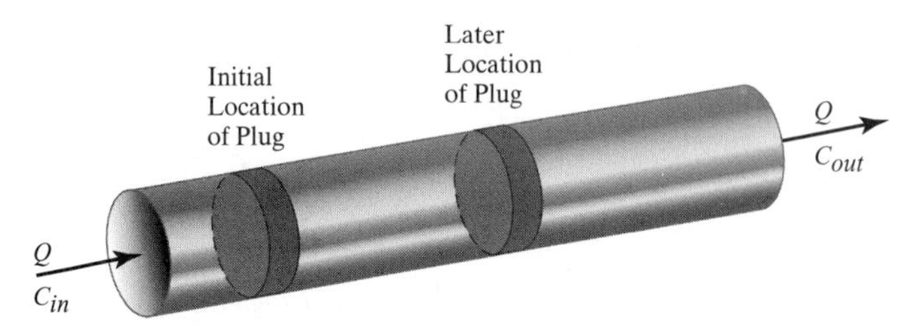

Figure 4-5. Schematic diagram of a plug-flow reactor.

or gases flow. Of course, it can even represent a pipe. A schematic diagram of a PFR is shown in Figure 4-5, and examples of PFRs in an engineered and a natural system are shown in Figure 4-6.

As fluid flows down the PFR, the fluid is mixed in the radial direction, but mixing does not occur in the axial direction. That is, each *plug* of fluid is considered a separate entity as it flows down the pipe. However, time passes as the plug of fluid moves downstream (or downwind). Therefore, there is an implicit time

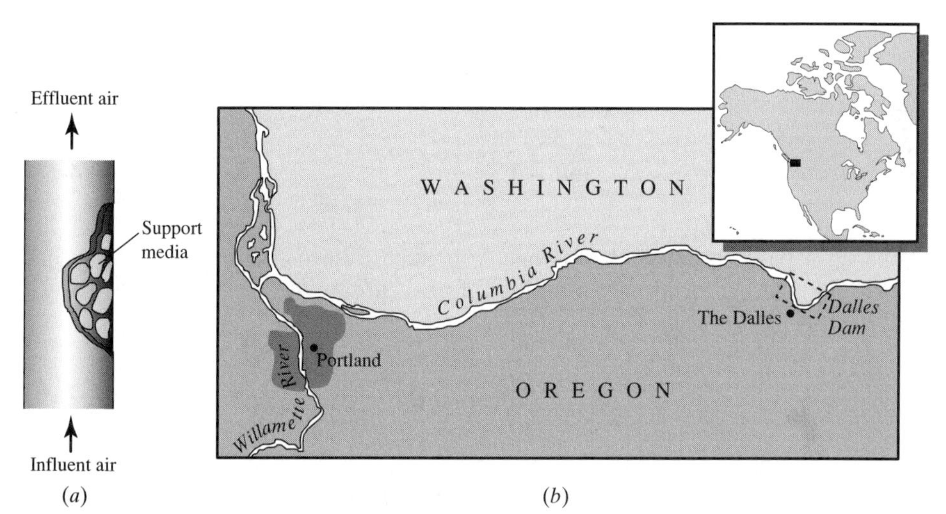

Figure 4-6. Examples of plug-flow reactors in an engineered and a natural system. (*a*) Packed-tower biofilters are used to remove organic compounds from wastewater and odorous air emissions, such as hydrogen sulfide (H_2S) from gas-phase emissions. Biofilters consist of a column packed with a support medium, such as rocks, plastic rings, or activated carbon, on which a biofilm is grown. Contaminated water or air is passed through the filter, and bacterial degradation results in the desired reduction of pollutant emissions. (*b*) The Columbia River flows 1,200 mi from its source in Canada to the Pacific Ocean. Prior to reaching the ocean, the Columbia River flows southward into the United States and forms the border between the states of Oregon and Washington. Shown is a stretch of the river near The Dalles, Washington, where the river once narrowed and spilled over a series of rapids, christened *les Dalles*, or the trough, by the voyageurs—early French explorers. A large dam has since been constructed near The Dalles. The section of the river downstream of the dam could be modeled as a plug-flow reactor.

dependence even in steady-state PFR problems. *However, because the velocity of the fluid* (v) *in the PFR is constant, time and downstream distance* (x) *are interchangeable, and* $t = x/v$. That is, it always takes an amount of time equal to x/v for a fluid plug to travel a distance x down the reactor. This observation can be used with the mass-balance formulations just given to determine how chemical concentrations vary during flow through a PFR.

To develop the equation governing concentration as a function of distance down a PFR, the evolution of concentration with time within a single fluid plug will be analyzed. The plug is assumed to be well mixed in the radial direction, but does not mix at all with the fluid ahead or behind it. As the plug flows downstream, chemical decay occurs, and concentration decreases. The mass balance for mass within this moving plug is then the same as that for a batch reactor:

$$\frac{dm}{dt} = \dot{m}_{\text{in}} - \dot{m}_{\text{out}} + \dot{m}_{rxn} \tag{4-15}$$

$$V\frac{dC}{dt} = 0 - 0 + V\left(\frac{dC}{dt}\right)_{\text{reaction only}} \tag{4-16}$$

where $\dot{m}_{\text{in}}$ and $\dot{m}_{\text{out}}$ are set equal to zero because there is no mass exchange across the plug boundaries.

Equation 4-16 can be used to determine concentration as a function of flow time within the PFR for any reaction kinetics. In the case of first-order decay, $V(dC/dt)_{\text{reaction only}} = -VkC$, and

$$V\frac{dC}{dt} = -VkC \tag{4-17}$$

which results in

$$\frac{C_t}{C_0} = \exp(-kt) \tag{4-18}$$

It is generally desirable to express the concentration at the outlet of PFR in terms of the inlet concentration and PFR length or volume, rather than time spent in the PFR. In a PFR of length L, each plug travels for a period $\theta = L/v = L \times A/Q$, where A is the cross-sectional area of the PFR and Q is the flow rate. The product of length and cross-sectional area is simply the PFR volume, so Equation 4-18 is equivalent to

$$\frac{C_{\text{out}}}{C_{\text{in}}} = \exp -\frac{kV}{Q} \tag{4-19}$$

Note that Equation 4-19 has no time dependence. Although concentration within a given plug changes over time as that plug flows downstream, the concentration at a given fixed location within the PFR is constant with respect to time, since all plugs reaching that location have spent an identical period in the PFR.

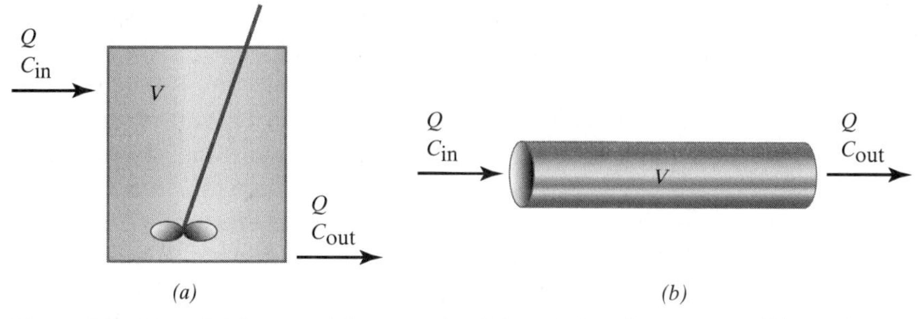

Figure 4-7. Completely mixed flow reactor (*a*) and plug-flow reactor (*b*) used to compare the two reactor types.

4.1.5.1 Comparison of the PFR to the CMFR

The ideal CMFR and the PFR are fundamentally different and thus behave differently. When a parcel of fluid enters the CMFR, it is immediately mixed throughout the entire volume of the CMFR. In contrast, each parcel of fluid entering the PFR remains separate during its passage through the reactor. These differences are highlighted using an example involving the continuous addition of a pollutant to each reactor, with destruction of the pollutant within the reactor according to first-order kinetics. The two reactors are depicted in Figure 4-7.

This example assumes that the incoming concentration (C_{in}), the flow rate (Q), and the first-order reaction rate constant (k) are known and are the same for both reactors. Then, two common problems are considered: (1) if the volume V is known (the same for both reactors), what is the resulting outlet concentration (C_{out}) exiting the CMFR and PFR? and (2) if an outlet concentration is specified,

Table 4-2. Comparison of CMFR and PFR Performance*

CMFR	PFR
Example 1. Determine C_{out}, given C_{in}, V, Q, and k.†	
$C_{out} = C_{in}/(1 + kV/Q)$	$C_{out} = C_{in} \exp(-kV/Q)$
$C_{out}/C_{in} = 0.50$	$C_{out}/C_{in} = 0.37$
Example 2. Determine V, given C_{in}, C_{out}, Q, and k.‡	
$V = (C_{in}/C_{out} - 1) \times (Q/k)$	$V = -(Q/k) \ln(C_{out}/C_{in})$
$V = 1{,}900\ L$	$V = 300\ L$

*Example 1 compares the effluent concentration (C_{out}) for a PFR and CMFR of the same volume; Example 2 compares the volume required for each reactor type if a given percent removal is required.

†$V = 100\ L$, $Q = 5.0\ L\ s^{-1}$, $k = 0.05\ s^{-1}$.

‡$C_{out}/C_{in} = 0.5$, $Q = 5.0\ L\ s^{-1}$, $k = 0.05\ s^{-1}$.

what volume of reactor is required for the CMFR and for the PFR? Table 4-2 summarizes the results of this comparison and lists the input variables.

The results shown in Table 4-2 indicate that, for equal reactor volumes, the PFR is more efficient than the CMFR and, for equal outlet concentrations, a smaller PFR is required. Why is this? The answer has to do with the fundamental difference between the two reactors—fluid parcels entering the PFR travel downstream without mixing, while fluid parcels entering the CMFR are immediately mixed with the low-concentration fluid within the reactor. Since the rate of chemical reaction is proportional to concentration, the result is that the rate of chemical reaction within the CMFR is reduced, relative to that within the PFR.

This effect is illustrated in Figure 4-8. The mass flux due to reaction is equal to ($-VkC$) in both reactors. However, in the PFR, concentration decreases exponentially as each plug passes through the PFR, as shown by the solid curve in Figure 4-8. The average mass flux due to reaction in the PFR is simply the average value of this curve—the value indicated by the dashed line in Figure 4-8. In

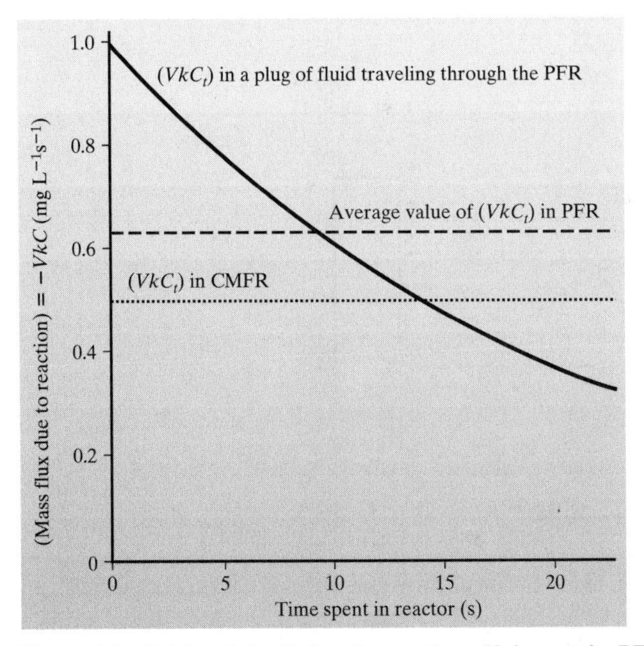

Figure 4-8. Origin of the higher destruction efficiency of a PFR under conditions of first-order decay. The rate of chemical destruction ($\dot{m}_{rxn} = -VkC$) is shown as a function of time spent in the reactor for a PFR (solid line) and CMFR (dotted line) for the conditions given in example 1 of Table 4-2. Since concentration changes as each plug passes through the PFR ($C = C_{\text{in}} \exp[-k\Delta t]$), the value of $\dot{m}_{rxn}$ changes. The average rate of destruction in the PFR is shown by a dashed line. The rate of chemical destruction is constant throughout the well-mixed CMFR and is equal to $-VkC$. Since the high inlet concentration is diluted immediately upon entering the CMFR, the rate of reaction is lower than that throughout most of the PFR, and is lower than the average rate of reaction in the PFR.

contrast, dilution as the incoming fluid is mixed into the CMFR immediately reduces the influent concentration to that within the CMFR, resulting in a reduced rate of destruction, indicated by the dotted line in Figure 4-8.

4.1.5.1.1 Response to Inlet Spikes. CMFRs and PFRs also differ in their response to spikes in the inlet concentration. In many pollution-control systems, inlet concentrations or flows are not constant. For example, flow into municipal wastewater-treatment plants varies dramatically over the course of each day. It is often necessary to ensure that a temporary increase in inlet concentration does not result in excessive outlet concentrations. This requires the use of CMFRs, as a result of the mixing that occurs within CMFRs but not within PFRs.

Consider the effect of a temporary doubling of the concentration entering a PFR and CMFR, each designed to reduce the influent concentration by the same amount, with the flow, first-order decay rate constant, and required degree of destruction equal to the values given in example 2 of Table 4-2.

The resulting changes in outlet concentration for the PFR and CMFR are shown in Figure 4-9. The concentration in fluid exiting the CMFR begins to rise

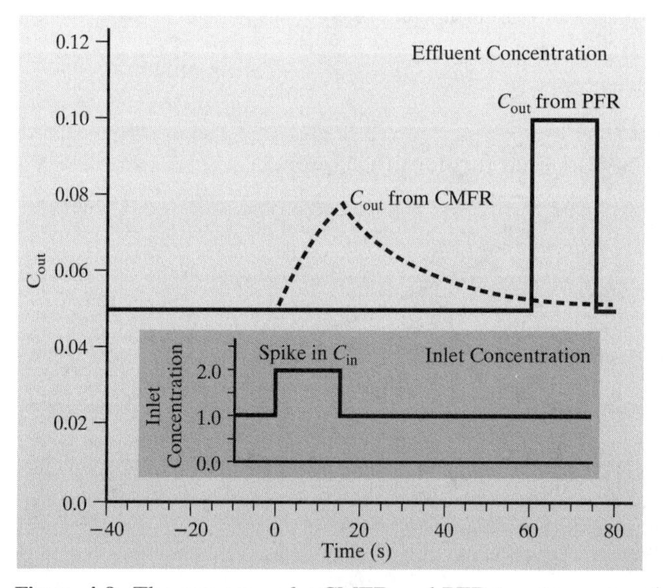

Figure 4-9. The response of a CMFR and PFR to a temporary increase in inlet concentration. The influent concentration, shown in the lower, inset figure, increases to 2.0 during the period t = 0–15 s. The resulting concentrations exiting the CMFR and PFR of example 2 of Table 4-2 are shown as a function of time before, during, and after the temporary doubling of inlet concentration. The concentration exiting the CMFR is shown with a dashed line; the concentration exiting the PFR is shown with a solid line. The maximum concentration reached in the CMFR effluent is less than that reached in the PFR effluent because the increased inlet concentration is diluted by the volume of lower-concentration fluid within the CMFR.

immediately after the inlet concentration increases, as the more concentrated flow is mixed throughout the CMFR. The outlet concentration does not immediately double in response to the doubled inlet concentration, however, as the higher-concentration influent flow is diluted by the volume of low-concentration fluid within the CMFR. The CMFR outlet concentration rises exponentially and would eventually double, but the inlet spike does not last long enough for this to occur. In contrast, the outlet concentration exiting the PFR does not change until enough time has passed for the first plug of higher-concentration fluid to traverse the length of the PFR. At that time, the outlet concentration doubles, and it remains elevated for a period equal to the duration of the inlet spike.

4.1.5.1.2 Selection of CMFR or PFR. Selection of a CMFR or PFR in an engineered system is based on the considerations just described—control efficiency as a function of reactor size, and response to changing inlet conditions. In many cases, a CMFR may be used to reduce sensitivity to spikes and be followed by a PFR for efficient use of resources.

In natural systems, the choice is based upon whether or not the system is mixed (in which case a CMFR would be used to model the system) or flows downstream without mixing (requiring use of a PFR). In some cases, it is necessary to use both the CMFR and PFR models. A common example of this occurs in cases involving effluent flow into a river. A CMFR is used to define a mixing problem, as was done in Example 4.2. This sets the inlet concentration for a PFR, which is used to model degradation of the pollutant as it flows further downstream. (This type of problem is investigated further in Chapter 5.)

EXAMPLE 4.6. REQUIRED VOLUME IN A PFR

Determine the volume required for a PFR to obtain the same degree of pollutant reduction as the CMFR in Example 4.3. Assume that the flow rate and first-order decay rate constant are unchanged ($Q = 50\ \text{m}^3\ \text{day}^{-1}$, $k = 0.216\ \text{day}^{-1}$).

SOLUTION

The CMFR in Example 4.3 achieved a pollutant decrease of $C_{\text{out}}/C_{\text{in}} = 32/100 = 0.32$. From Equation 4-19,

$$\frac{C_{\text{out}}}{C_{\text{in}}} = e^{-(kV/Q)}$$

or

$$0.32 = \exp -\frac{0.216\ \text{day}^{-1}V}{50\ \text{m}^3\ \text{day}^{-1}}$$

Solving for V,

$$V = \ln(0.32) \times \frac{50 \text{ m}^3 \text{ day}^{-1}}{-0.216 \text{ day}^{-1}}$$

$$= 264 \text{ m}^3$$

As expected, this volume is smaller than the 500 m^3 required for the CMFR in Example 4.3.

4.1.6 Retention Time and Other Expressions for *V/Q*

A number of terms, including *retention time*, *detention time*, and *residence time*, are used to refer to the average period spent in a given control volume, θ. The retention time is given by

$$\theta = \frac{V}{Q} \quad \textbf{(4-20)}$$

where V is the volume of the reactor, and Q is the total volumetric flow rate exiting the reactor.

Typical retention times in some engineered systems are shown in Table 4-3. Examples 4.7 and 4.8 illustrate the calculation and application of retention time.

Table 4-3. Typical Retention Times in Unit Processes Used for Treating Drinking Water and Wastewater

Unit Operation	Used for	Approximate Retention Time
Wastewater Treatment		
Grit removal	Removal of large particles (grit)	30 min
Primary settling	Removal of large solids	$\leq$ 1 h
Secondary Settling	Removal of smaller solids	$\leq$ 2 h
Activated sludge	Removal of organic matter using microorganisms and oxygen	4–8 h
Anaerobic digester	Stabilization of organic matter in sludge in absence of oxygen	15–30 days
Drinking-water Treatment		
Rapid-mix tank	Blending of chemical coagulants with water prior to treatment	< 1 min
Flocculator	Gentle mixing to promote flocculation of small particles	30 min
Disinfection	Destruction of pathogens	< 15 min

EXAMPLE 4.7. RETENTION TIME IN A CMFR AND A PFR

Calculate the retention times in the CMFR of Example 4.3 and the PFR of Example 4.6.

SOLUTION

For the CMFR,

$$\theta = \frac{V}{Q} = \frac{500\ \text{m}^3}{50\ \text{m}^3\ \text{day}^{-1}} = 10\ \text{days}$$

For the PFR,

$$\theta = \frac{V}{Q} = \frac{264\ \text{m}^3}{50\ \text{m}^3\ \text{day}^{-1}} = 5.3\ \text{days}$$

EXAMPLE 4.8. RETENTION TIMES FOR THE GREAT LAKES

The Great Lakes region is shown in Figure 4-10. Calculate the retention times for Lake Michigan and Lake Ontario using the data provided in Table 4-4.

SOLUTION

For Lake Michigan,

$$\theta = \frac{4{,}900 \times 10^9\ \text{m}^3}{36 \times 10^9\ \text{m}^3\ \text{yr}^{-1}} = 136\ \text{yr}$$

For Lake Ontario,

$$\theta = \frac{1{,}634 \times 10^9\ \text{m}^3}{212 \times 10^9\ \text{m}^3\ \text{yr}^{-1}} = 8\ \text{yr}$$

These values mean that Lake Michigan changes its water volume completely once every 136 years and Lake Ontario once every 8 years. The higher flow and smaller volume of Lake Ontario result in a significantly shorter retention time. This means that pollutant concentrations can increase in Lake Ontario much more quickly than they can in Lake Michigan, but it also means that concentrations will drop much more quickly in Lake Ontario if a pollutant source is eliminated, provided that flow out of the lakes is the dominant pollutant sink.

These values of θ can be used to determine whether it would be appropriate to model the lakes as CMFRs in a mass-balance problem. As discussed in Chapter

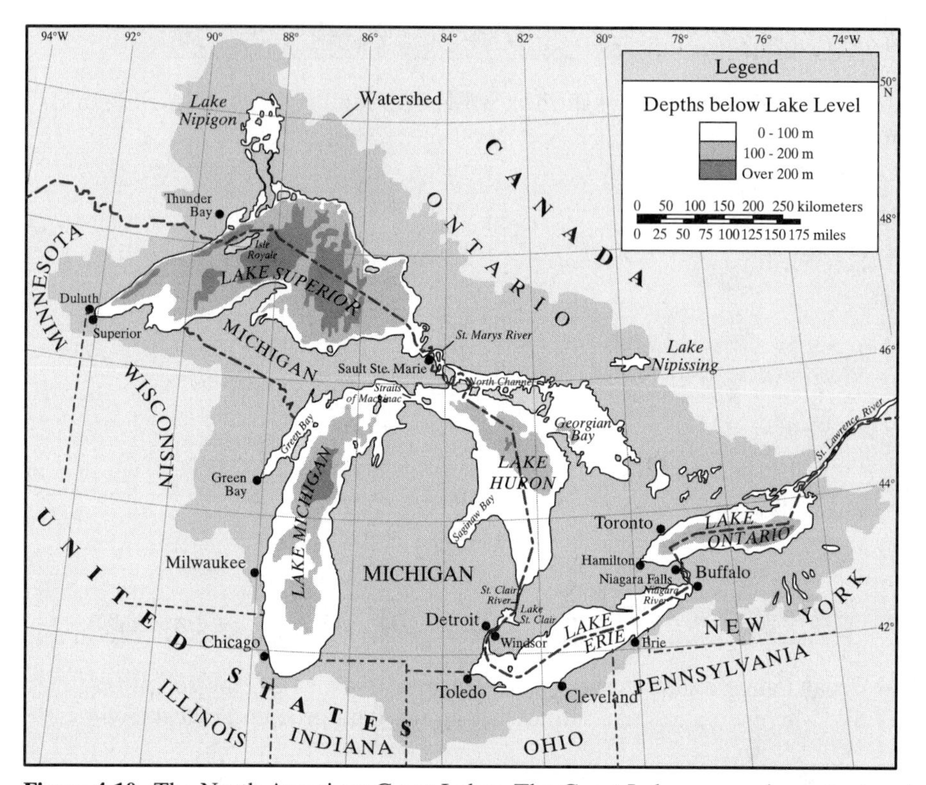

Figure 4-10. The North American Great Lakes. The Great Lakes are an important part of the physical and cultural heritage of North America. The Great Lakes contain approximately 18% of the world's supply of freshwater, making them the world's largest system of available surface freshwater (only the polar ice caps contain more freshwater). The first humans arrived in the area approximately 10,000 years ago. Around 6,000 years ago, copper mining began along the south shore of Lake Superior and hunting/fishing communities were established throughout the area. Population in the region in the sixteenth century is estimated at between 60,000 and 117,000, a level that resulted in few human disturbances. Today, the combined Canadian and U.S. population in the region exceeds 33 million. Increases in settlement and exploitation over the past two hundred years have caused many disturbances to the ecosystem. The outflow from the Great Lakes is less than 1% per year. Therefore, pollutants that enter the lakes by air, direct discharge, or from nonpoint pollution sources may remain in the system for a long period of time, as demonstrated in Example 4.8.

5, temperate lakes generally are mixed twice per year. Therefore, over the period required for water to flush through Lakes Michigan and Ontario, the lakes would be mixed many times. It would therefore be appropriate to model the lakes as CMFRs in mass balances involving pollutants that do not decay significantly in less than approximately 1 year.

Table 4-4. Volume and Flows for the Great Lakes

Lake	Volume 10^9 m^3	Outflow 10^9 m^3 yr^{-1}
Superior	12,000	67
Michigan	4,900	36
Huron	3,500	161
Erie	468	182
Ontario	1,634	211

Source: Chapra and Reckhow, 1983.

4.2 ENERGY BALANCES

Modern society is dependent on the use of energy. Such use requires transformations in the form of energy and control of energy flows. For example, when coal is burned at a power plant, the chemical energy present in the coal is converted to heat, which is then converted in the plant's generators to electrical energy. Eventually, the electrical energy is converted back into heat for warmth or used to do work. However, energy flows and transformation can also cause environmental problems. For example, thermal heat energy from electrical power plants can result in increased temperatures in rivers used for cooling water, "greenhouse" pollutants in the atmosphere alter the energy balance of the Earth and may cause significant increases in global temperatures, and burning of fossil fuels to produce energy is associated with emissions of pollutants.

The movement of energy and changes in its form can be tracked using *energy balances*, which are analogous to mass balances. The first law of thermodynamics states that energy can neither be produced nor destroyed. Conservation of energy provides a basis for energy balances, just as the law of conservation of mass provides a basis for mass balances. However, all energy balances are treated as conservative; as long as all the possible forms of energy are considered (and in the absence of nuclear reactions), there is no term in energy balances that is analogous to the chemical-reaction term in mass balances.

4.2.1 Forms of Energy

The forms of energy can be divided into two types: *internal energy* and *external energy*. Energy that is a part of the molecular structure or organization of a given substance is internal. Energy that results from the location or motion of the substance is external. Examples of external energy include *gravitational potential energy* and *kinetic energy*. Gravitational potential energy is the energy gained when a mass is moved to a higher location above the Earth. Kinetic energy is the energy that results from the movement of objects. When a rock thrown off of a cliff accelerates toward the ground, the sum of kinetic and potential energy is

Table 4-5. Some Common Forms of Energy Encountered in Environmental Engineering and Science

	Representation for Energy or Change in Energy
Heat internal energy	ΔE = (mass)$c\Delta T$
Chemical internal energy	$\Delta E = \Delta H_{rxn}$ at constant volume
Gravitational potential	ΔE = (mass)(Δheight)
Kinetic	E = (mass)(velocity)^2/2
Electromagnetic	E = (Planck's constant)(photon frequency)

conserved (neglecting friction): as the rock falls, it loses potential energy, but increases in speed, gaining kinetic energy. Examples of some common forms of energy are provided in Table 4-5.

Heat is a form of internal energy—it results from the random motions of atoms. Heat is thus really a form of kinetic energy, although it is considered separately, because the motion of the atoms cannot be seen. When a pot of water is heated, energy is being added to the water. That energy is stored in the form of internal energy, and the change in internal energy of the water is given by

$$\text{Change in internal energy} = (\text{mass of } H_2O) \times c \times \Delta T \qquad \textbf{(4-21)}$$

where c is the heat capacity or specific heat of the water, with units of energy mass^{-1} temperature^{-1}. Heat capacity is a property of a given material. For water, the heat capacity is 4,184 J kg^{-1} C^{-1} (1 BTU lb^{-1} F^{-1}).

Chemical internal energy reflects the energy in the chemical bonds of a substance. This form of energy is composed of two parts:

1. ***The strengths of the atomic bonds in the substance.*** When chemical reactions occur, if the sum of the internal energies of the products is less than that for the reactants, a reduction in chemical internal energy has occurred. As a result of the conservation of energy, this leftover energy must show up in a different form. Usually, the energy is released as heat. The most common example of this is the combustion of fuel, in which hydrocarbons and oxygen react to form carbon dioxide and water. The chemical bonds in carbon dioxide and water are much lower in energy than are those in hydrocarbons, and therefore combustion releases a significant amount of heat.
2. ***The energy in the interactions between molecules.*** Solids and liquids form as a result of interactions between adjacent molecules. These bonds are much weaker than are the chemical bonds between atoms in molecules, but are still important in many energy balances. The energy required to break these bonds is referred to as *latent heat.* Values of latent heat are tabulated for various substances for the phase changes from solid to liquid and from liquid to gas. The *latent heat of condensation* for a given substance is equal

to the heat released when a unit of mass of the substance condenses to form a liquid. (An equal amount of energy is required for evaporation.) The *latent heat of fusion* is equal to the heat released when a unit of mass solidifies. (Again, an equal amount of energy is required to melt the substance.)

4.2.2 Conducting an Energy Balance

In analogy with the mass-balance equation (Equation 4-3), the following equation can be used to conduct energy balances:

$$\begin{pmatrix} \text{Change in} \\ \text{internal plus} \\ \text{external energy} \\ \text{per unit time} \end{pmatrix} = (\text{energy flux in}) - (\text{energy flux out})$$

or

$$\frac{dE}{dt} = \dot{E}_{in} - \dot{E}_{out} \qquad \textbf{(4-22)}$$

The use of this relationship is illustrated in Examples 4.9 and 4.10.

EXAMPLE 4.9. HEATING WATER

A 40-gallon electric water heater is used to heat tap water (temperature 10°C). The heating level is set to the maximum level while several people take consecutive showers. If, at the maximum heating level, the heater uses 5 kW of electricity, and the water use rate is a continuous 2 gal min^{-1}, what is the temperature of the water exiting the heater? Assume that the system is at steady state and that the heater is 100% efficient; that is, it is perfectly insulated and all of the energy used goes to heat the water.

SOLUTION

The control volume is the water heater. Because the system is at steady state, dE/dt is equal to zero. The energy flux added by the electric heater is used to heat water entering the water heater to the temperature at the outlet. The energy balance is thus

$$\frac{dE}{dt} = 0 = \dot{E}_{in} - \dot{E}_{out}$$

The energy flux into the water heater comes from two sources: the heat content of the water entering the heater and the electrical heating element. The heat content of the water entering the heater is the product of the water-mass flux, the heat capacity, and the inlet temperature. The energy added by the heater is given as 5 kW.

The energy flux out of the water heater is just the internal energy of the water leaving the system ($\dot{m}_{H_2O} \times c \times T_{out}$). There is no net conversion of other forms of energy. Therefore, the energy balance may be rewritten as

$$0 = (\dot{m}_{H_2O}\, cT_{in} + 5 \text{ kW}) - \dot{m}_{H_2O}\, cT_{out}$$

Each term of this equation is an *energy flux*, and has the units of (energy time^{-1}). To solve, each term must be placed in the same units, in this case watts (1 W equals 1 J s^{-1}). In addition, the water flow rate (gallons min^{-1}) needs to be converted to units of mass of water per unit time using the density of water. Combining the first and third terms

$$0 = \dot{m}_{H_2O}\, c(T_{out} - T_{in}) + 5\ kW \qquad \textbf{(4-23)}$$

$$0 = \frac{2 \text{ gal } H_2O}{\text{min}} \times \frac{3.785 \text{ L}}{\text{gal}} \times \frac{1.0 \text{ kg}}{\text{L}} \times \frac{4{,}184\ J}{\text{kg} \times {}^\circ\text{C}} \times (T_{in} - T_{out}) + \frac{5{,}000\ J}{\text{s}} \times \frac{60 \text{ s}}{\text{min}}$$

$$= 3.16 \times 10^4 \frac{J}{\text{min} \times {}^\circ\text{C}} \times (T_{in} - T_{out}) + 3.00 \times 10^5 \frac{J}{\text{min}}$$

or

$$T_{out} = T_{in} + 9.5^\circ\text{C}$$

$$= (10 + 9.5) = 19.5^\circ\text{C}$$

This is a cold shower! This makes sense; many people have had the experience of taking a shower after the hot water in the tank was used up by previous showers.

EXAMPLE 4.10. HEATING WATER

Example 4.9 showed that it is necessary to wait until the water in the tank is reheated before a hot shower can be taken. How long would it take the temperature to reach 54°C if no hot water were used during the heating period and the water temperature started at 20°C?

SOLUTION

In this case, the only energy input is the electrical heat, and there is no energy leaving the tank. Therefore, the rate of increase in internal energy is equal to the rate at which electrical energy is used:

$$\frac{dE}{dT} = \dot{E}_{in} - \dot{E}_{out} = \dot{E}_{in} - 0$$

From Table 4.5, $\Delta E = (\text{mass})c\Delta T$, so

$$\frac{dE}{dT} = \frac{(\text{mass of } H_2O) \times c \times \Delta T}{\Delta\text{time}}$$

and

$$\frac{(\text{mass of } H_2O) \times c \times \Delta T}{\Delta\text{time}} = \dot{E}_{in} = 5{,}000\ J\ s^{-1}.$$

This expression can be solved for Δtime, given that ΔT is equal to $(54 - 20) = 34°C$.

$$\Delta\text{time} = \frac{(\text{mass of } H_2O) \times c \times \Delta T}{5{,}000\ J\ s^{-1}}$$

$$= \frac{\left(40 \text{ gal } H_2O \times \frac{3.785 \text{ L}}{\text{gal}} \times \frac{1.0 \text{ kg}}{\text{L}}\right)\left(4{,}184 \frac{J}{\text{kg} \times °\text{C}}\right)(130 - 20°\text{C})}{5{,}000\ J\ \text{s}^{-1}}$$

$$= 4.3 \times 10^3 \text{ s} = 1.2\ h$$

In Examples 4.9 and 4.10, energy balances were conducted for the heating of water in an intentional and desirable energy transfer process. The following two examples describe unintentional or undesirable impacts of energy balances.

Example 4.11 deals with the impacts of waste heat. The second law of thermodynamics states that *heat energy cannot be converted to work with 100% efficiency*. As a result, a significant fraction of the heat released in electrical power plants is lost as waste heat; in modern large power plants, this loss accounts for 65–70% of the total heat released from combustion.

Example 4.12 deals with the energy balance of the Earth as a whole. This energy balance is being increasingly altered by human activities, mainly through the addition to the atmosphere of carbon dioxide from fuel combustion.

EXAMPLE 4.11. THERMAL POLLUTION

A typical large coal-fired electric power plant produces 1,000 MW of electricity by burning fuel with an energy content of 2,800 MW. Three hundred and forty MW are lost as heat up the smokestack, leaving 2,460 MW to power turbines that drive a generator to produce electricity. However, the thermal efficiency of the turbines is only 42%. That is, 42% of this power goes to drive the generator, but the rest (58% of 2,460 = 1,430 MW) is waste heat that must be removed by cooling water.

Assume that cooling water from an adjacent river, which has a total flow rate of 100 $m^3\ s^{-1}$, is used to remove this waste heat. How much will the temperature of the river rise as a result of the addition of this heat?

SOLUTION

This problem is very similar to Example 4.9, in that a specified amount of heat is being added to a flow of water, and the resulting temperature rise must be determined. An energy balance can be written over the region of the river to which the heat is added. Here T_{in} represents the temperature of the water upstream, and T_{out} represents the temperature after heating.

$$\frac{dE}{dt} = \dot{E}_{in} - \dot{E}_{out}$$

$$0 = \begin{pmatrix}\text{1,430 MW of heat} \\ \text{from power plant}\end{pmatrix} + (\dot{m}_{H_2O} \times c \times T_{H_2O,\,in}) - (\dot{m}_{H_2O} \times c \times T_{H_2O,\,out}).$$

Rearranging,

$$\dot{m}_{H_2O} \times c \times (T_{out} - T_{in}) = 1{,}430 \text{ MW}$$

The remainder of this problem is essentially unit conversion. To obtain $\dot{m}_{H_2O}$ requires multiplication of the given river volumetric flow rate by the density of water (approximately 1,000 kg m^{-3}). The heat capacity of water, c = 4,184 J kg^{-1} C^{-1}, is also required. Thus,

$$\left(100\ \frac{m^3}{s} \times 1{,}000\ \frac{kg}{m^3}\right) \times \left(4{,}184\ \frac{J}{kg \times {}^\circ C}\right) \times \Delta T = 1{,}430 \times 10^6\ J\ s^{-1}$$

or

$$\Delta T = 3.4^\circ C$$

Consideration of this temperature increase is important, as the Henry's law constant for oxygen changes with temperature (see Section 3.4.2). This results in a reduced dissolved-oxygen concentration in the river in warmer water, which may be harmful to aquatic life. Oxygen exchange in rivers is discussed further in Section 5.5.

The global average surface temperature of the Earth is determined by a balance between the energy provided to the Earth by the Sun and the energy radiated away by the Earth to space. The energy radiated by the Earth to space is emitted in the form of infrared radiation. As illustrated in Figure 4-11, some of

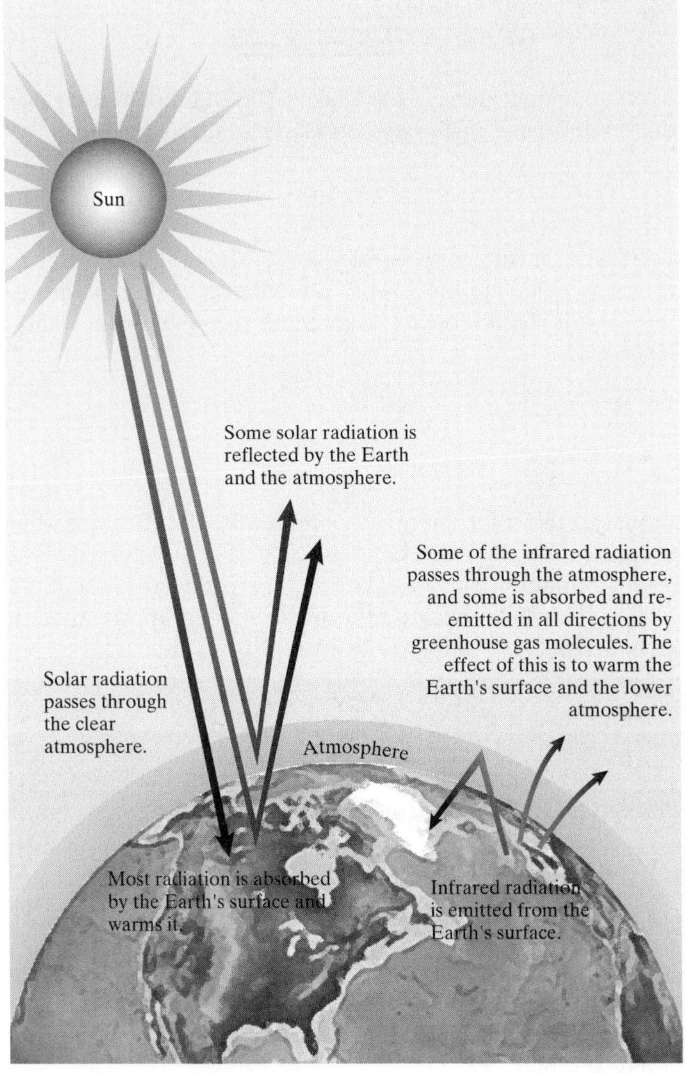

Figure 4-11. The greenhouse effect. (From "Our Changing Planet: The FY 1996 U.S. Global Change Research Program," A Report by the Subcommittee on Global Change Research, Committee on Environment and Natural Resources Research of the National Science and Technology Council, A Supplement to the President's Fiscal Year 1996 Budget.)

the infrared radiation emitted by the Earth is absorbed in the atmosphere. The gases that are responsible for this absorption are called greenhouse gases, and without them the Earth would not be habitable, as demonstrated in the following example.

EXAMPLE 4.12. EARTH'S ENERGY BALANCE AND THE GREENHOUSE EFFECT

Calculate the global average temperature of the Earth without greenhouse gases and show the effect that greenhouse gases have on Earth's energy balance.

SOLUTION

An energy balance can be written with the control volume being the entire Earth. For this system, the goal is to calculate the Earth's annual average temperature. Over time periods of ≥ 1 yr, it is reasonable to assume that the system is at steady state. The energy balance is

$$\frac{dE}{dt} = 0 = \dot{E}_{\text{in}} - \dot{E}_{\text{out}}$$

The energy flux in is equal to the solar energy intercepted by the Earth. At the Earth's distance from the Sun, the Sun's radiation is 342 W m^{-2}, referred to as S. The Earth intercepts an amount of energy equal to S times the cross-sectional area of the Earth: $S \times \pi R_e^2$. However, because the Earth reflects approximately 30% of this energy back to space, $\dot{E}_{\text{in}}$ equals only 70% of this value:

$$\dot{E}_{\text{in}} = 0.7 S \pi R_e^2$$

The second term, $\dot{E}_{\text{out}}$, is equal to the energy radiated to space by the Earth. The energy emitted per unit surface area of the Earth is given by Boltzmann's law:

$$\begin{pmatrix}\text{Energy flux per} \\ \text{unit area}\end{pmatrix} = \sigma T^4$$

where σ is Boltzmann's constant, equal to $5.67 \times 10^{-8} W\ m^{-2}\ K^{-4}$. To obtain $\dot{E}_{\text{out}}$, this value is multiplied by the total surface area of the Earth, $4\pi R_e^2$. (The total surface area of the sphere is used here because energy is radiated away from the Earth during both day and night.)

$$\dot{E}_{\text{out}} = 4\pi R_e^2 \sigma T^4$$

The energy balance can be solved by setting $\dot{E}_{\text{in}}$ equal to $\dot{E}_{\text{out}}$:

$$4\pi R_e^2 \sigma T^4 = 0.7 S \pi R_e^2$$

or

$$T^4 = \frac{0.7S}{4\sigma}$$

Plugging in the values for S and σ yields an average annual temperature of the Earth of $T = 255°$ K, or $-18°$C. This is too cold! In fact, the globally averaged temperature at the surface of the Earth is much warmer: 287° K. The reason for the difference is the presence of gases in the atmosphere that absorb the infrared radiation emitted by the Earth and prevent it from reaching space. These gases include water vapor, CO_2, CH_4, and N_2O, and were neglected in the initial energy balance. Their influence can be included by adding a new term in the energy balance, the energy flux absorbed and retained by these gases. If the impact of greenhouse gas absorption is given by $E_{greenhouse}$, then the corrected $\dot{E}_{out}$ term is

$$\dot{E}_{out} = 4\pi R_e^2 \sigma T^4 - E_{greenhouse}$$

The reduction in $\dot{E}_{out}$ that results from greenhouse gas absorption is sufficient to cause the higher observed surface temperature. Clearly, this is largely a natural phenomenon, since surface temperatures were well above 255° K long before humans began burning fossil fuels.

However, human activities—primarily the burning of fossil fuels—are changing the atmospheric composition to a significant extent, and are increasing the magnitude of the greenhouse effect. Atmospheric concentrations of carbon dioxide, the most important greenhouse gas, are shown in Figure 4-12. Increasing atmospheric concentrations of carbon dioxide, as well as those of methane, nitrous oxide, chluorofluorocarbons, and tropospheric ozone, which have occurred as a result of human activities, increase the value of $E_{greenhouse}$. This enhanced greenhouse effect, termed the *anthropogenic greenhouse effect*, is currently equivalent to an increase in the energy flux to the Earth of approximately 2 W m^{-2}.

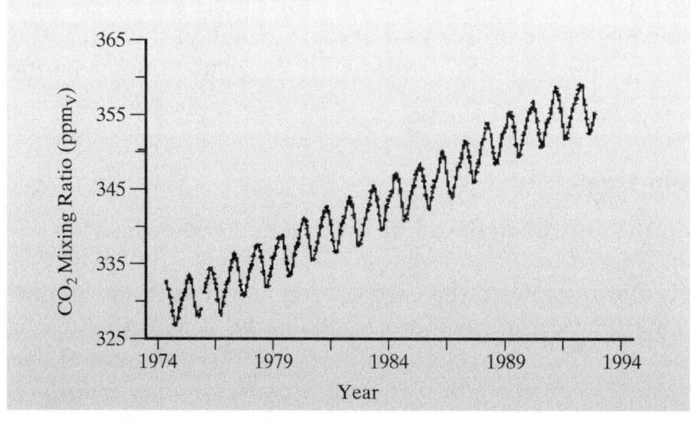

Figure 4-12. The global-average carbon dioxide concentration trend. Measurements of CO_2 made at the Mauna Loa, Hawaii, observatory by the National Oceanic and Atmospheric Administration. The annual increase of approximately 0.4%/year is attributed to fossil-fuel combustion and deforestation; the annual cycle is the result of photosynthesis and respiration, which result in a drawdown of CO_2 during the summer growing season and an increase during winter. (From: http://cdiac.esd.ornl.gov/cdiac/)

Projections indicate that the increase could be as high as 5 W m^{-2} over the next 50 years.

As the energy absorbed by greenhouse gases increases, some other term in the energy balance must respond to maintain steady state. If the solar radiation absorbed by the Earth remains constant, then the Earth's average temperature must increase. The magnitude of the resulting temperature increase depends upon the response of the complex global climate system, including changes in cloudiness and ocean circulation. Current global climate models predict that the anthropogenic greenhouse effect will cause a global-average temperature increase (relative to 1990) of 1–3.5°C by the year 2050 (Houghton et al., 1996). Resulting alterations to global and regional climate are predicted to include increased rainfall and increased frequency of severe storms, although some regions of the Earth may experience increased frequency of drought or even regional cooling, as a result of changes in atmospheric and oceanic circulation patterns.

4.3 MASS-TRANSPORT PROCESSES

Processes that move chemicals through the air, surface water, subsurface environment, or engineered systems (for example, treatment reactors) are of particular interest to environmental engineers and scientists. Transport processes move pollutants from the location at which they are generated, resulting in impacts that can be distant from the pollution source. In addition, environmental engineers make use of transport processes in the design of emission-control systems. In this section we discuss some of the processes that transport pollutants in the environment and in engineered systems. The goals of this discussion are twofold: to provide an understanding of the processes that cause pollutant transport, and to present and apply the mathematical formulas used to calculate the resulting pollutant fluxes.

4.3.1 Advection and Dispersion

Transport processes in the environment can be divided into two categories: *advection* and *dispersion*. *Advection refers to transport with the mean fluid flow.* For example, if the wind is blowing toward the east, advection will carry any pollutants present in the atmosphere toward the east. Similarly, if a bag of dye is emptied into the center of a river, advection will carry the resulting spot of dye downstream. In contrast, *dispersion refers to the transport of compounds through the action of random motions.* Dispersion works to eliminate sharp discontinuities in concentration and results in smoother, flatter concentration profiles. Advective and dispersive processes can usually be considered independently. For the spot of dye in a river, while advection moves the center of mass of the dye downstream, dispersion spreads out the concentrated spot of dye to a larger, less concentrated region.

4.3.1.1 *Definition of the Mass-Flux Density*

The term *mass flux* ($\dot{m}$, with units of mass time^{-1}) was used in Section 4.1 to calculate the rates at which mass was transported into and out of a control volume in mass balances. Because mass-balance calculations are always made with reference to a specific control volume, it was clear that this value referred to the rate at which mass was transported *across the boundary of the control volume.* However, in calculations of advective and dispersive fluxes, a specific, well-defined control volume will not be created. Instead, the *flux density* across an imaginary plane oriented perpendicular to the direction of mass transfer will be determined. The resulting mass-flux density is defined as the rate of mass transferred across the plane per unit time *per unit area.* The symbol J will be used to represent the flux density:

J represents the mass-flux density, expressed as the rate per unit area at which mass is transported across an imaginary plane. J has units of (mass length^{-2} time^{-1}).

The total mass flux across a boundary ($\dot{m}$) can be calculated from the flux density simply by multiplying J by the area of the boundary:

$$\dot{m} = J \times A \tag{4-24}$$

The mass-transfer process that J describes can result from advection, dispersion, or a combination of both processes.

4.3.1.2 *Calculation of the Advective Flux*

The advective flux refers to the movement of a compound along with flowing air or water. The advective-flux density depends simply on concentration and flow velocity:

$$J = C \times v \tag{4-25}$$

The fluid velocity, v, is a vector quantity. It has both magnitude and direction, and the flux J refers to the movement of pollutant mass in the same direction as the fluid flow. The coordinate system is generally defined so that the x-axis is oriented in the direction of fluid flow. In this case, the flux J will reflect a flux in the x-direction, and the fact that J is really a vector quantity will be ignored.

EXAMPLE 4.13. CALCULATION OF THE ADVECTIVE-FLUX DENSITY

Calculate the average flux density J of phosphorus downstream of the wastewater discharge of Example 4.2. The cross-sectional area of the river is 30 m^2.

SOLUTION

In Example 4.2, the following conditions downstream of the spot where a pipe discharged to a river were determined: volumetric flow rate $Q = 26\ \text{m}^3\ \text{s}^{-1}$ and downstream concentration $C_d = 0.20\ \text{mg L}^{-1}$ as phosphorus. The average river velocity is $v = Q/A = (26\ \text{m}^3\ \text{s}^{-1})/(30\ \text{m}^2) = 0.87\ \text{m s}^{-1}$. Using the definition of flux density (Equation 4-25):

$$J = \left[(0.20\ \text{mg L}^{-1}) \times \frac{10^3\ L}{\text{m}^3}\right] \times (0.87\ \text{m s}^{-1})$$

$$= 174\ \text{mg m}^{-2}\ \text{s}^{-1} \text{ or } 0.17\ \text{g m}^{-2}\ \text{s}^{-1}$$

4.3.1.3 Dispersion

Dispersion results from random motions of two types: the random motion of molecules, and the random eddies that arise in turbulent flow. Dispersion from the random molecular motion is termed *molecular diffusion*; dispersion that results from turbulent eddies is called *turbulent dispersion* or *eddy dispersion.*

4.3.1.3.1 Fick's Law. Fick's law is used to calculate the dispersive flux density. It can be derived by analyzing the mass transfer that results from the random motion of gas molecules.[1] The purpose of this derivation is to provide a qualitative and intuitive understanding of the reason that diffusion occurs, and the derivation is useful only for that purpose. In problems where it is necessary to actually calculate the diffusive flux, Fick's law (Equation 4-35 below) will be used.

Consider a box that is initially divided into two parts, as shown in Figure 4-13. Each side of the box has a height and depth of one unit, and a width of length Δx. Initially, the left portion of the box is filled with 10 molecules of gas x and the right side is filled with 20 molecules of gas y, as shown in the top half of Figure 4-13. What happens if the divider is removed?

Molecules are never stationary. All of the molecules in the box are constantly moving around, and at any moment they have some probability of crossing the imaginary line at the center of the box. Assume that the molecules on each side are counted every Δt seconds. The probability that a molecule crosses the central line during the period between observations can be defined as k, which is assumed to equal 20% (any value would do for the present purpose). The first time the box is checked, after a period Δt, 20% of the molecules that were originally on the left will have moved to the right, and 20% of the molecules that were originally on the right will have moved to the left. When the molecules on each side are then counted, the situation shown in the bottom of Figure 4.13 will be found. Eight "x" molecules remain on the left, and 2 have moved to the right, while 16 "y" molecules remain on the right, 4 having moved to the left.

[1]This derivation is based closely on one presented by Fischer et al. (1979).

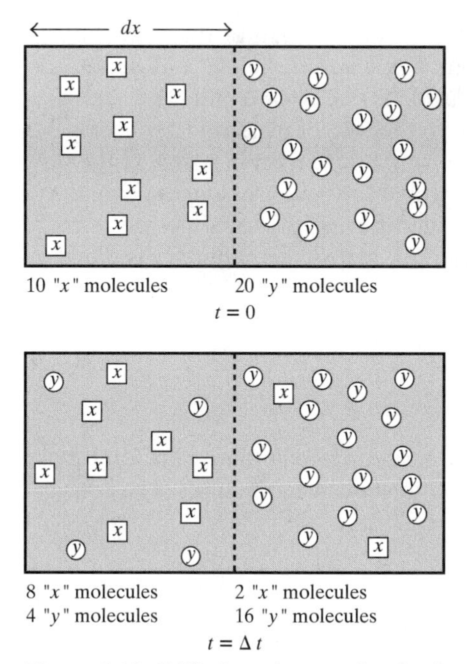

Figure 4-13. Diffusion of gas molecules in a box. A box is divided into two regions of equal size. Ten gas molecules of one type (x) are added to the left side, while 20 gas molecules of another type (y) are added to the right side. Although they are distinguishable, the two types of molecules are identical in every physical respect. At time $t = 0$, the divider separating the two regions is removed. As a result of random motion, each molecule within the box has a 20% probability of moving to the opposite side of the box during each time interval of duration Δt. The result after one time interval is shown in the bottom figure. (From Fischer et al., 1979. Reprinted by permission of Academic Press, Inc.)

Since the boxes are equal in size, the concentration within each box is proportional to the number of molecules within it. Therefore, the random motion of molecules in the boxes has resulted in a reduction in the concentration difference between the boxes—from (20–0) to (16–4) for the "y" molecules, and from (10–0) to (8–2) for the "x" molecules. This result leads to a fundamental property of dispersive processes.

Dispersion moves mass from regions of high concentration to regions of low concentration and acts to reduce concentration gradients.

The flux density, J, can also be derived for the two-box experiment. For this calculation, the situation shown in Figure 4-13 is used again, with the probability of any molecule crossing the central boundary during a period Δt equal to k. Since each molecule can be considered independently, the movement of a single molecule type, say molecule "y," can be analyzed.

Let M_L be the total mass of molecule "y" in the left half of the box, and M_R equal the mass in the right half. Since our box has unit height and depth, the area perpendicular to the direction of diffusion is one square unit. Thus, the flux density, the flux per unit area, is just equal to the rate of mass transfer across the boundary. The amount of mass transferred from left to right in a single time step is equal to kM_L, since each molecule has a probability k of crossing the boundary, while the amount transferred from right to left during the same period is kM_R. Thus, the *net rate* of mass flux from left to right across the boundary is equal to $(kM_L - kM_R)$ divided by Δt, or

$$J = \frac{k}{\Delta t}(M_L - M_R) \tag{4-26}$$

Since it is more convenient to work with concentrations than with total mass values, Equation 4-26 needs to be converted to concentration units. The concentration in each half of the box is given by

$$C_L = \frac{M_L}{\Delta x \times (\text{height}) \times (\text{depth})} \tag{4-27}$$

$$= \frac{M_L}{\Delta x} \text{ (height and depth are both equal to 1)} \tag{4-28}$$

and

$$C_R = \frac{M_R}{\Delta x} \tag{4-29}$$

Thus, substituting $C\Delta x$ for the mass in each half of the box, the flux density is equal to

$$J = \frac{k}{\Delta t}(C_L\Delta x - C_R\Delta x) \tag{4-30}$$

$$= \frac{k}{\Delta t}(\Delta x)(C_L - C_R) \tag{4-31}$$

Finally, note that as $\Delta x \to 0$, $(C_R - C_L)/\Delta x \to dC/dx$, and therefore if we multiply Equation 4-31 by $(\Delta x/\Delta x)$

$$J = \frac{k}{\Delta t}(\Delta x)(C_L - C_R)\frac{\Delta x}{\Delta x} \tag{4-32}$$

$$= \frac{k}{\Delta t}(\Delta x)^2 \frac{(C_L - C_R)}{\Delta x} \tag{4-33}$$

we obtain

$$J = -\frac{k}{\Delta t}(\Delta x)^2 \frac{dC}{dx} \quad \textbf{(4-34)}$$

(Note that the negative sign in this equation is simply a result of the convention that flux is positive when it flows from left to right, while the derivative is positive when concentration increases toward the right.)

This equation states that the flux of mass across an imaginary boundary is proportional to the concentration gradient at the boundary. Since the resulting flux cannot depend on arbitrary values of Δt or Δx, the factor $k(\Delta x)^2/\Delta t$ must be constant.

This product is the value called the *diffusion coefficient*, D. Rewriting Equation 4-34 results in *Fick's Law*

$$J = -D\frac{dC}{dx} \quad \textbf{(4-35)}$$

The units of the diffusion coefficient are clear from an analysis of the units of Equation 4-35 or from the units of the parameters in Equation 4-34; the diffusion coefficient has the same units as $k(\Delta x)^2/\Delta t$. Since k is a probability, and thus has no units, the units of D are (length^2 time^{-1}). Diffusion coefficients are commonly reported in units of cm^2 s^{-1}.

Note the form of Equation 4-35:

$$\text{Flux density} = (\text{constant}) \times (\text{concentration gradient}) \quad \textbf{(4-36)}$$

This form of equation will also appear later when Darcy's law is covered. Darcy's law governs the rate at which water flows through porous media as in groundwater flow. The same equation also governs heat transfer, with the concentration gradient replaced with a temperature gradient.

4.3.1.3.2 Molecular Diffusion. The molecules-in-a-box analysis used earlier is essentially an analysis of molecular diffusion. Purely molecular diffusion is relatively slow. Table 4-6 reports some typical values for D. Typical values of the

Table 4-6. Selected Molecular Diffusion Coefficients in Water and Air

Compound	Temperature (°C)	Diffusion Coefficient (cm^2 s^{-1})
Methanol in H_2O	15	1.26×10^{-5}
Ethanol* in H_2O	15	1.00×10^{-5}
Acetic acid in H_2O	20	1.19×10^{-5}
Ethylbenzene in H_2O	20	8.1×10^{-6}
CO_2 in air	20	0.151

*Note that of the two similar compounds methanol and ethanol, the less massive compound, methanol, has the higher diffusion coefficient.

diffusion coefficient are approximately 10^{-2} to 10^{-1} cm^2 s^{-1} for gases, and much lower, around 10^{-5} cm^2 s^{-1} for liquids. The difference in diffusion coefficient between gases and liquids is understandable because gas molecules are free to move much greater distances before being stopped by bumping into another molecule.

The diffusion coefficient also varies with temperature and the molecular weight of the diffusing molecule. This is because the average speed of the random molecular motions is dependent on the kinetic energy of the molecules. As heat is added to a material and temperature increases, the thermal energy is converted to random kinetic energy of the molecules, and the molecules move faster. This results in an increase in the diffusion coefficient with increasing temperature. If molecules of differing molecular weight are compared, however, it is found that at a given temperature a heavier molecule moves more slowly, and thus the diffusion coefficient decreases with increasing molecular weight.

EXAMPLE 4.14. MOLECULAR DIFFUSION

The transport of polychlorinated biphenyls (PCBs) from the atmosphere into the Great Lakes is of concern because of health impacts to the aquatic life and to people and wildlife that eat fish from the lakes. PCB transport is limited by molecular diffusion across a thin stagnant film at the surface of the lake, as shown in Figure 4-14. Calculate the flux density J and the total annual amount of PCBs deposited into Lake Superior if the transport is by molecular diffusion, the PCB concentration in the air just above the lake's surface is 100×10^{-12} g m^{-3}, and the concentration at a height of 2.0 cm above the water surface

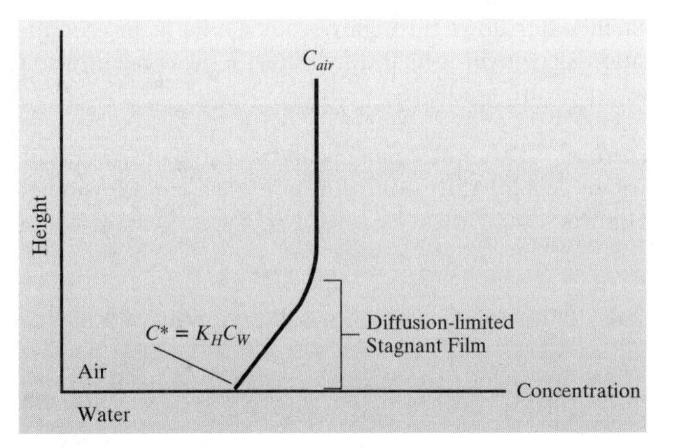

Figure 4-14. Variation of PCB concentration with height above Lake Superior. C_{air} is the PCB concentration in the atmosphere above the lake, and C^* is the concentration at the air–water interface, which is determined by Henry's law equilibrium with the dissolved PCB concentration. The flux of PCBs into the lake is determined by the rate of diffusion across a stagnant film above the lake.

is $450 \times 10^{-12} = \text{g m}^{-3}$. The diffusion coefficient for PCBs is equal to 0.044 cm^2 s^{-1}, and the surface area of Lake Superior is 8.2×10^{10} m^2. (The PCB concentration in the air at the air–water interface is determined by Henry's law equilibrium with dissolved PCBs. Henry's law is described in Chapter 3.)

SOLUTION

To calculate the flux density, first determine the concentration gradient. Assume that concentration changes linearly with height between the surface and 2.0 cm, as no concentration information was provided between those two heights. The gradient is then

$$\frac{dC}{dz} = \frac{450 \times 10^{-12} \text{ g m}^{-3} - 100 \times 10^{-12}}{2.0 \text{ cm} - 0} \times \frac{10^2 \text{ cm}}{\text{m}}$$

$$= 1.8 \times 10^{-8} \text{ g m}^{-4}$$

Fick's law (Equation 4-35) can be used to calculate the flux density:

$$J = -D\frac{dC}{dz}$$

$$= -(0.044 \text{ cm}^2 \text{ s}^{-1}) \times 1.8 \times 10^{-8} \text{ g m}^{-4} \times \frac{\text{m}^2}{10^4 \text{ cm}^2} \times \frac{3.15 \times 10^7 \text{ s}}{\text{yr}}$$

$$= -2.4 \times 10^{-6} \text{ g m}^{-2} \text{ yr}^{-1}$$

where the negative sign indicates that the flux is downward. (However, it's not necessary to pay attention to the sign to determine that. Remember that diffusion always transports mass from higher concentration to lower concentration regions.)

The total depositional flux is given by $\dot{m} = J \times A$:

$$\dot{m} = -2.4 \times 10^{-6} \text{ g m}^{-2} \text{ yr}^{-1} \times 8.2 \times 10^{10} \text{ m}^2$$

$$= -2.0 \times 10^5 \text{ g yr}^{-1}$$

Thus, the total annual input of PCBs into Lake Superior from the atmosphere is approximately 200 kg. Although this is a small annual flux for such a large lake, PCBs do not readily degrade in the environment, and they bioaccumulate, or concentrate, in the fish, resulting in unhealthy levels. Bioaccumulation is discussed further in Chapter 5.

4.3.1.3.3 Turbulent Dispersion. In turbulent dispersion, mass is transferred through the mixing of *turbulent eddies* within the fluid. This is fundamentally different from the processes that determine molecular diffusion. In turbulent dispersion, it is the random motion of the *fluid* that does the mixing, while in molecular diffusion it is the random motion of the pollutant molecules that is important.

Random motions of the fluid are generally present in the form of whorls, or eddies. These are familiar in the form of eddies or whirlpools in rivers, but occur in all forms of fluid flow. The size of turbulent eddies is several orders of magnitude larger than the mean free path of individual molecules, so turbulence moves mass much faster than does molecular diffusion. As a result, turbulent, or eddy-dispersion coefficients used in Fick's law are generally several orders of magnitude larger than are molecular diffusion coefficients.

The value of the turbulent dispersion coefficient depends on properties of the fluid flow. It is not dependent on molecular properties of the compound being dispersed (as was the molecular diffusion coefficient) because in turbulence the molecules are simply carried along with the macroscale flow. For flow in pipes or streams, the most important flow property determining the turbulent dispersion coefficient is the flow velocity. Turbulence is only present at flow velocities above a critical level, and the degree of turbulence is correlated with velocity. (More precisely, the presence or absence of turbulence depends on the *Reynolds number*, a dimensionless number that depends on velocity, width of the river or pipe, and the viscosity of the fluid.) In addition, the degree of turbulence depends on the material over which the flow is occurring, so that flow over bumpy surfaces will be more turbulent than flow over a smooth surface, and the increased turbulence will cause more rapid mixing. In lakes and in the atmosphere, buoyant mixing that results from temperature-induced density gradients can also cause turbulent mixing, even in the absence of currents.

Except in the case of transport across a boundary, such as at the air–water interface considered in Example 4.14, turbulent dispersion almost always entirely dominates molecular diffusion. This is so because even an occasional amount of weak turbulence will cause more mixing than would several days of molecular diffusion.

Fick's law applies to turbulent dispersion just as it does to molecular diffusion. Thus, flux-density calculations are the same for both processes; only the magnitude of the dispersion coefficient is different.

4.3.1.3.4 Mechanical Dispersion. The final dispersion process considered in this chapter is similar to turbulence, in that it is a result of variations in the movement of the fluid that carries a chemical. In *mechanical dispersion* these variations are the result of (a) variations in the flow pathways taken by different fluid parcels that originate in nearby locations, or (b) variations in the speed at which fluid travels in different regions.

Dispersion in groundwater flow provides a good example of the first process. Figure 4-15 shows a magnified depiction of the pores through which groundwater flows within a subsurface sample. (Note that, as shown in Figure 4-15, groundwater movement is not a result of underground rivers or creeks, but rather is caused by the flow of water through the pores of the soil, sand, or other material

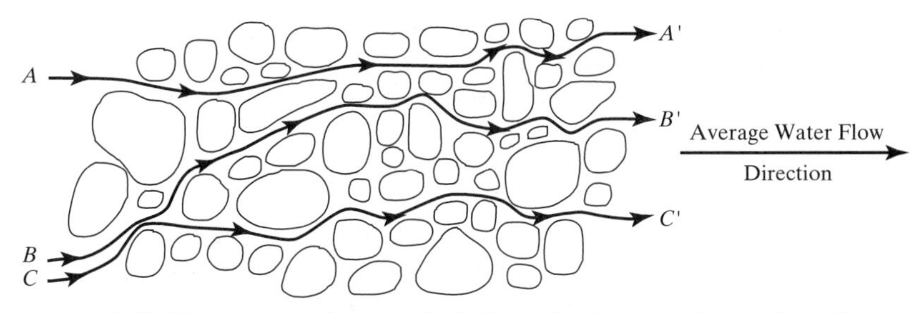

Figure 4-15. The process of mechanical dispersion in groundwater flow. Two fluid parcels starting near each other at locations B and C are dispersed to locations farther apart (B' and C') during transport through the soil pores, while parcels from A and B are brought closer together, resulting in mixing of water from the two regions. (From Hemond, H. F., and E. J. Fechner, 1994. Reprinted by permission of Academic Press, Inc.)

underground.) Because transport through the soil is limited to the pores between soil particles, each fluid particle takes a convoluted path through the soil and, as it is transported horizontally with the mean flow, it is displaced vertically a distance that depends on the exact flow path it took. The great variety of possible flow paths results in a random displacement in the directions perpendicular to the mean flow path. Thus, a spot of dye introduced into the groundwater flow between points B and C in Figure 4-15 would be spread out, or *dispersed* into the region between points B' and C' as it flows through the soil.

The second type of mechanical dispersion results from differences in flow speed. Anywhere that a flowing fluid contacts a stationary object, the speed at which the fluid moves will be slower near the object. For example, the speed of water flowing down a river is fastest in the center of a river, and can be very slow near the edges. Thus, if a line of dye were somehow laid across the river at one point, it would be stretched out in the upstream/downstream direction as it flowed down the river, with the center part of the line moving faster than the edges. This type of dispersion spreads things out in the *longitudinal* direction in the direction of flow. This is in contrast to mechanical dispersion in groundwater, which spreads things out in the direction *perpendicular* to the direction of mean flow.

4.3.2 The Movement of a Particle in a Fluid: Stokes' Law

This section analyzes the forces that determine the movement of a particle suspended in a fluid. Suspensions of particles are important in a variety of environmental applications, from the cleaning of particulate pollutants from coal-fired power plants and settling out of suspended particles in wastewater treatment plants or lakes to the removal of turbidity during drinking-water treatment. In order to design or analyze these processes, it is necessary to understand the movement of particles within the air or water fluid.

The movement of a particle in a fluid is determined by a balance of the viscous drag forces resisting the particle movement with gravitational or other forces that cause the movement. In this section, a force balance on a particle is used to derive the relationship between particle size and settling velocity known as Stokes' law, and Stokes' law is used in examples involving particle-settling chambers.

4.3.2.1 *Gravitational Settling*

Consider the settling particle shown in Figure 4-16. To determine the velocity at which it falls (the settling velocity) a force balance will be conducted. There are three forces acting on the particle: the downward gravitational force, an upward buoyancy force, and an upward drag force.

The gravitational force F_g is equal to the gravitational constant g times the mass of the particle, m_p. In terms of particle density, ρ_p, and diameter, D_p, m_p is equal to $(\rho_p \pi/6D_p^3)$. Therefore,

$$F_g = \rho_p \frac{\pi}{6} D_p^3 g \tag{4-37}$$

The buoyancy force, F_B, is a net upward force that results from the increase of pressure with depth within the fluid. The buoyancy force is equal to the gravitational constant times the mass of the fluid displaced by the particle:

$$F_B = \rho_f \frac{\pi}{6} D_p^3 g \tag{4-38}$$

where ρ_f is equal to the fluid density.

The only remaining force to determine is the drag force, F_D. The drag force is the result of frictional resistance to the flow of fluid past the surface of the particle. This resistance depends upon the speed at which the particle is falling through the fluid, the size of the particle, and the *viscosity*, or resistance to shear, of the fluid. (Viscosity is essentially what one would qualitatively call the "thickness" of the fluid—honey has a high viscosity, while water has a relatively low viscosity, and the viscosity of air is much lower yet.) Over a wide range of con-

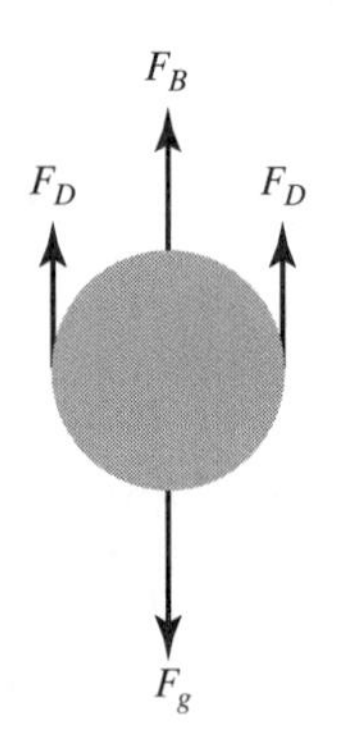

Figure 4-16. The forces acting on a particle settling through air or water. The gravitational force F_g is in the downward direction, and is counteracted by the buoyancy force F_B and the drag force F_D.

ditions, the friction force can be correlated with the *Reynolds number* discussed in fluid dynamics and hydromechanic texts.

Most particle-settling situations involve "creeping flow" conditions (Reynolds number less than 1). In this case, the Stokes' drag force can be used:

$$F_D = 3\pi\mu D_p v_r \tag{4-39}$$

where μ is the fluid viscosity (units of $g\ cm^{-1}\ s^{-1}$) and v_r is the velocity of the particle relative to the fluid (the settling velocity).

The net downward force acting on the particle is equal to the vector sum of all forces acting on the particle and equals

$$F_{\text{down}} = F_g - F_B - F_D \tag{4-40}$$

$$= \rho_p \times \frac{\pi}{6} D_p^3 \times g - \rho_f \times \frac{\pi}{6} D_p^3 \times g - 3\pi\mu D_p v_r \tag{4-41}$$

$$= (\rho_p - \rho_f) \times \frac{\pi}{6} D_p^3 \times g - 3\pi\mu D_p v_r \tag{4-42}$$

The particle will respond to this force according to Newton's second law, which states that force equals mass times acceleration. Thus,

$$F_{\text{down}} = m_p \times \text{acceleration} \tag{4-43}$$

$$= m_p \times \frac{dv_r}{dt} \tag{4-44}$$

This differential equation can be solved to determine the time-varying velocity of a particle that is initially at rest. The solution indicates that, in almost all cases of environmental interest, the period of time required before the particle reaches its final *settling velocity* is very short (much less than 1 s). For this reason, in this text only the final, or terminal settling velocity is considered.

When the particle has reached terminal velocity, it is no longer accelerating, so $dv/dt = 0$. Thus, from Equation 4-44, $F_{\text{down}} = 0$. Setting F_{down} equal to zero and noting that v_r is equal to the settling velocity v_s at terminal velocity, Equation 4-42 can be rearranged to yield

$$(\rho_p - \rho_f) \frac{\pi}{6} D_p^3 g = 3\pi\mu D_p v_s \tag{4-45}$$

or

$$v_s = \frac{g(\rho_p - \rho_f)}{18\mu} D_p^2 \tag{4-46}$$

Equation 4-46 is referred to as Stokes' law. The resulting settling velocity is often called the Stokes' velocity.

Stokes' law, so called because it is based upon the Stokes' drag force, is the fundamental equation used to calculate terminal settling velocities of particles in both air and water. It is used in the design of treatment systems to remove particles from exhaust gases and from wastewater, as well as in analyses of settling particles in lakes and in the atmosphere.

An important implication of Stokes' law is that the settling velocity increases as the *square* of the particle diameter, so larger particles settle much faster than do smaller particles. This result is used in drinking-water treatment: coagulation and flocculation are used to get small particles to come together and form larger particles, which can then be removed by gravitational settling in a reasonable amount of time. This process results in reduction in the turbidity (i.e., increase in the clarity) of the water. In contrast, particles with very small diameters settle extremely slowly. As a result, atmospheric particles with diameters less than 1–10 μm generally fall more slowly than the speed of turbulent eddies of air, with the result that they are not removed by gravitational settling.

Note that particle–particle interactions have been ignored in this derivation. Thus, Stokes' law is valid for *discrete particle settling*. In some situations in which particle concentration is extremely high, particles form agglomerations or mats, and Stokes' law may no longer be valid.

EXAMPLE 4.15. DISCRETE SETTLING DURING WASTEWATER TREATMENT

A rectangular tank, called a *grit chamber*, is used to settle out large particles (e.g., sand or grit) prior to subsequent wastewater treatment. These large particles can be abrasive and damaging to plant equipment and piping, and are removed before beginning the other wastewater treatment processes. To design the grit chamber in a sufficient size, the settling velocity of the particles must first be determined. If the particles are assumed to be spheres with diameter D_p = 100 μm and density ρ_p = 2.65 g cm^{-3}, what would be their settling velocity? (Note that the viscosity of water is 0.01185 g cm^{-2}s^{-1} and the density of water is 1.00 g cm^{-3}.)

SOLUTION

The steady-state solution based on Stokes' law, Equation 4-46, is

$$v_s = \frac{(2.65 \text{ g cm}^{-3} - 1.00 \text{ g cm}^{-3})(980 \text{ cm s}^{-2})}{18 \times 0.01185 \text{ g cm}^{-1} s^{-1}} (100 \times 10^{-4} \text{ cm})^2$$

$$= 0.76 \text{ cm s}^{-1}, \text{ or } 27 \text{ m h}^{-1}$$

(Note that 1 μm = 10^{-6}m = 10^{-4}cm.) The settling velocity is relatively slow. For these very small particles with correspondingly low mass, the drag force is significant in comparison to the gravitational force.

EXAMPLE 4.16. DISCRETE SETTLING IN AIR

Calculate the gravitational settling velocity for two atmospheric particles having diameters of 0.1 μm and 100 μm, respectively, and with density $\rho_p = 1.0\ \text{g cm}^{-3}$. Based on the result, determine whether removal of these particles in a settling chamber of 1-m depth would be a realistic proposition. The viscosity of air is $\mu_{\text{air}} = 1.695 \times 10^{-4}\ \text{g cm}^{-1}\text{s}^{-1}$. The density of air is extremely small, and thus the buoyancy force may be neglected.

SOLUTION

Use Equation 4-46 to determine the settling velocity. For the 0.1-μm particle, we obtain

$$
\begin{aligned}
v_s &= \frac{(1.0\ \text{g cm}^{-3})(980\ \text{cm s}^{-2})}{18 \times 1.695 \times 10^{-4}\ \text{g cm}^{-1}\text{s}^{-1}} (0.1 \times 10^{-4}\ \text{cm})^2 \\
&= 3.2 \times 10^{-5}\ \text{cm s}^{-1}
\end{aligned}
$$

and, for the 100-μm particle,

$$
\begin{aligned}
v_s &= \frac{(1.0\ \text{g cm}^{-3})(980\ \text{cm s}^{-2})}{18 \times 1.695 \times 10^{-4}\ \text{g cm}^{-1}\text{s}^{-1}} (100 \times 10^{-4}\ \text{cm})^2 \\
&= 32\ \text{cm s}^{-1}
\end{aligned}
$$

The 0.1-μm particle would take about 900 h to settle a distance of 1 m. However, because the settling velocity is proportional to the square of the particle size, the 100-μm particle settles 10^6 times faster, and would settle a distance of 1 m in only 3 s. It would clearly be realistic to construct a chamber with a residence time of >3 s. However, a much larger chamber would be required to reach a residence time of 900 h, and such a chamber would not be economical.

Note that the 100-μm particle in this example settles with a velocity that is much greater than that of the 100-μm particle settling through water in Example 4.15. This is due to the much lower viscosity of air relative to water, which results in reduced drag.

4.3.2.2 Calculation of the Minimum Particle Size Removed

It is often necessary to determine the minimum particle size removed in a settling chamber. To do this, one must first determine the minimum settling velocity that

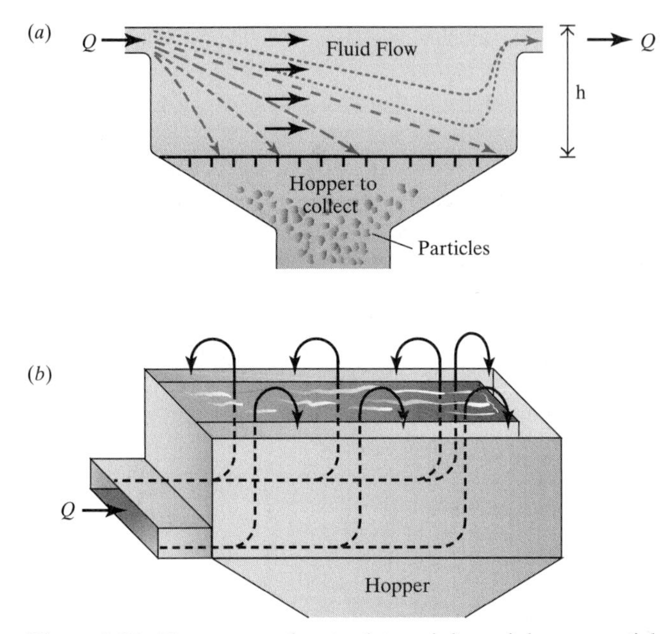

Figure 4-17. Two approaches to determining minimum particle size removed in a nonturbulent chamber. (*a*) The fluid (water or air) travels smoothly from left to right across the settling chamber. Large particles fall rapidly while traveling through the chamber, resulting in a steep trajectory, as indicated by dashed lines; small particles fall slowly, resulting in a shallow trajectory, as indicated by dotted lines. (*b*) The inflow enters near the bottom of the chamber and flows upward, exiting by overflowing out of the open top. Particles that fall faster than the water rises have a net downward velocity and eventually reach the hopper. Particles with settling velocities less than the upward water velocity rise, and are carried out with the effluent flow.

is required for a particle to be removed, and then use Stokes' law (Equation 4-46) with this settling velocity to solve for the particle diameter, D_p.

For settling chambers with smooth (nonturbulent) flow, there are two ways that the minimum settling velocity can be determined. These two approaches are based upon different flow patterns within the settling chamber, as illustrated in Figure 4-17. The two methods (described below) yield the same result, so choice of which method to use may be based on convenience rather than on flow geometry.

4.3.2.2.1 Distance Settled during Retention Period. Consider the settling chamber shown in Figure 4-17*a*. Liquid (or air) enters the chamber on the left side, and flows from left to right across the chamber. (Baffles, not shown, are present to distribute the flow throughout the depth of the chamber.) To be removed from the flow, a particle must fall into the hopper during the amount of time it spends in the chamber. The farthest any particle has to fall is the entire height of the chamber, h. The particle must fall this distance within the retention time of the chamber, $\theta = V/Q$. Therefore, the required vertical settling velocity of the particle must be

$$v_s \geq \frac{h}{\theta} \tag{4-47}$$

which is equivalent to

$$v_s \geq \frac{hQ}{V}. \tag{4-48}$$

A fraction of the particles with settling velocities less than this critical value may also be removed. To see why this is so, consider a particle entering the left-hand side of the settling chamber at a location $0.5h$ above the hopper. This particle would have to fall only $0.5h$ to be removed, and would thus be removed if $v_s = 0.5hQ/V$. Particles with this settling velocity that enter at or below a height of $0.5h$ would be removed; those entering above $0.5h$ would not.

4.3.2.2.2 Settling Velocity Greater than the Overflow Rate. In the settling chamber shown in Figure 4-17*b*, water flows vertically upward and overflows out of the top. The vertical velocity of the water is referred to as the overflow rate (OFR). Since the settling velocity is the speed a particle falls *relative to the fluid*, the actual downward velocity of a particle in this chamber is equal to $(v_s - \text{OFR})$. The particle will settle downward as long as the settling velocity is greater than the overflow rate.

The overflow rate—the upward velocity of the water—is equal to Q/A_{top}, where A_{top} is the surface area of the top of the chamber. Therefore, the particle will be removed if

$$v_s \geq \text{OFR} \tag{4-49}$$

which is equivalent to

$$v_s \geq \frac{Q}{A_{\text{top}}} \tag{4-50}$$

In municipal wastewater treatment, *clarifiers*, which are used to settle out suspended particles and thereby to increase the clarity of the treated wastewater, generally use the approach shown in Figure 4-17*b*. Typical overflow rates for municipal wastewater-treatment clarifiers are shown in Table 4-7.

4.3.2.2.3 Equivalence of the Two Methods. In the settling chamber of Figure 4-17*a*, the settling velocity must exceed hQ/V, while in the chamber of Figure 4-17*b*, settling velocity must exceed the overflow rate, Q/A_{top}. Since the chamber volume is equal to hA_{top}, $hQ/V = Q/A_{\text{top}}$. Therefore, the same required settling

Table 4-7. Typical Overflow Rates Used in the Design of Municipal Wastewater Primary and Secondary Clarifiers*

Unit Operation	Overflow Rate ($m^3m^{-2}d^{-1}$)†	Overflow Rate ($cm\ s^{-1}$)‡
Primary clarifier	>40	>0.046
Secondary clarifier	>20	>0.023

*The suspended solids removed in secondary clarifiers are smaller than those removed in the primary clarifier; this difference results in the requirement of a lower overflow rate in the secondary clarifier.

†Overflow rate expressed in units of Q/A_{top}, where Q is expressed in m^3d^{-1} and A_{top} is expressed in m^2.

‡Overflow rate expressed in units of $cm\ s^{-1}$, units commonly used to describe particle-settling velocity.

velocity will be obtained, regardless of which method is used, and either method can be successfully used with chambers of either type.

EXAMPLE 4.17. CALCULATION OF THE MINIMUM DIAMETER REMOVED BY A SETTLING CHAMBER

A grit chamber is used to remove particles from the influent of a municipal wastewater-treatment plant. The chamber is 2 m deep, and the residence time (retention time) of water in the chamber is 1 h. What is the minimum-size particle that would be completely removed by settling to the bottom of the chamber during passage through the chamber? The density of sand particles is 2.65 g cm^{-3}, and the viscosity of water is 0.01185 g $cm^{-1}\ s^{-1}$. Assume that any particle that settles to the bottom of the chamber is removed.

SOLUTION

Since only those particles that reach the bottom of the chamber are removed, 100% removal requires that the distance settled during passage through the chamber is equal to the chamber depth. This results in a minimum settling velocity required for 100% removal:

$$v_s > \frac{2\ \text{m}}{1\ \text{h}} \times \frac{\text{h}}{3600\ \text{s}} \times \frac{100\ \text{cm}}{\text{m}} = 5.6 \times 10^{-2}\ \text{cm s}^{-1}$$

Plugging into the Stokes' law equation for settling velocity (Equation 4-46),

$$v_s = \frac{(980\ \text{cm s}^{-1})(2.65 - 1.00\ \text{g cm}^{-3})}{18(0.01185\ \text{g cm}^{-1}\ \text{s}^{-1})} D_p^2 > 5.6 \times 10^{-2}\ \text{cm s}^{-1}$$

Solving for D_p^2,

$$D_p^2 > 7.39 \times 10^{-6}$$

or

$$D_p > 2.7 \times 10^{-3}\ \text{cm} = 27\ \mu\text{m}$$

Thus, the minimum-size particle removed with 100% efficiency has a diameter of 27 μm.

EXAMPLE 4.18. DETERMINATION OF HYDRAULIC RETENTION TIME AND OVERFLOW RATE DURING SEDIMENTATION OF WASTEWATER SOLIDS

Suspended solids that are found in municipal wastewater are typically removed during conventional wastewater treatment by sedimentation in primary and secondary sedimentation tanks (called clarifiers). Assume that the average wastewater flow of a municipal wastewater plant is 3.2 MGD (1.2×10^4 m^3/day) and the average maximum weekly flow is 8.5 MGD (3.2×10^4 m^3/day). The total volume of the two primary clarifiers, which are operated in parallel, is 2,040 m^3 (total surface area of both clarifiers equals 600 m^2). The total volume of the three secondary clarifiers which are operated in parallel is 4,600 m^3 (total surface area equals 1,070 m^2).

What is the hydraulic retention time (in hrs) and overflow rate (m^3/m^3-day) in each clarifier at the maximum weekly flow rate? Assume all the clarifiers are in operation.

For the primary clarifier, the retention time at the maximum weekly flow equals

$$\theta = \frac{V}{Q} = [2{,}080\ \text{m}^3/(3.2 \times 10^4\ \text{m}^3\ \text{day}^{-1})]\ \frac{24\ \text{hr}}{\text{day}} = 1.5\ \text{hr}$$

The overflow rate equals

$$\frac{Q}{A} = \frac{3.2 \times 10^4\ \text{m}^3\ \text{day}^{-1}}{600\ \text{m}^2} = 53\ \text{m}^3\ \text{m}^{-2}\ \text{day}^{-1}$$

For the secondary clarifier, the retention time at the maximum weekly flow equals

$$\frac{V}{Q} = [4{,}600\ \text{m}^3/(3.2 \times 10^4\ \text{m}^3\ \text{day}^{-1})]\ \frac{24\ \text{hr}}{\text{day}} = 3.4\ \text{hr}$$

The overflow rate equals

$$\frac{Q}{A} = \frac{3.2 \times 10^4 \text{ m}^3 \text{ day}^{-1}}{1{,}070 \text{ m}^2} = 30 \text{ m}^3 \text{ m}^{-2} \text{ day}^{-1}$$

Note several interesting things about this solution. Municipal wastewater treatment plants are designed to handle variable flow, which changes daily and seasonally. Thus, the hydraulic retention time of individual unit processes will change with that variable flow. Also, notice that the secondary clarifiers are designed with a larger retention time (and lower overflow rate) than the primary clarifiers. This is because the primary clarifiers are designed to remove larger particles than the secondary clarifiers. In fact, the main purpose of secondary clarifier is to separate and then concentrate microorganisms from the wastewater which can then be recycled back into the biological reactor. Thus, because the flow is the same in both types of clarifiers, the secondary clarifiers must be designed with a greater total volume.

4.3.2.3 Other Applications of Force Balances on a Particle

Analyses similar to the one used to calculate the gravitational settling velocity can be used in a number of other applications. Essentially, in any situation where a particle moves in response to an applied force, the applied force can be balanced against the drag force to determine the particle's terminal velocity. In air pollution control, devices called *electrostatic precipitators* are used to remove smaller particles by applying an electric force to the particles, which causes them to move out of the airstream and onto charged collection plates. Other air-pollution control devices, called *cyclones*, are based on the use of inertial forces to remove particles. To determine the particle-removal efficiency in these devices, it is necessary to balance the drag force against a centrifugal force.

4.3.3 Flow of Water Through a Porous Medium: Darcy's Law for Groundwater Flow

The final problem considered in this chapter is the calculation of the rate at which water flows through a porous medium. The example of groundwater flow is used here. However, the discussion and resulting equations can be applied to a variety of other systems, such as the flow of oil underground, the flow of water through a water-treatment filter composed of an activated-carbon-packed cylinder, or the flow of air through an air filter.

This problem is similar in some respects to the particle-settling problem previously discussed. In both situations, a fluid and a particle or particles are moving relative to one another, and the relative velocity results from a balance of the drag forces with an applied force. For a settling particle, the applied force is the downward gravitational force acting on the particle, and the drag force is due to the movement of a solid particle through a stationary fluid. In groundwater flow, the drag force is due to the movement of water (the fluid) past stationary pore surfaces. The applied force that causes groundwater movement is a *pressure force*.

Groundwater is water that fills the cracks and pore spaces of underground soil and rock (Figure 4-18). Accessible groundwater constitutes the world's largest freshwater resource, containing about 40 times the freshwater in all the world's lakes and rivers, and more than twice as much freshwater as the world's glaciers (excluding the polar icecaps, which are not readily available for freshwater use). Groundwater is used extensively as a water source for agricultural and industrial use, and about half of the U.S. population uses groundwater resources for drinking water. In fact, some regions of the United States and other countries are entirely dependent on groundwater supplies.

However, groundwater flow is extremely slow in comparison to surface water. In combination with the large volume of groundwater reservoirs, this results in slow pollutant transport and very long residence times in the groundwater aquifer. For this reason, if a region of groundwater is contaminated, the contaminated water will move quite slowly. However, for the same reason—that large volumes and slow flow velocities are involved—once groundwater is contaminated it can be very difficult to clean up.

This section presents and applies the equations that govern the rate at which groundwater moves through the subsurface environment. First, however, it is

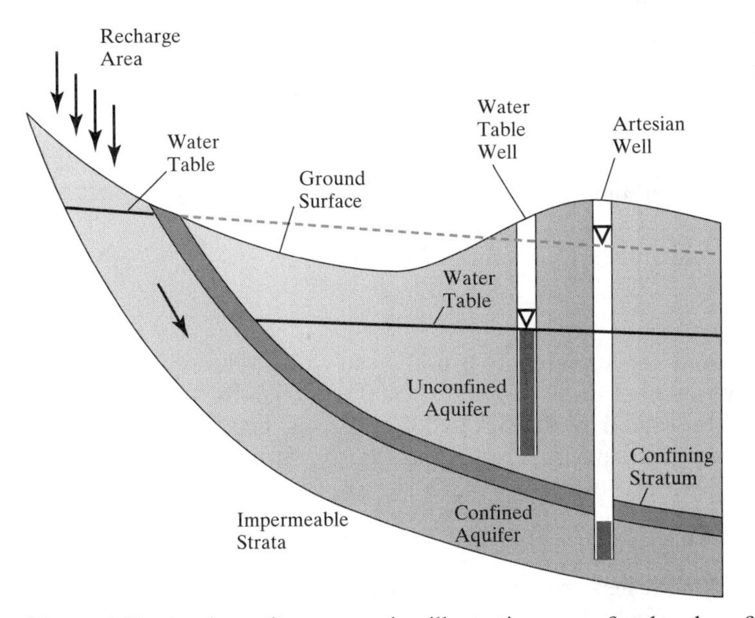

Figure 4-18. A schematic cross-section illustrating unconfined and confined groundwater aquifers. (From D. K. Todd, Groundwater Hydrology, Wiley, New York, 1980. Reprinted by permission of John Wiley and Sons, Inc.)

necessary to define some terms used in groundwater discussions: *head*, *hydraulic gradient*, and *porosity*.

4.3.3.1 Head and Hydraulic Gradient

The term *head* is used to refer to the height to which water rises within a well. In an unconfined aquifer, such as the one depicted in Figure 4-18, the head within the well is equal to the height of the surrounding water table. In cases in which the groundwater is confined below an impermeable layer (*confined aquifers*), pressure forces generally result in the head rising above the surrounding water table, as shown by the well extending into the confined aquifer in Figure 4-18.

Water flows from one point toward another as a result of differences in pressure between the two points. *The magnitude of this difference is expressed by the hydraulic gradient—the rate at which head changes with horizontal distance, or dh/dx.* The hydraulic gradient along a line between two wells can be calculated using Equation 4-51:

$$\text{Hydraulic gradient} = \frac{dh}{dx} \tag{4-51}$$

(The hydraulic gradient is actually a vector quantity, having a direction as well as a magnitude. Equation 4-51 results in calculation of the component of the hydraulic gradient in the x-direction. Fully characterizing the hydraulic gradient requires three wells.)

4.3.3.2 Porosity

Porosity is defined as the fraction of the total volume of soil or rock that is empty pore space capable of containing water or air. Porosity is given the symbol eta (η):

$$\eta = \frac{\text{volumes of pores}}{\text{total volume}} \tag{4-52}$$

Since the volume units cancel, porosity is unitless. Typical values of porosity in the subsurface environment range from 5% to 30% for sandstone rock formations, 25% to 50% for sand deposits, 5% to 50% for Karst limestone formations, and 40% to 70% for clay deposits (Freeze and Cherry, 1979).

4.3.3.3 Darcy's Law

The pressure force that drives groundwater flow is proportional to the hydraulic gradient. The velocity at which groundwater moves is determined by a balance between this driving force and resistive forces impeding the flow. These drag forces are proportional to the flow velocity, just as was the case in the Stokes' drag (Equation 4-39). Without considering the specific values of the proportionality factors relating force to hydraulic gradient and resistive force to flow velocity, a force balance can be conducted to determine the expected form of the relationship between flow velocity and hydraulic gradient. Setting the drag force equal to the resistive forces results in

$$\text{(Constant one)} \times \frac{dh}{dx} = \text{(constant two)} \times \text{velocity} \tag{4-53}$$

This equation can be rearranged to give

$$\text{Velocity} = \text{(constant three)} \times \frac{dh}{dx} \tag{4-54}$$

This analysis indicates that groundwater flow velocity should be proportional to the hydraulic gradient.

Henri Darcy (1803–1858), a French civil engineer, conducted experiments in the 1800s to determine the constant in this relationship, as part of a study of the water supply for Dijon, France. Figure 4-19 shows the experimental setup used. Darcy used a variety of materials to fill the flow tube, and found that *the total volumetric flow through the tube, Q, is related to the hydraulic gradient by the following equation*:

$$Q = -KA \frac{dh}{dx} \tag{4-55}$$

in which K is termed the *hydraulic conductivity*. Equation 4-55 is referred to as *Darcy's law*.

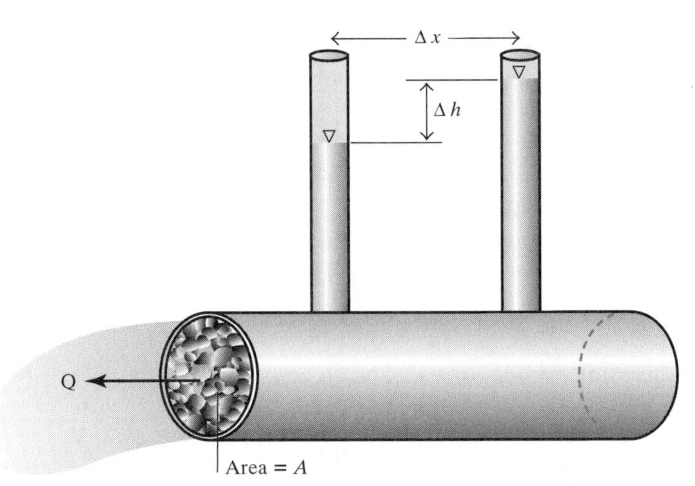

Figure 4-19. Experimental setup for a Darcy's law experiment.

The hydraulic conductivity is a property of the medium through which flow is occurring and of the fluid (e.g., water or oil). K is high for materials through which it is easy to push water, such as sand or gravel, and is low for materials that greatly restrict flow, such as clays. It also depends on the fluid viscosity, just as the drag force in Stokes' settling was a function of fluid viscosity.

Hydraulic conductivity is measured in the field by performing a "pumping test." Values can range over 13 orders of magnitude for common geologic materials (10^{-11} to 100 cm s^{-1}). Representative values of hydraulic conductivity are 0.2 to 0.5 cm s^{-1} for gravel (which has large pores that are easy for water to flow through); 3×10^{-3} to 5×10^{-2} cm s^{-1} for sand; and 2×10^{-7} cm s^{-1} for clay (which is composed of very fine particles and thus has very small, tight pores that are difficult to force water through). In the construction of solid and hazardous waste landfills, liners are placed under the waste material. The liners are usually composed of compacted nonswelling clay materials (K approximately 10^{-7} cm s^{-1}) and/or synthetic membranes, such as high-density polyethylene, which can have hydraulic conductivity less than 10^{-11} cm s^{-1}.

4.3.3.4 *The Darcy Velocity and the True Horizontal Velocity*

Darcy's law relates the volumetric flow rate of groundwater, Q (units of volume water per unit time), to the cross-sectional area A of the soil or rock through which the flow occurs. The ratio of these two quantities gives the *Darcy velocity*:

$$\text{Darcy velocity } (v_d) = \frac{Q}{A} = -K\left(\frac{dh}{dx}\right) \qquad \textbf{(4-56)}$$

However, this velocity does not reflect the true speed at which groundwater moves through the subsurface environment, because *the cross-sectional area of the pores through which groundwater flows is smaller than the total soil or rock cross-sectional area.*

The effect of limited pore area can be seen through a simple experiment. Consider a pipe of length L through which water flows at a volumetric flow rate of Q, as shown in Figure 4-20. If the cross-sectional area of the pipe is A, the velocity of water through the pipe is equal to Q/A, and the retention time within the pipe is given by $\frac{V}{Q}$, or $(A \times L)/Q$.

If the flow rate Q is held constant while the pipe is filled with rocks, it will require more pressure to force the water through, and water will squirt out of the pipe at a much greater velocity. The new velocity through the pipe is equal to L divided by the residence time in the pipe, θ. This residence time is still equal to $\frac{V}{Q}$, but the volume available for water flow is reduced; it is now equal to the porosity times the total volume, or ηV. The velocity through the pipe is thus given by

$$\text{Velocity} = \frac{L}{\frac{(\eta V)}{Q}} = \frac{QL}{\eta V} \qquad \textbf{(4-57)}$$

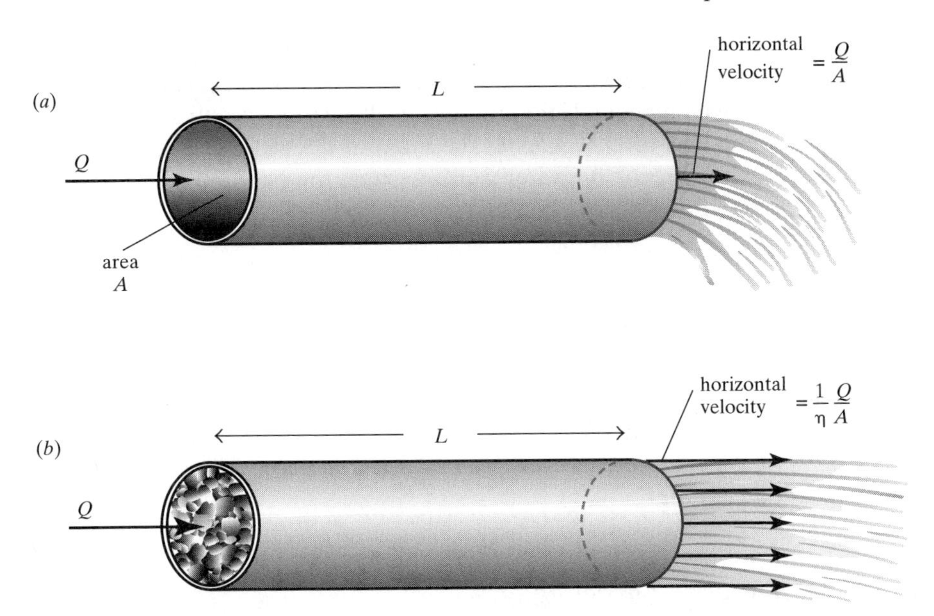

Figure 4-20. The effect of porosity on flow velocity. (*a*) Water flows through a pipe at a volumetric flow rate Q. The water velocity is equal to Q/A. (*b*) The pipe is filled with rocks having a porosity η, while the flow rate Q is maintained. The water velocity through the pipe is increased by the factor $1/\eta$.

Since V/L is equal to the cross-sectional area A,

$$\text{Velocity} = \frac{1}{\eta}\frac{Q}{A} \tag{4-58}$$

This example demonstrates that the true velocity through a porous volume is increased, relative to Q/A, by a factor of $1/\eta$. Since the Darcy velocity v_d is equal to Q/A, this means that

$$v_{\text{true}} = \frac{1}{\eta} v_d \tag{4-59}$$

In problems requiring calculations of how quickly groundwater travels from one point to another, the true velocity should be used. If, however, one wishes to calculate the amount of water flowing through a given cross-sectional area of an aquifer, Darcy's law should be used to directly determine Q/A.

EXAMPLE 4.19. USE OF DARCY'S LAW IN DETERMINING THE TRANSPORT TIME BETWEEN TWO WELLS

An underground storage tank has discharged diesel fuel into groundwater. A drinking-water well is located 200 m downgradient from the fuel spill. To ensure the safety of the drinking-water supply, a monitoring well is drilled halfway between the drinking-water well and the fuel spill. The difference in hydraulic head between the drinking-water well and the monitoring well is 40 cm (with the head in the monitoring well higher). If the porosity is 39% and hydraulic conductivity is 45 m day^{-1}, how long after it reaches the monitoring well would the contaminated water reach the drinking water well? Assume the pollutants move at the same speed as the groundwater.

SOLUTION

First, determine the true velocity of the groundwater between the two wells. The time for travel between the two wells will then be $\Delta t = (100 \text{ m})/v_{true}$. The hydraulic gradient is equal to $dh/dx = (0.40 \text{ m})/(100 \text{ m}) = 0.0040$. The Darcy velocity is given by (Equation 4-56)

$$v_d = K\frac{dh}{dx} = 45 \text{ m day}^{-1} \times 0.0040 = 0.18 \text{ m day}^{-1}$$

The true velocity is equal to this value divided by the porosity (Equation 4-59):

$$v_{\text{true}} = \frac{v_d}{\eta} = \frac{0.18 \text{ m day}^{-1}}{0.39} = 0.46 \text{ m day}^{-1}$$

Thus, the period for flow from the monitoring well to the drinking water well is $\Delta t = (100 \text{ m})/(0.46 \text{ m day}^{-1}) = 217$ days. This result is typical of groundwater flow speeds. That is, groundwater transport is usually very slow relative to surface waters.

Note that this solution is based on the assumption that the chemical of concern travels at the same velocity as the groundwater. In reality, the chemical's movement may be slowed by sorption to solids in the aquifer, a phenomenon discussed in Section 3.4.5.2.

PROBLEMS

4-1. A pond is used to treat a dilute municipal wastewater before the liquid is discharged into a river. The inflow to the pond has a flow rate of $Q =$ 4,000 m^3 day^{-1} and a BOD concentration of $C_{\text{in}} = 25$ mg L^{-1}. The volume of the pond is 20,000 m^3. The purpose of the pond is to allow time for the decay of BOD to occur before discharge into the environment. BOD decays in the pond with a first-order rate constant equal to 0.25 day^{-1}. What is the BOD concentration at the outflow of the pond, in units of mg L^{-1}?

4-2. For each of the following problems, answer the following questions. (You do not need to actually solve the problems.) (i) What control volume would you use to solve the problem? (ii) Are conditions or concentrations in the problem changing with time? (iii) Is the problem steady state or non-steady state, and why? (iv) Is the chemical being produced or destroyed by chemical reaction within the control volume, or not? (v) Is the chemical conservative or nonconservative, and why?

(a) An accident has resulted in the release of a pollutant inside a chemical manufacturing plant. The spill released the pollutant into a lake near the chemical plant. The lake flows into a small stream. How long would it take the pollutant, which is inert, to reach a safe level in the lake and in the stream leaving the lake?

(b) Carbon dioxide (CO_2) emissions from fossil-fuel burning are mixed throughout the atmosphere. Assume that these emissions are mixed immediately throughout the entire atmosphere, and that CO_2 does not degrade chemically. If the total emission rate of carbon dioxide and the volume of the atmosphere is known, what would be the rate of increase of atmospheric carbon dioxide levels in ppm yr^{-1}?

(c) An air freshener emits perfume into a room at a constant rate. The perfume is mixed throughout the room and diluted by the room ventilation flow, which brings fresh air into the room at a flow rate Q. What is the resulting concentration of perfume in the room? (Note that the first-order decay rate constant of the perfume is very slow relative to the amount of time it takes to mix fresh air through the room.)

4-3. A mixture of two gas flows is used to calibrate an air-pollution measurement instrument. The calibration system is shown in Figure 4-21. If the calibration gas concentration C_{cal} is 4.90 ppm$_V$, the calibration gas flow rate Q_{cal} is 0.010 L min^{-1}, and the total gas flow rate Q_{total} is 1.000 L min^{-1}, what is the concentration of calibration gas after mixing ($C_{d/s}$)? (Assume that the concentration upstream of the mixing point is zero.)

4-4. Consider a house into which radon is emitted through cracks in the basement. The total volume of the house is 650 m^3 (assume that it is well mixed

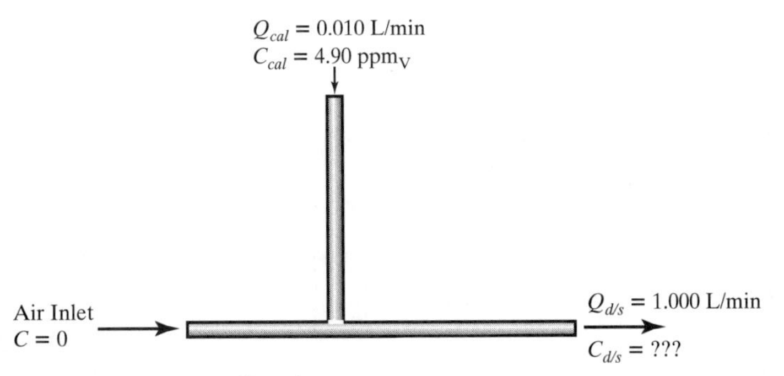

Figure 4-21. Gas calibration system.

throughout). The radon source emits 250 pCi s^{-1} (pCi, or picoCuries, is a unit proportional to the amount of radon gas, and indicates the amount of radioactivity of the gas). Air inflow and outflow can be modeled as a flow of clean air into the house of 722 m^3 h^{-1}, and an equal air flow out. Radon can be considered to be conservative in this problem.

(a) What is the retention time of the house?

(b) What is the steady-state concentration of radon in the house? Express your answer in units of pCi/L.

4-5. You are in an old spy movie, and have been locked into a small room (volume 1,000 ft^3). You suddenly realize that a poison gas has just started entering the room through a ventilation duct. Recognizing the type of poison from its smell, you know that if the gas reaches a concentration of 100 mg m^{-3}, you will die instantly, but that you are safe as long as the concentration is less than 100 mg m^{-3}. If the ventilation air flow rate in the room is 100 ft^3 min^{-1} and the incoming gas concentration is 200 mg m^{-3}, how long do you have to escape?

4-6. In the simplified depiction of an ice rink with an ice resurfacing machine operating shown in Figure 4-22, points 1 and 3 represent the ventilation air intake and exhaust for the entire ice rink, and point 2 is the resurfacing machine's exhaust. Conditions at each point are (C indicates the concentration of carbon monoxide, CO):

Point 1	$Q_1 = 3.0$ m^3/s
	$C_1 = 10$ mg/m^3
Point 2	Emission rate = 8 mg/s of CO. Note that CO is nonreactive.
Point 3	Q_3, C_3 unknown
Ice rink volume	$V = 5.0 \times 10^4$ m^3

(a) Define a control volume as the interior of the ice rink. What is the mass flux of CO into the control volume, in units of mg/s?

(b) Assume that the resurfacing machine has been operating for a very long time, and that the air within the ice rink is well mixed. What is the concentration of CO within the ice rink, in units of mg/m^3?

Figure 4-22. Schematic diagram of an ice resurfacing machine in an ice rink.

4-7. An ice rink, depicted in Figure 4-22, has the following conditions:

Point 1	$Q_1 = 3.0\ m^3/s$
	$C_1 = 0.0\ mg/m^3$
Point 3	$Q_3 = 3.0\ m^3/s$, $C_3 = 50\ mg/m^3$
Ice rink volume	$V = 5.0 \times 10^4\ m^3$

at the time the resurfacing machine is turned off.

(a) What is the mass flux of CO out of the control volume at this moment, in units of mg/s?

(b) The U.S. ambient air-quality standard for CO is 40 mg/m^3. How long would it take before the CO concentration in the well-mixed ice rink air dropped to 40 mg/m^3? Express your answer in units of hours.

4-8. Poorly treated municipal wastewater is discharged to a stream. The river flow rate upstream of the discharge point is $Q_{u/s} = 8.7\ m^3\ s^{-1}$. The discharge occurs at a flow of $Q_d = 0.9\ m^3\ s^{-1}$ and has a BOD concentration of 50.0 mg L^{-1}. Assuming that the upstream BOD concentration is negligible

(a) What is the BOD concentration just downstream of the discharge point?

(b) If the stream has a cross-sectional area of 10 m^2, what would the BOD concentration be 50 km downstream? (BOD is removed with a first-order decay rate constant equal to 0.20 day^{-1}.)

4-9. Two towns, located directly across from each other, operate municipal wastewater-treatment plants that are situated along a river. The river flow is 50 million gallons per day (50 MGD). Coliform counts are used as a measure to determine a water's ability to transmit disease to humans. The coliform count in the river upstream of the two treatment plants is 3 coliforms/100 mL. Town 1 discharges 3 MGD of wastewater with a coliform count of 50 coliforms/100 mL, and town 2 discharges 10 MGD of wastewater with a coliform count of 20 coliforms/100 mL. Assume that the state requires that the downstream coliform count not exceed 5 coliforms/100 mL.

(a) Is the state water-quality standard being met downstream? (Assume that coliforms do not die by the time they are measured downstream.)

(b) If the state standard downstream is not met, the state has informed town 1 that they must treat their sewage further so the downstream standard is met. Use a mass-balance approach to show that the state's request is unfeasible.

4-10. In the winter, a stream flows at 10 m^3 s^{-1} and receives discharge from a pipe that contains road runoff. The pipe has a flow of 5 m^3 s^{-1}. The stream's chloride concentration just upstream of the pipe's discharge is 12 mg L^{-1} and the runoff pipe's discharge has a chloride concentration of 40 mg L^{-1}. Chloride is a conservative substance.

(a) Does wintertime salt usage on the road elevate the downstream chloride concentration above a 20 mg L^{-1} water-quality standard?

(b) What is the maximum daily mass of chloride (tons d^{-1}) that can be discharged through the road runoff pipe without exceeding the water-quality standard?

4-11. A small lake with volume 20.0×10^3 m^3 has no natural pollutant sources. However, a sewage pipe discharges sewage containing a pollutant into the lake. Conditions in the sewage pipe are $Q_s = 0.5$ m^3/s, and $C_s =$ 100.0 mg/L. The pollutant is destroyed in the lake with a first-order rate constant of 0.20/day. The flow rate out of the lake is 5.0 m^3/s. Assume that the lake is well-mixed.

(a) Calculate the steady-state pollutant concentration in the lake.

(b) What is the retention time for this lake?

4-12. **(a)** Calculate the hydraulic residence times (the retention time) for Lake Superior and for Lake Erie using data in Table 4-4.

(b) Assume that both lakes currently are polluted with the same compound at a concentration that is 10 times the maximum acceptable level. If all sources of the compound are removed, how long will it take the concentration to reach acceptable levels in each lake? Assume that the pollutant does not decay chemically.

(c) Comment on the significance of your answers.

4-13. The total flow at a wastewater-treatment plant is 600 m^3 day^{-1}. Two biological aeration basins are used to remove BOD from the wastewater and are operated in parallel. They each have a volume of 25,000 L. What is the aeration period of each tank in hours?

4-14. You are designing a reactor that uses chlorine in a PFR or CMFR to destroy pathogens in water. A minimum contact time of 30 min is required to reduce the pathogen concentration from 100 pathogens L^{-1} to below 1 pathogen L^{-1} through a first-order decay process. You plan on treating water at a rate of 1,000 gal min^{-1}.

(a) What is the first-order decay rate constant?

(b) What is the minimal size (in gallons) of the reactor required for a plug flow reactor?

(c) What size (in gallons) of CMFR would be required to reach the same outlet concentration?

(d) Which type of reactor would you select if your treatment objective stated that "no discharge can ever be greater than 1 pathogen L^{-1}?" (Explain your reasoning.)

(e) If the desired chlorine residual in the treater water after it leaves the reactor is 0.20 mg L^{-1}, and the chlorine demand used during treatment is 0.15 mg L^{-1}, what must be the daily mass of chlorine added to the reactor in grams?

4-15. What is the most effective way to wash laboratory glassware? Assume that you have a 1.0 L flask that contains 5 mL of residual droplets of a 1.0 M NaCl solution. You can rinse the flask with a total volume of 1.0 L of pure water ([NaCl] = 0.000 M). Consider the following options: (a) Fill the flask up completely once, adding the full 1.0 L of pure water. Shake the flask to mix well, and empty it. (b) Add one-third of the water, shake to mix, empty, and repeat two more times.

> *In each case, assume that 5 mL of the resulting mixture of clean and dirty water is left after* each *rinse.*

For your answer, calculate the concentration of NaCl that would result in case (a) and in case (b) if the remaining NaCl in the flask were diluted to 1.0 L with pure water. Discuss your results.

4-16. Derive the four solutions shown in Table 4-2, starting with Equation 4-3 for the CMFR and Equation 4-18 for the PFR.

4-17. The concentration of BOD (an indicator of organic pollution) in a river just downstream of a sewage-treatment plant's effluent pipe is 75 mg/L^{-1}. If the BOD is destroyed through a first-order reaction with rate constant equal to 0.05/day^{-1}, what is the BOD concentration 50 km downstream? The velocity of the river is 15 km/day^{-1}.

4-18. A 1.0×10^6 gallon reactor is used in a sewage-treatment plant. The influent concentration is 100 mg/L^{-1}, the effluent concentration is 25 mg/L^{-1}, and the flow rate through the reactor is 500 gal/min^{-1}.

(a) What is the first-order rate constant for decay of BOD in the reactor? Assume that the reactor can be modeled as a CMFR. Report your answer in units of h^{-1}.

(b) Assume that the reactor should be modeled as a PFR with first-order decay, *not* as a CMFR. In that case, what must the first-order decay rate constant be within the PFR reactor?

(c) It has been determined that the outlet concentration is too high, and that as a result, the residence time in the reactor must be doubled. Assuming that all other variables remain constant, what must be the volume of the new CMFR?

4-19. How many watts of power would it take to heat 1 L of water (weighing 1.0 kg) by 10°C in 1.0 h? Assume that no heat losses occur, so that all of the energy expended goes into heating the water.

4-20. The concentration of a pollutant along a quiescent water-containing tube is shown in Figure 4-23. The diffusion coefficient for this pollutant in water is equal to 10^{-5} cm^2 s^{-1}.

(a) What is the initial pollutant flux density in the x-direction at the following locations: $x = 0.5, 1.5, 2.5, 3.5$, and 4.5?

(b) If the diameter of the tube is 3 cm, what is the initial flux of pollutant mass in the x-direction at the same locations?

(c) As time passes, this diffusive flux will change the shape of the concentration profile. Draw a sketch of concentration in the tube versus x-axis location showing what the shape at a later time might look like. (It is

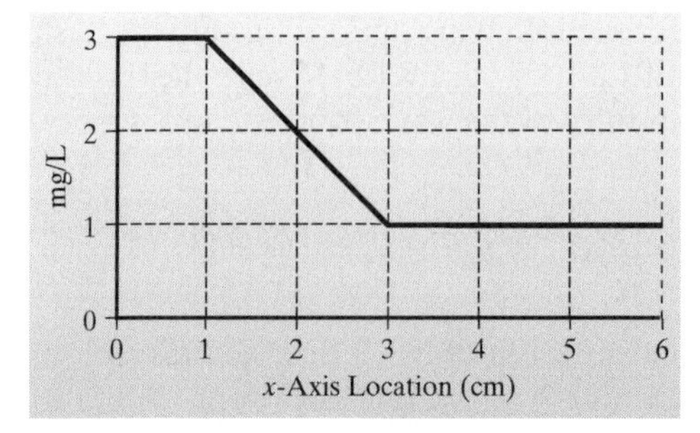

Figure 4-23. Hypothetical concentration profile in a closed pipe.

not necessary to do any calculations to draw this sketch.) Assume that the concentration at $x = 0$ is held at 3 mg/L and the concentration at $x = 6$ is held at 1 mg/L.

(d) Describe, in one paragraph, why the concentration profile changed in the way that you sketched in your solution to part (c).

4-21. The tube in Problem 4-20 is connected to a source of flowing water, and water is passed through the tube at a rate of 100 $cm^3\ s^{-1}$. If the pollutant concentration in the water is constant at 2 mg L^{-1}, what is

(a) The mass flux density of the pollutant through the tube due to advection?

(b) The total mass flux through the tube due to advection?

4-22. A treatment plant has two primary clarifiers, each with a depth of 5 m and with a combined volume of 5,000 m^3. The clarifiers were designed to have a detention time greater than or equal to 2 h.

(a) What is the maximum flow the plant is designed to accept based upon the design detention time?

(b) What is the minimum settling velocity (in units of m s^{-1}) required so that 100% of the particles entering the chamber are removed? (Particle density is 2,000 kg m^{-3}, liquid density is 1,000 kg m^{-3}, particle diameter is 0.0001 m, and liquid viscosity is 0.01 kg $m^{-1}\ s^{-1}$.)

4-23. A settling chamber is used to remove particles from an air flow. Settling chamber height is 3 m and residence time of air in the chamber is 10 s.

(a) What minimum particle-settling velocity is required, if all particles are to be removed?

(b) What is the minimum-size particle that the chamber would remove?

The particle density is 1 g cm^{-3}, air viscosity is 1.69×10^{-4} g $cm^{-1}\ s^{-1}$, and the air density is small enough that the buoyancy force may be ignored. Express your answer as particle diameter in μm.

4-24. A wastewater-treatment plant uses a grit chamber to settle out sand particles. Water flows at a rate of 100 gal/min through the chamber, which is

2 m in height and has a volume of 100 m^3. Calculate the minimum-size sand particle that would be removed in the chamber. To be removed, the particles must settle the depth of 2 m during the period the wastewater spends in the chamber ($\Delta t = V/Q$). (Sand density is 2.65 g m^{-3}, water density is 1.0 g cm^{-3}, and the viscosity of water is 0.01185 g cm^{-1} s^{-1}.)

4-25. Cyclones are air-pollution control devices that remove particles by using centrifugal force to throw the particles out of the air stream and into a wall. The force on the particle is given by

$$\frac{V_g^2}{R} \times m_p$$

where V_g is the gas flow velocity in the cyclone, R is the cyclone radius, and m_p is the particle mass, equal to particle density (assume the density is equal to 1 g/cm^3) times particle volume, $(\pi/6)D_p^3$. If a cyclone is designed such that V_g^2/R equals 1.6×10^6 cm s^{-2}, what is the resulting terminal velocity of a 15 μm (15×10^{-4} cm) particle exposed to this centrifugal force? (Gravitational settling and buoyancy are negligible in a cyclone. Express your answer in units of cm s^{-1}.)

4-26. Electrostatic precipitators (ESPs) are used to clean particles from gas flows such as the exhaust from a power plant boiler or a municipal waste incinerator. They work by applying a charge to the particles, which then move at their terminal velocity in response to the electrostatic force that results from an applied voltage. In this problem, you are to do the following (conditions and equations that apply to the entire problem are given at the end of the problem):

(a) Calculate the terminal velocity, for which it is first necessary to clearly identify the forces acting on the particle. Draw a picture similar to that in Figure 4-16 and identify the forces acting on a particle passing through the ESP.

(b) Calculate the terminal drift velocity (w) of particles in response to the electrostatic force as a function of particle diameter. Report your solution as an equation that gives the w in units of cm/s, and specify the units of D_p that you are using. Note: Neglect gravity in this calculation—assume that the only external force acting on the particle is the electrostatic force. Since buoyancy is a result of the gravitational force, neglect buoyancy also.

(c) Calculate the terminal drive velocity for the following particle sizes: 0.5 μm and 10.0 μm.

(d) For comparison, calculate the gravitational settling velocity for the same particle sizes.

(e) By comparing your answer for part (a) to the Stokes' settling velocity equation, determine the particle size for which the gravitational settling velocity is equal to the drift velocity in the ESP. Express your answer in units of μm.

(f) For particles *smaller* than your answer to part (e), which would be larger: the gravitational settling velocity or the ESP drive velocity?

Conditions and equations:

$$\text{Electrostatic force} = q \times E$$

where the electrostatic force is given in Newtons (kg $\times$ m $\times$ s^{-2}), q is the charge on the particle (coulombs, C), and E is the electric-field strength (V m^{-1}). The charge on each particle is dependent on particle diameter, and is approximately given by

$$q = 3 \times 10^{-9}\, ED_p^2,$$

where q is given in coulombs, E is the electric field (V m^{-1} as before), and D_p is the particle diameter in m.

Assume that the electric field strength is 10^5 V m^{-1}, and the particle density is 1,000 kg m^{-3}. Assume that the particles are suspended in air at 298° K and 1 atm, which has a viscosity of 1.695×10^{-5} kg m^{-1} s^{-1}. (Note that particle density is *much* greater than the density of air.)

4-27. A settling chamber is used to remove sand and grit from the influent water to a wastewater-treatment plant. Conditions are:

Chamber height	3 m
Chamber width	2 m
Chamber length	7 m
Water flow rate	0.5 m^3/min
Particle density	2.65 g/cm^3

(a) What is the minimum settling velocity required for a particle to be removed by this settling chamber? Express your answer in units of cm s^{-1}.

(b) Assume that this or another settling chamber removes all particles having settling velocities greater or equal to 0.04 cm s^{-1}. What is the minimum particle diameter that would be entirely removed by such a settling chamber? Express your answer in units of μm.

4-28. Two groundwater wells are located 100 m apart in permeable sand and gravel. The water level in well 1 is 50 m below the surface, and in well 2 the water level is 75 m below the surface. The hydraulic conductivity is 1 m day^{-1}, and porosity is 0.60. What is

(a) The Darcy velocity, $v_d = Q/A$?

(b) The true velocity of the groundwater flowing between wells?

(c) The period it takes water to travel between the two wells, in days?

4-29. The hydraulic gradient of groundwater in a certain location is 2 ft/100 ft. Groundwater in this location flows through sand, with a high hydraulic conductivity equal to 40 m day^{-1} and a porosity of 0.5. An oil spill has caused the pollution of the groundwater in a small region beneath an industrial site. How long would it take the polluted water from that location to reach a drinking water well located 100 m downgradient? Assume no retardation of the pollutant's movement.

REFERENCES

Chapra, S. C. and K. H. Reckhow, 1983. Engineering Approaches for Lake Management, Vol. 2, Mechanistic Modeling. Butterworth Publishers, Boston.

H. B. Fischer, E. J. List, J. Imberger, and N. H. Brooks, 1979. Mixing in Inland and Coastal Waters. Academic Press, New York.

Freeze and Cherry, 1979. Groundwater, Prentice-Hall Inc., Englewood Cliffs.

Hemond, H. F., and E. J. Fechner. 1994. Chemical Fate and Transport in the Environment. Academic Press. San Diego.

J. J. Houghton, L. G. Meiro Filho, B. A. Callander, N. Harris, A. Kattenberg, and K. Maskell, Eds., Climate Change 1995: The Science of Climate Change Contribution of Working Group I to the Second Assessment Report of the Intergovernmental Panel on Climate Change. Cambridge Univ. Press, New York, 584 pages.

Todd, D. K. 1980. Groundwater Hydrology, John Wiley, New York.

Chapter 5

Biology

Martin T. Auer
Michael R. Penn
James R. Mihelcic

In this chapter, the reader will be introduced to the fundamental biological principles governing the ecosystems of the world, with special attention to those processes that mediate the fate of chemical substances in natural and engineered systems. The chapter begins with a discussion of ecosystem structure and function, including a description of population dynamics, that is, the simulation of organism growth and attendant resource consumption. Production and consumption in ecosystems is then examined, leading to a description of trophic structure and energy flow. Attention then turns to the demand for oxygen associated with the mineralization of organic matter. The significance of this phenomenon is considered in relation to waste treatment and oxygen dynamics in flowing waters that receive inputs of organic waste. The chapter continues with an introduction to material flow in ecosystems, focusing on key biogeochemical cycles (e.g., oxygen, carbon, nitrogen, sulfur, and phosphorus) and the related hydrologic cycle. Building on the concept of material cycling, the role of nutrients in governing water-quality conditions in lakes is described and engineered approaches for lake and reservoir management are presented. Additional materials on photosynthesis, aerobic and anaerobic respiration, toxicity, bioconcentration, and pathogenic organisms round out the chapter.

Biology is defined as the scientific study of life and living things, often taken to include their origin, diversity, structure, activities, and distribution. *Biotic* effects, that is, those produced by or involving organisms, are important in many phases of environmental engineering. This chapter's exploration of environmental biology will focus on biological activities—the ways that organisms are affected by and have an effect on the environment. These include:

- Effects on humans (e.g., infectious disease);
- Impacts on the environment (e.g., introduction of exotic species);
- Impacts by humans (e.g., threats to endangered species);
- Mediation of chemical transformations in the environment (e.g., breakdown of toxic chemicals);
- Application in the treatment of contaminated air, water, and soil.

5.1 ECOSYSTEM STRUCTURE AND FUNCTION

The Earth can be conceptualized as being composed of "great spheres" of living and nonliving material (Figure 5-1). The *atmosphere* (air), *hydrosphere* (water), and *lithosphere* (soil) constitute the *abiotic* or nonliving component. The *biosphere* contains all of the living things on Earth. Any intersection of the biosphere with the nonliving spheres, that is, living things and their attendant abiotic environment, constitutes an ecosystem. Examples include both natural (lake, forest, and desert) and engineered (activated sludge—a waste-treatment system where microorganisms are used to decompose organic wastes) ecosystems. Taken together, all of the ecosystems of the world make up the *ecosphere*. *Ecology* is the study of structure and function in nature: interactions between living things and their nonliving (abiotic) environment or habitat.

Engineers often envision the biotic component of their work as being rather indistinct in form (a "black box") and for convenience reduce it to a surrogate or composite, that is, grams of organic carbon as opposed to numbers of individual organisms. However, the biology of natural and engineered systems can be extremely complex, and such simplifications can lead to critical engineering oversights. It is valuable then to begin with a brief overview of the levels of organization in nature, recognizing the tremendous diversity in form and function represented in the biosphere. Engineers and scientists must understand how to operate across this entire spectrum, from biochemical transformations at the molecular level to the processes that influence the global climate at the ecosphere level.

Although the field of taxonomy (classification) is highly dynamic and hotly debated, biologists have traditionally placed living things within one of five *kingdoms*, differentiated by the organization of their nuclear material and by their feeding strategies. *Procaryotic* organisms have their nuclear material distributed throughout the cell, while *eucaryotic* organisms utilize a membrane to segregate the nuclear material, that is, a distinct nucleus is present. Feeding strategies include *absorption* (uptake of dissolved nutrients), *photosynthesis* (fixation of light energy in simple carbohydrates), and *ingestion* (intake of particulate nutrients). The five kingdoms are Monera (unicellular procaryotes that obtain nutrients strictly by absorption), Protista (mostly unicellular eucaryotes that obtain food by absorption, photosynthesis, or ingestion), Fungi (mostly multicellular eucary-

Figure 5-1. The Earth's Great Spheres of living and nonliving material. The atmosphere, hydrosphere, and lithosphere are the nonliving components and the biosphere contains all the living components. The ecosphere is the intersection of the abiotic spheres and the biotic component. (Adapted from *Environmental Science* by Kupchella/Hyland, © 1986, by permission of Prentice-Hall, Inc., Upper Saddle River, NJ.)

otes that obtain food by absorption), Plantae (multicellular eucaryotes that obtain food by photosynthesis), and Animalia (multicellular eucaryotes that obtain food by ingestion). This system of classification is not without exceptions. For example, the cyanobacteria have no distinct nuclear structure (a Moneran characteristic), but obtain food through photosynthesis (a Protistan characteristic). Once referred to as blue-green algae, these "borderline" forms are today firmly placed among the Monera. More recently, a three-domain system has been proposed in an attempt to resolve this and other difficulties with the five-kingdom approach.

Each kingdom can be further subdivided into phyla, classes, orders, families, genera, and species. A *species* is a group of individuals that possess a common gene pool and that can successfully interbreed. Each species is assigned a scientific name (genus–species), in Latin, to avoid the confusion associated with common names. Under this system of binomial nomenclature, *Stizostedion vitreum*, is the scientific name for the fish species commonly referred to as walleye, walleye pike, pike, pike perch, pickerel, yellow pike, yellow pickerel, yellow pike perch, or yellow walleye. All of the members of a species in a given area make up a *population*, for example, the walleye population of a lake. All of the populations (of different species) that interact in a given system make up the *community*, for example, the fish community of a lake. Finally, all of the communities, plus the abiotic factors, make up the ecosystem (Figure 5-2) and the ecosystems, the ecosphere.

5.1.1 Major Organism Groups

Environmental engineers encounter a variety of populations and communities in both *natural* (e.g., rivers, wetlands, soil) and *engineered systems* (e.g., landfills, wastewater-treatment plants). Certain groups of organisms are especially impor-

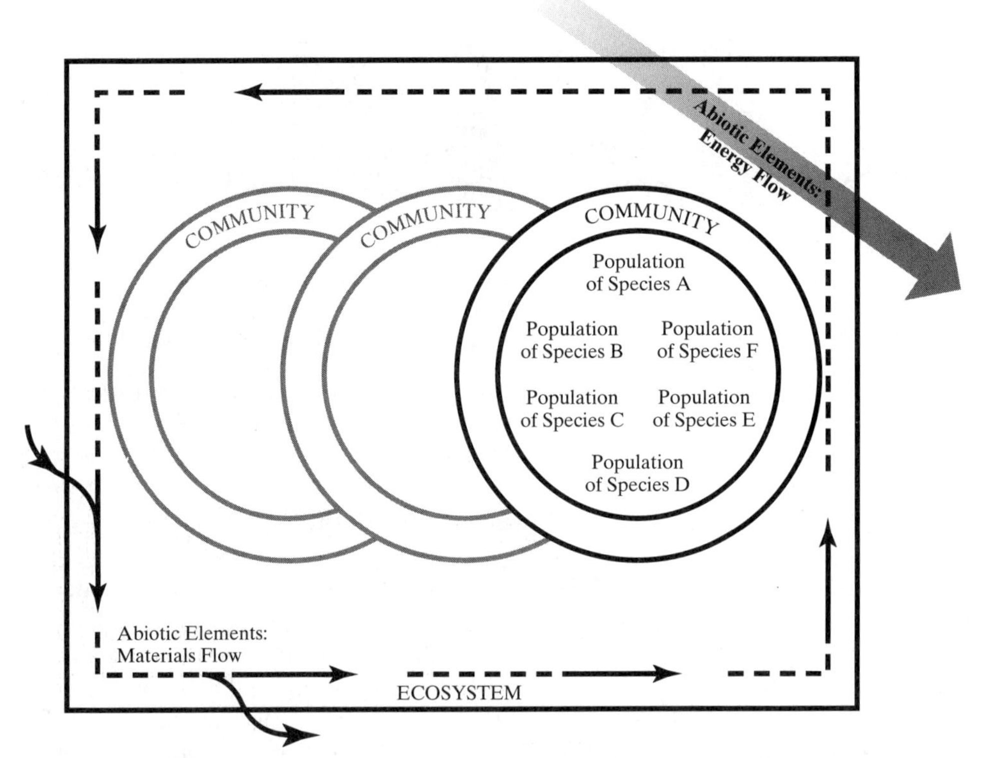

Figure 5-2. The biotic component of an ecosystem can be organized according to species, populations, and communities. Note in this schematic that energy flows through and chemicals cycle largely within the ecosystem. Both natural and engineered environments may be considered as ecosystems. The various biological unit processes employed for wastewater treatment (activated sludge, anaerobic digester, trickling filter, oxidation pond) have communities composed of a variety of microorganism populations. The nature of the ecosystem is determined by the physical design of the unit process and by the chemical and biological character of the wastewater entering the system.

tant in environmental engineering. Figures 5-3*a* to 5-3*j* present illustrations of some of these, discussed in more detail below.

Viruses: Submicroscopic particles (there is some argument as to whether they are very complex biochemicals or very simple organisms, i.e., are they truly alive?) ranging in size from 0.02 μm to 0.3 μm, composed of a nucleic acid core and a protein coat and containing all of the hereditary material required for reproduction; all are parasitic, depending on a host for protein and the energy needed to reproduce; all are pathogenic, causing a variety of diseases: notably AIDS and, in water, hepatitis, polio, and gastroenteritis; because of public health concerns, viruses are of particular importance to engineers involved in water and wastewater treatment. Other noncellular agents of disease include the viroids, consisting only of small RNA molecules (lacking a protein coat) that infect plants and the prions, protein units that infect animals, causing scrapie in sheep and goats and mad cow disease.

Bacteria: Monerans; 0.1–10 μm in size; typically reproduce by fission (splitting); acquire nutrients by absorption; many are pathogenic, causing tuberculosis, diphtheria, strep throat, whooping cough, Lyme disease, tetanus and, in water, cholera and typhoid, thus also of importance in water and wastewater disinfection. Although some bacteria depend on sunlight as a source of energy, most use chemicals for this purpose, and thus play an important role in mediating various biochemical transformations, for example, decomposition of organic matter, oxidation of ammonia to nitrate, and reduction of sulfate to sulfide. Bacteria are of major importance in cycling material and energy in natural and engineered systems, for example, hydrocarbon-degrading bacteria have received significant attention for their ability to break down toxic chemicals (e.g., gasoline, solvents), thus aiding in the remediation of contaminated soil and water environments. In soils, the unsaturated zone typically contains 10^5 to 10^8 viable bacterial cells per gram of soil, and the saturated zone (i.e., groundwater) typically contains 10^2 to 10^7 viable bacterial cells per gram of soil–water mixture. Bacteria are important in the production of foods, especially fermented milks (cheese, yogurt, buttermilk) and vegetables (pickles, olives, sauerkraut), antibiotics, enzymes, and industrial solvents.

Algae: Protistans (unicellular; 1–100 μm) and nonvascular plants (multicellular; ranging to several meters); obtain nutrition through photosynthesis; reproduce asexually (simple division with no exchange of genetic material) and/or sexually (with exchange of genetic material). Algae play an important role in the cycling of materials and energy in aquatic ecosystems, and together with macrophytes, are the major sources of organic matter in lakes and reservoirs. Excessive algal growth can lead to taste and odor problems in water supplies, a reduction in water clarity in lakes, and depletion of oxygen reserves as algae settle to the bottom of lakes and decompose. The free-floating algae of lakes are termed *phytoplankton* (i.e., plants dependent on currents and eddies for transport).

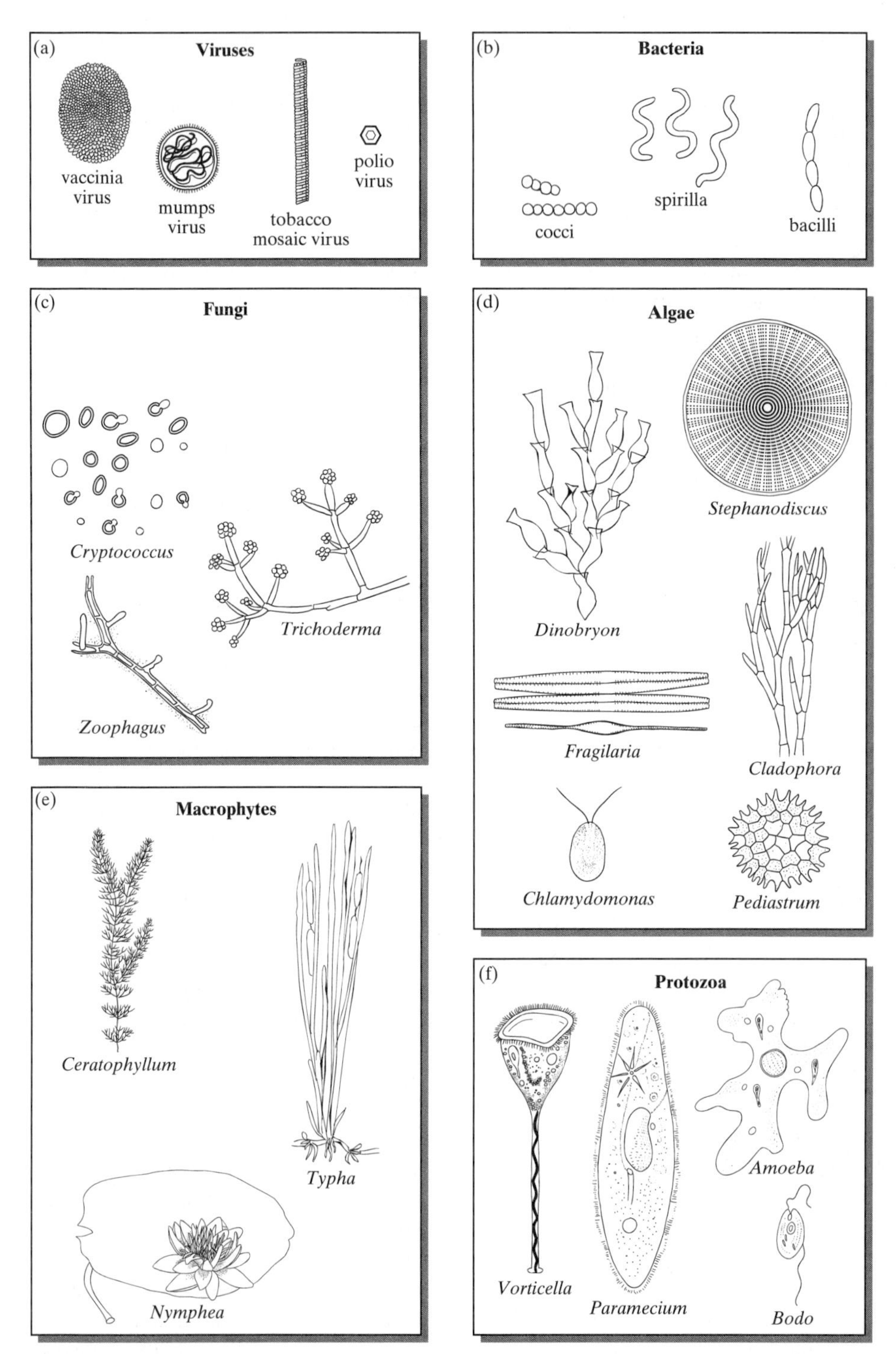

Figure 5-3. Major organism groups with representative members important in environmental science and engineering: (*a*) viruses, (*b*) bacteria, (*c*) fungi, (*d*) algae, (*e*) macrophytes, (*f*) protozoa.

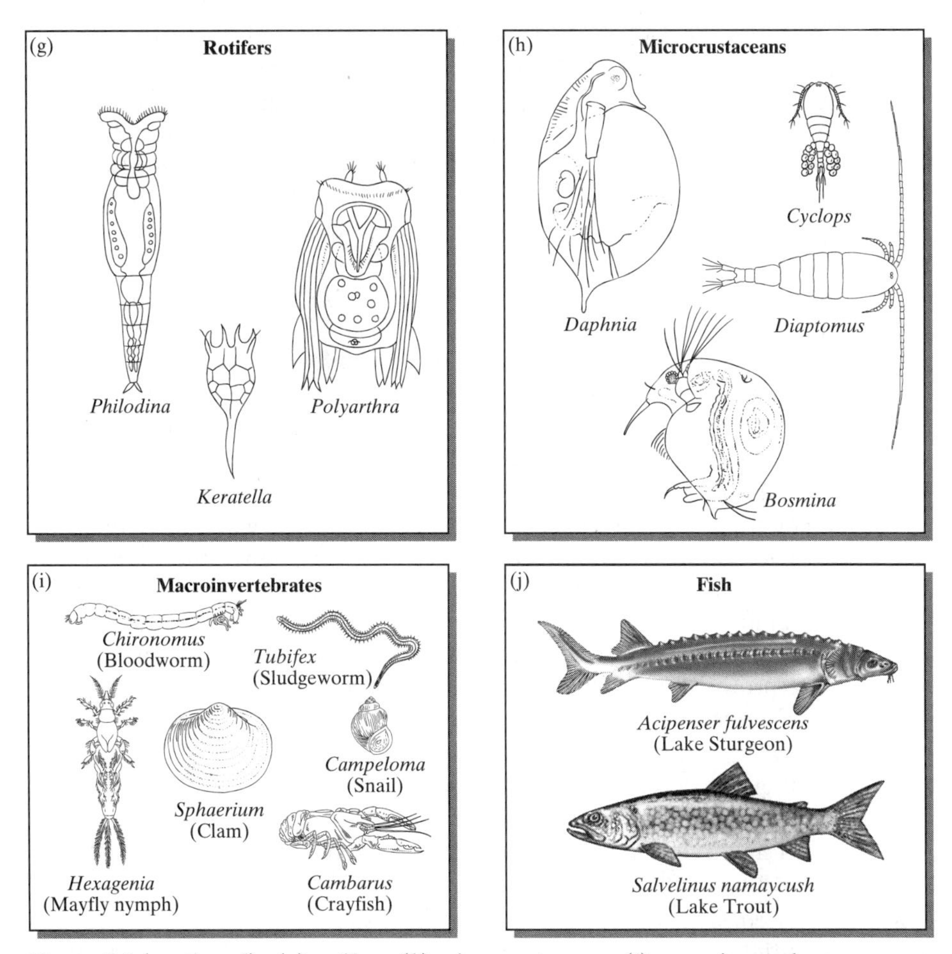

Figure 5-3 (continued)**.** (*g*) rotifers, (*h*) microcrustaceans, (*i*) macroinvertebrates, and (*j*) fish.

Fungi: Unicellular (yeasts) or multicellular (molds) Fungi; range in size from a few μm to several cm (some filamentous soil fungi may cover hectares of land area); reproduce asexually (budding, spores) or sexually (spores); lack chlorophyll and feed by absorption. In tribute to their role in cycling organic matter in soil, water, and wastewater, fungi are sometimes called "the great decomposers." Fungi are important in the pharmaceutical (antibiotics) and food (alcoholic beverages, cheese, soy sauce) industries, during composting, and are responsible for a variety of diseases: ringworm, athlete's foot, and toxic shock syndrome.

Protozoa: Protistans; 10–300 μm in size; reproduce asexually by fission (splitting) and budding or sexually; some form "resting" cysts to weather hostile environmental conditions. Protozoa are considered to be "animal-like" because they lack chlorophyll, are motile, and ingest dead particulate matter or living cells, for example, bacteria, algae, or other protozoa; however, some feed by absorption. They are important in the decomposition process in wastewater treatment and in lakes, as they solubilize particulate organic matter, producing the dissolved substrates required by bacteria and fungi. This group includes the well-known genera *Amoeba* and *Paramecium*, and the pathogenic genera *Giardia* and *Cryptosporidium* are of concern to drinking water supply engineers because they produce cysts which are resistant to disinfection.

Rotifers: Microscopic animals; 100–1000 μm in size, with one or more rings of cilia or hairs at the head of the body that aid in locomotion and in the drawing in of food. The feeding strategy that rotifers utilize is similar to that of protozoa, ingesting living and dead particles and excreting soluble organic matter useful to bacteria and fungi. Rotifers are thus important in recycling energy and material in wastewater-treatment plants and in natural systems.

Microcrustaceans: Microscopic animals; 1–10 mm in size; commonly represented by the copepods and cladocerans (*Daphnia*, the water flea); relatives of crabs, lobster, and shrimp; feed on bacteria, algae, and other particles in lakes. A primary food source for many species of fish, microcrustaceans are important in energy and material transfer in aquatic systems, but rarely exist in biological wastewater treatment. Taken together, the free-floating animals of lakes (protozoa, rotifers, and microcrustaceans) are termed the *zooplankton* (i.e., animals dependent on currents and eddies for transport).

Macrophytes: Large, vascular plants; grow submerged, floating, or emergent in lakes and rivers. Macrophytes provide important habitat, for example, nursery areas, but can reach nuisance proportions in rivers and lakes enriched with nutrients, creating problems with recreational use and negatively impacting dissolved oxygen budgets.

Macroinvertebrates: Higher animals lacking a spine or backbone, usually inhabiting the bottom muds of lakes and rivers. Macroinvertebrates include worms, clams, snails, and the early life stages of insects. They are important in processing dead organic matter in aquatic ecosystems and are a major food source for fish. Because of their relative lack of mobility, macroinvertebrates are often exposed to and accumulate toxic chemicals, and thus serve as indicators ("canary in the coal mine") of ecosystem health.

Fish: Much could be said about this group of animals that both influence and are influenced by the environment. As a result of their tendency to bioconcentrate hydrophobic organic chemicals and mercury in their tissues, fish can impact the health of humans and other animals that feed on them. The public perception of water quality is clearly linked to the presence of an abundant, diverse, and healthy fish community.

Interaction among the various organism groups results in highly dynamic communities in both natural and engineered systems. Seasonal cycles and natural and anthropogenic perturbations of the environment often lead to dramatic shifts in population size and community structure. For example, the transparency or clarity of lakes varies with the quantity of soil particles delivered from terrestrial sources by tributary streams and with the abundance of algae in the water column. The abundance of algae may, in turn, fluctuate with the size of the microcrustacean populations which graze upon them and with the availability of nutrients introduced from the watershed. In some lakes, water clarity can go from "crystal clear" to "pea soup" to "crystal clear" over a matter of days as algal and microcrustacean populations wax and wane.

A similar dynamic is observed in the activated sludge process in wastewater treatment. The microbial community ideally comprises protozoans and rotifers, which act to "polish" the wastewater by removing bacteria and abiotic particulates. A sudden spike of industrial waste or a dramatic shift in temperature can disrupt the system, allowing less desirable organisms to proliferate. A shift to filamentous bacteria (which are not easily removed by protozoans and rotifers) or microcrustaceans (which prey on protozoans and rotifers) can reduce the efficiency of particulate removal in sedimentation tanks and lead to a treated wastewater of poor quality, that is, high in suspended solids.

5.2 POPULATION DYNAMICS

This section focuses on population growth and the development of models that simulate the dynamics of population change, including the regulation of those dynamics by environmental factors. Population dynamics play a role in many areas of environmental engineering, including the fate of fecal bacteria discharged to surface waters, the efficiency of microbes in biological waste treatment, and substrate/organism interactions in the cleanup of contaminated soils. Other applications include nuisance algae growth in lakes, biomanipulation as a management approach for surface water quality, and the transfer of toxic chemicals through the food chain.

An engineer's ability to manage and protect the environment can be enhanced through an understanding of population dynamics, for example, by simulation or modeling of the response of populations to environmental stimuli. Models can provide a wealth of information about "what makes the garden grow," and thus help us to be better stewards of creation.

5.2.1 Units of Expression for Population Size

Although it is the individual that is born, the individual that reproduces, and the individual that dies, the environmental effect of these events is best appreciated by examining the community at large. While it is possible to characterize individual populations through direct enumeration (e.g., number of walleye), the communities typical of natural or engineered systems (e.g., the algae community, the bacterial community) comprise many populations of various sizes and forms, and thus are difficult to describe in this way. An alternative approach involves the use of a surrogate parameter (e.g., dry weight—mgDW/L, or suspended solids—mg SS/L) to represent all living material or biomass. This concept, reporting concentrations as a common constituent, was introduced previously in Section 2.5.2. For example, in the activated sludge reactor of a wastewater-treatment plant, total suspended solids (TSS, see Section 2.5.3) and volatile suspended solids (VSS, see Section 2.5.3) can provide measures of biomass. Particulate organic carbon and (for the algal community) chlorophyll are surrogate parameters commonly applied in natural systems.

5.2.2 Models of Population Growth

A physical model is defined as a small object, built to scale, that represents another often larger object (*American Heritage Dictionary*) and that can be used to examine the response of an entity to perturbation. Environmental engineers use mathematical models to this end, such as those developed in Chapter 4 where a mass-balance approach was applied in assessing environmental behavior. Here the mass balance is applied to living organisms.

Consider the case of the algal or bacterial community of a lake or river, or the community of microorganisms in a waste-treatment system. The mass balance on biomass may be written as:

$$V \frac{dX}{dt} = Q\, X_{\text{in}} - Q\, X \pm \text{Reaction} \qquad \textbf{(5-1)}$$

where X is biomass (mg/L), V is volume (L), Q is flow (L/day), t is time (days), and "Reaction" refers to all the kinetic processes describing the growth or death of the organisms. Note that in this "mass balance" each term in Equation 5-1 ($V \times dX/dt$, $Q \times X$, and Reaction) has units of mass per time (mg/d). To simplify the conceptual development of the models that follow, the flow terms will be ignored here (i.e., $Q = 0$, a batch reactor). Assuming that first-order kinetics adequately describe the reaction term (in this case, population growth), Equation 5-1 can be rewritten as:

$$V \frac{dX}{dt} = V\, k\, X \qquad \textbf{(5-2)}$$

Dividing Equation 5-2 by V results in:

$$\frac{dX}{dt} = k\, X \qquad \textbf{(5-3)}$$

where k is the first-order rate coefficient (time^{-1}). Because the reaction term is describing growth, the right side of Equation 5-3 is positive. Equation 5-3 will be used to develop realistic, but not overly complex models to simulate the rates of organism growth. Three models are introduced here describing unlimited (exponential), space-limited (logistic), and resource-limited (Monod) growth.

5.2.2.1 Exponential or Unlimited Growth

The dynamics of many organisms, from bacteria to (surprisingly) humans, can be described using a simple expression termed the *exponential growth model*:

$$\frac{dX}{dt} = \mu X \tag{5-4}$$

This relationship is identical to that presented previously as Equation 5-3, with μ, the *specific growth-rate coefficient* (day^{-1}) being a special case of the first-order rate constant, k. Equation 5-4 can be integrated to yield:

$$X_t = X_0 \, e^{(\mu \, t)} \tag{5-5}$$

where X_t is the biomass at some time t, and X_0 is the initial biomass, reported as numbers or as a surrogate concentration, for example, mg DW/L. Example 5.1 illustrates calculations relating to exponential growth.

EXAMPLE 5.1. EXPONENTIAL GROWTH AND EFFECT OF SPECIFIC GROWTH RATE ON THE RATE OF GROWTH

Consider a population or community with an initial biomass (X_0) of 2 mg DW/L and a specific growth rate (μ) of 1 day^{-1}. Determine the biomass concentration (mg DW/L) over a time period of ten days.

SOLUTION

Assume exponential growth. The biomass at any time is given by (Equation 5-5):

$$X_t = X_0 \, e^{(\mu \, t)} \qquad \text{and} \qquad X_t = 2 \, e^{(1 \, t)}$$

The table below and accompanying Figure 5-4*a* show biomass as a function of time. The form of this curve (J-shaped) is typical of exponential growth. The steepness of the curve is determined by the value of the specific growth-rate coefficient. The influence of the value of μ on the shape of the growth-rate curve is shown in Figure 5-4*b*.

The table below presents calculations of biomass as a function of time for the exponential growth model using the initial biomass and specific growth-rate coefficient just provided.

Time (days)	Biomass (mg DW/L)	Time (days)	Biomass (mg DW/L)
1	5	6	807
2	15	7	2,193
3	40	8	5,692
4	109	9	16,206
5	297	10	50,015

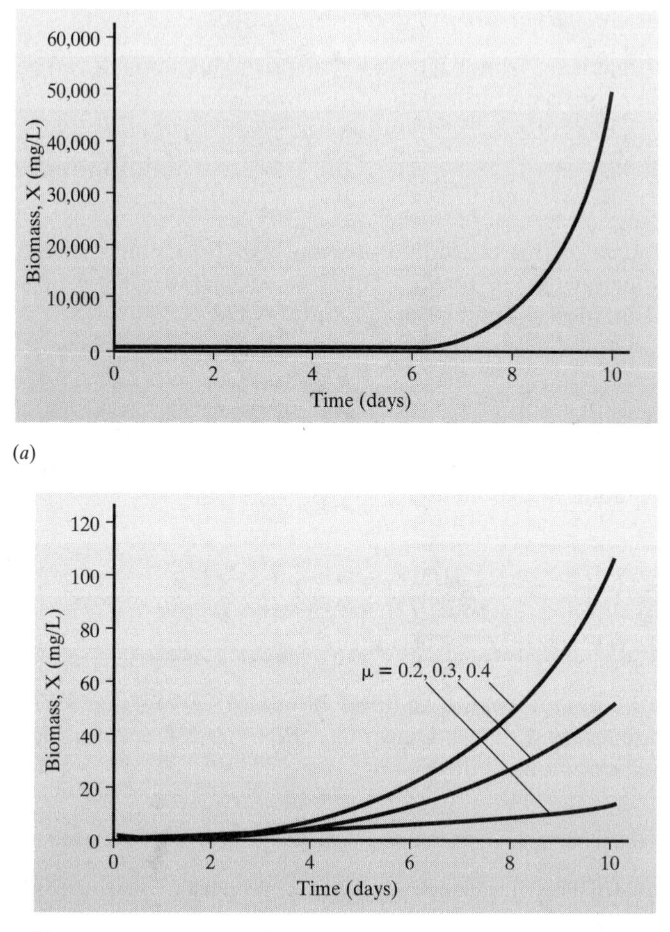

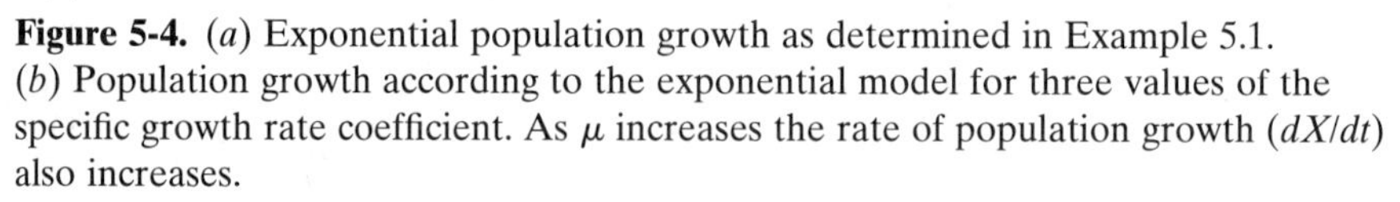

Figure 5-4. (*a*) Exponential population growth as determined in Example 5.1. (*b*) Population growth according to the exponential model for three values of the specific growth rate coefficient. As μ increases the rate of population growth (dX/dt) also increases.

Note the similarity between the exponential growth model (Equation 5-4), applied here for living organisms and the expression for first-order decay (Equation 3-16) introduced in Chapters 3 and 4 for application to chemical losses. Compare:

$$\frac{dX}{dt} = \mu X$$

with:

$$\frac{dC}{dt} = -k C$$

Both of these are first-order expressions, that is, both rates are a direct function of either the organism concentration or chemical concentration. In typical applications of these first-order reactions, however, organism concentrations increase exponentially (growth) while chemical concentrations decrease exponentially (decay).

5.2.2.2 Logistic Growth: The Effect of Carrying Capacity

If the predictions generated by the exponential growth model were examined a bit further along in time some interesting biomass levels would be observed. For example, in 100 days, the biomass simulated in Example 5.1 would reach 5.4×10^{43} mg/L! Is there something wrong with this? No wonder the exponential growth model is sometimes called *unlimited growth*: there are no constraints or upper bounds on biomass.

In many situations the exponential growth model has value. However, the *logistic growth model* provides a framework that is more in tune with our concept of how populations and communities of organisms behave, that is, the size of a population or community is limited by the *carrying capacity* of the environment to sustain growth. Carrying capacity refers to an upper limit to population or community size (biomass) imposed through environmental resistance. Figure 5-5 illustrates this. In nature, that resistance is related to the availability of renewable (e.g., food) and nonrenewable (e.g., space) resources as they impact biomass

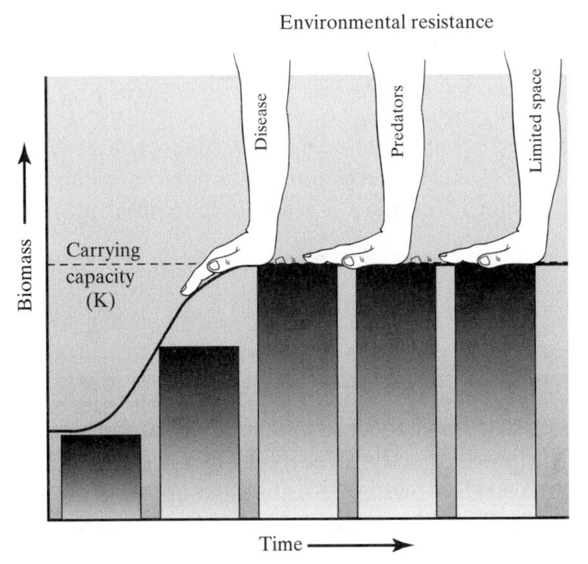

Figure 5-5. Effect on biomass of limitation by nonrenewable (space-related) resources as manifested through carrying capacity. According to the logistic growth model, environmental resistance (represented by the downward pressure of the hand) reduces the growth rate. At some time the population reaches some carrying capacity which the population cannot exceed. (Adapted from Enger et al., 1983.)

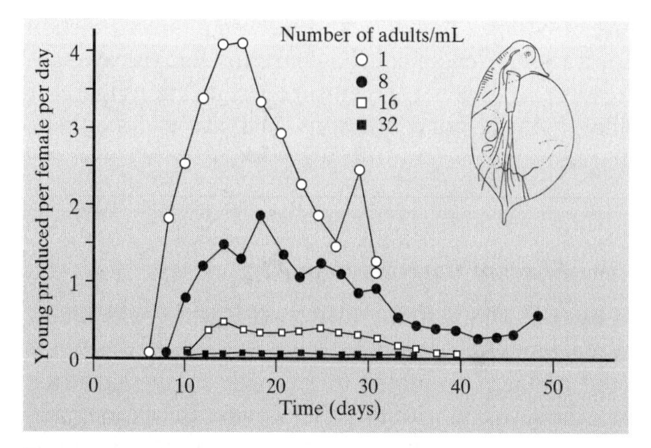

Figure 5-6. The reproductive response to space limitation in *Daphnia* (water flea), a microcrustacean common to many aquatic environments. (Adapted from Ricklefs, 1983.)

through reproduction, growth, and survival. Figure 5-6 illustrates the reproductive response to space limitation for a population of *Daphnia* (water flea), a microcrustacean common to many aquatic environments.

Here, we limit application of the logistic growth model to nonrenewable resources and their effect on growth. A companion approach (termed the Monod model and introduced in the subsequent section) treats limitation by renewable resources. The logistic growth model is developed by modifying the maximum specific growth rate to account for carrying-capacity effects:

$$\mu = \mu_{max}\left(1 - \frac{X}{K}\right) \tag{5-6}$$

where μ_{max} is the *maximum specific growth-rate coefficient* (day^{-1}), and K is the *carrying capacity* (mg DW/L), that is, the maximum sustainable population biomass. Substituting Equation 5-6 into Equation 5-4 (the exponential model, $dX/dt = \mu X$) yields:

$$\frac{dX}{dt} = \mu_{max}\left(1 - \frac{X}{K}\right)X \tag{5-7}$$

The reader should examine the behavior of the term in parentheses in Equation 5-7 to appreciate the way in which carrying capacity mediates the rate of population growth. Note that as the carrying capacity is reached, the specific growth rate approaches zero, causing the overall growth rate (dX/dt) to approach zero. Equation 5-7 can be integrated to yield:

$$X_t = \frac{K}{1 + \left[\left(\frac{K - X_0}{X_0}\right)e^{(-\mu_{max}\,t)}\right]} \tag{5-8}$$

Equation 5-8 permits calculation of biomass as a function of time according to the logistic growth model. Example 5.2 illustrates calculations relating to exponential growth.

EXAMPLE 5.2. LOGISTIC GROWTH

Consider the population from Example 5.1 ($X_0 = 2$ mg DW/L; $\mu_{max} = 1$ day^{-1}), but with a carrying capacity (K) of 5,000 mg DW/L. Determine the population biomass over a time period of 10 days.

SOLUTION

Use the carrying-capacity term, K, to apply the logistic growth model. Equation 5-8 can be used to solve for the biomass concentration over the 10-day period. The table below and accompanying Figure 5-7*a* show the population biomass over time. Note that in this example, the specific growth rate begins to decrease

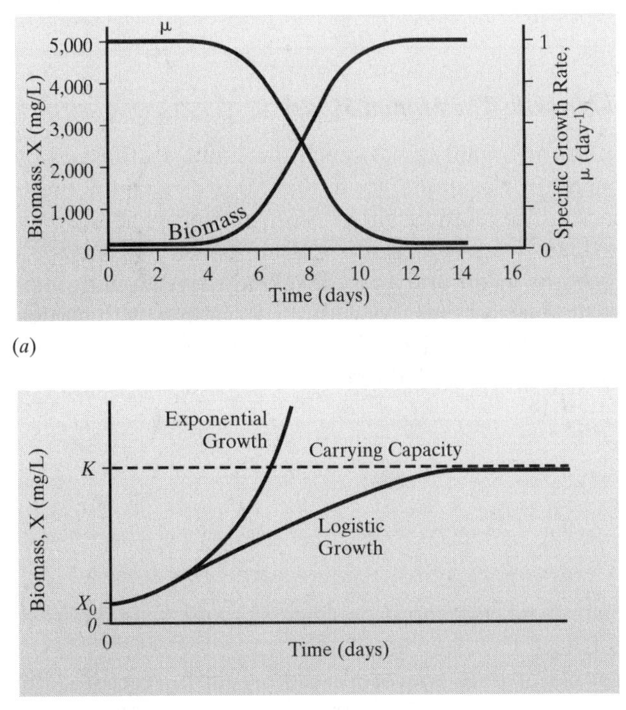

Figure 5-7. (*a*) Population biomass and specific growth rate according to the logistic model as determined in Example 5.2. (*b*) A comparison of the exponential and logistic growth models. Both predict a similar population response in the early stages when population size is small. However, the exponential model predicts that unlimited growth will continue, while the logistic model predicts an approach to the carrying capacity.

after several days and approaches zero. Also, the biomass concentration levels off over time as the carrying capacity (in this case, 5,000 mg DW/L) is approached. Figure 5-7*b* compares the exponential and logistic growth models. Note that both models predict the same population behavior at low population numbers. This suggests that the exponential model may be appropriately applied under certain conditions.

Time (days)	μ (day^{-1})	Biomass Concentration (mg DW/L)	Time (days)	μ (day^{-1})	Biomass Concentration (mg DW/L)
0	1.0	2	6	0.9	695
1	1.0	5	7	0.7	1,525
2	1.0	15	8	0.5	2,720
3	1.0	40	9	0.2	3,821
4	1.0	107	10	0.1	4,491
5	1.0	280			

5.2.2.3 *Resource Limited Growth: The Monod Model*

In nature, it is more common for organisms to reach the limits established by reserves of renewable resources, for example, food, than to approach the limits established by carrying capacity. The relationship between nutrients and the population or community growth rate can be described using the *Monod model*, a conceptual framework important in natural (e.g., lakes and rivers) and engineered (e.g., wastewater treatment) systems. As with the logistic growth model, the maximum specific growth rate is modified to account for the effects of limitation, in this case by renewable resources. The Monod model, graphically illustrated in Figure 5-8*a*, is written as:

$$\mu = \mu_{max} \frac{S}{K_s + S} \tag{5-9}$$

where S is the nutrient or *substrate* concentration (mg S/L) and K_s is the *half-saturation coefficient* (mg S/L).

Substrate may be either a major food source for the organism (e.g., organic matter in biological waste treatment) or a growth-rate-limiting micronutrient (e.g., phosphorus in lakes). The *half-saturation constant*, K_s, is defined as the substrate concentration at which the growth rate is one-half its maximum, that is, $\mu = \mu_{max}/2$. The value of K_s reflects the ability of an organism to acquire resources (substrate). Organisms with a small K_s approach the maximum specific

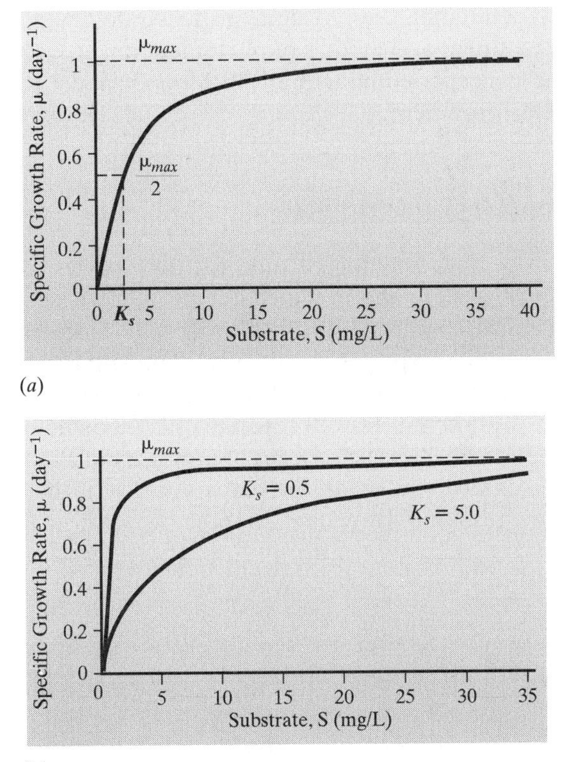

Figure 5-8. (*a*) The Monod model, illustrating the relationship between the specific growth rate (μ) and substrate (S) concentration. At high substrate concentrations ($[S] \gg K_s$), μ approaches its maximum value μ_{max} and growth is essentially independent of substrate concentration (i.e., zero-order kinetics). At low substrate concentrations ($[S] \ll K_s$), μ is directly proportional to substrate concentration (i.e., first-order kinetics). (*b*) Application of the Monod model illustrating the effect of variation in K_s. Organisms with a low K_s approach their maximum specific growth rate at lower substrate concentrations and thus may have a competitive advantage.

growth rate (μ_{max}) at comparatively low substrate concentrations, while those with high K_s values require more abundant substrate levels to achieve the same level of growth. Figure 5-8*b* illustrates the effect of variability in the half-saturation constant on growth rate. The physiological basis for this phenomenon lies in the role of enzymes in catalyzing biochemical reactions; low half-saturation constants reflect a strong affinity of the enzyme for substrate.

The Monod model (Equation 5-9) can be substituted into Equation 5-4 (the exponential model) to yield:

$$\frac{dX}{dt} = \mu_{max} \frac{S}{K_s + S} X \tag{5-10}$$

Equation 5-10 is the expression commonly used to simulate substrate-limited growth in biological wastewater treatment and models of phytoplankton growth in lakes. Example 5.3 applies the concepts embodied in the Monod model to important ecological issues in the environment.

EXAMPLE 5.3. RESOURCE-LIMITED GROWTH

Figure 5-9 shows population density as a function of time for two species of *Paramecium* (a protozoan) grown separately and in mixed culture. Grown separately, both species do well, acquiring substrate and achieving high biomass densities. In mixed culture, however, one species dominates, eliminating the other species. Organisms with a small K_s have a competitive advantage because they can reach a high growth rate at lower substrate levels. This can be demonstrated by inspection of the Monod model. A basic concept of ecology, the *Principle of Competitive Exclusion*, states that two organisms cannot coexist if they depend on the same growth-limiting resource. How then do these two species of

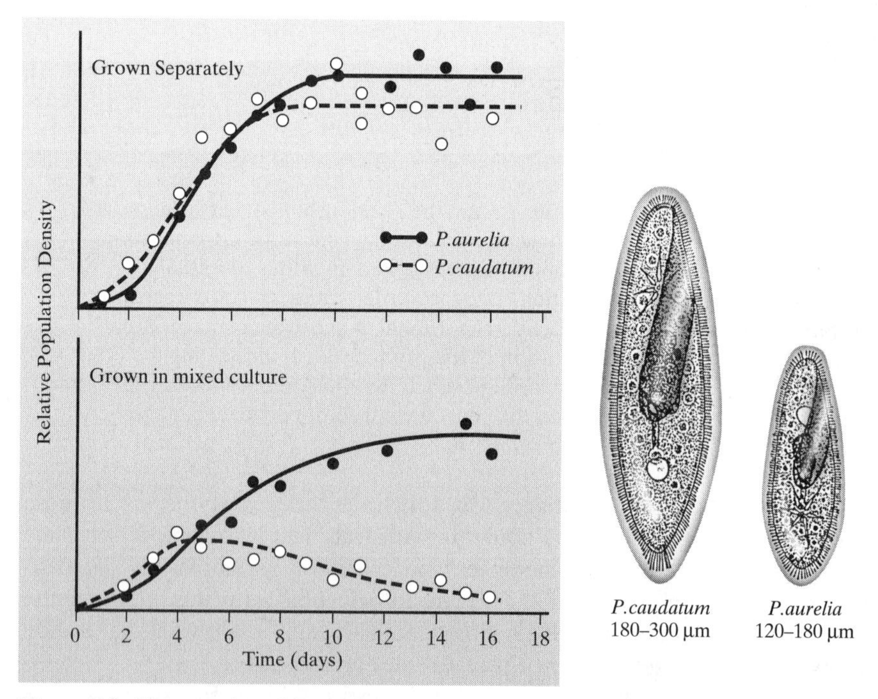

Figure 5-9. Two species of *Paramecium* grown separately and in mixed culture (adapted from Ricklefs, 1983). In separate culture both species do well, acquiring substrate and achieving high biomass densities. In mixed culture, however, one species dominates and eliminates the other species. This phenomenon is discussed in more detail in Example 5.3.

Paramecium manage to coexist in the natural world? Why isn't the poor competitor extinct?

SOLUTION

The answer lies in another ecological principle, niche separation. The term *niche* refers to the unique functional role or "place" of an organism in the ecosystem. Organisms that are poorly competitive from a purely kinetic perspective (e.g., K_s), can survive by exploiting a time or place where competition can be avoided.

5.2.2.4 The Yield Coefficient: Relating the Rate of Growth to the Rate of Substrate Utilization

While attention has been largely devoted here to tracking biomass, substrate fate may be of more interest in many engineering applications (e.g., bioremediation, wastewater treatment). The capability to model substrate concentrations (or to relate substrate consumption to organism growth) is gained through application of the *yield coefficient*, defined as the quantity of organisms produced per unit substrate consumed:

$$Y = \frac{\Delta X}{\Delta S} \tag{5-11}$$

In Equation 5-11, Y is the yield coefficient (mass of biomass produced per mass of substrate consumed). A yield coefficient value of $Y = 0.2$ indicates that 20 mg of biomass are produced for every 100 mg of substrate consumed. Note that Y for organic carbon is always <1 because organisms are not 100% efficient in converting substrate to biomass and because some energy must be expended for cell maintenance. *The yield coefficient is also commonly applied to relate the rate of substrate utilization (dS/dt) to the rate of organism growth (dX/dt):*

$$\frac{dS}{dt} = -\frac{1}{Y}\frac{dX}{dt} \tag{5-12}$$

Substituting the Monod growth-limitation model (Equation 5-10) for dX/dt in Equation 5-12 results in

$$\frac{dS}{dt} = -\frac{1}{Y}\frac{\mu_{max}\, S}{K_s + S}X \tag{5-13}$$

an expression used in engineering applications, for example, to develop the mass balances on organism growth and substrate utilization that support waste-treatment facility design and operation. Example 5.4 illustrates such an application.

EXAMPLE 5.4. USE OF THE YIELD COEFFICIENT

The organic matter present in municipal wastewater is removed in an aerated biological reactor, part of the activated sludge process at waste-treatment plants. The organic matter is removed at a rate of 25 mg BOD_5/L-h. As defined later in Section 5.4, BOD (biochemical oxygen demand) refers to the amount of oxygen consumed in oxidizing a given amount of organic matter; here, a representation by effect of substrate concentration.

Use the yield coefficient to compute the mass of microorganisms (measured as volatile suspended solids, VSS) produced daily due to the consumption of organic matter by microorganisms in the aeration basin. Assume that the aeration basin has a volume of 1500 m^3 (4.6 m deep by 9 m wide by 36 m long) and that the yield coefficient, Y, equals 0.6 mg VSS/mg BOD_5.

SOLUTION

The yield coefficient, Y, relates the rate of substrate (in this case, organic matter) disappearance to the rate of cell growth. This relationship (Equation 5-12) is written as:

$$\frac{dS}{dt} = -\frac{1}{Y}\frac{dX}{dt}$$

Therefore,

$$Y\frac{dS}{dt} = \frac{dX}{dt}$$

Substitute the values for Y and the rate of substrate depletion that were given:

$$\frac{0.6 \text{ mg VSS}}{\text{mg BOD}_5} \times \frac{25 \text{ mg BOD}_5}{\text{L-h}} = \frac{15 \text{ mg VSS}}{\text{L-h}}$$

Convert the tank volume to liters:

$$1500 \text{ m}^3 \times \frac{10^3 \text{ L}}{\text{m}^3} = 1.5 \times 10^6 \text{ L}$$

Next, convert this value to a mass per day basis:

$$\frac{15 \text{ mg VSS}}{\text{L-h}} \times 1.5 \times 10^6 \text{ L} \times \frac{24 \text{ h}}{\text{day}} = \frac{5.4 \times 10^8 \text{ mg VSS}}{\text{day}} = \frac{540 \text{ kg VSS}}{\text{day}}$$

Note that a lot of biological solids are produced at a wastewater-treatment plant each day (more than half a ton in this example, which is for a small plant). This explains why engineers spend so much time designing and operating facilities to handle and dispose of the residual solids (i.e., sludge) generated at a waste-treatment facility.

5.2.2.5 *Respiration: The Decay Coefficient*

Substrate acquired by organisms provides the energy required for maintenance of metabolic function, growth, and reproduction. This energy is released through the process of respiration, described in more detail in Section 5.3.1. The models developed to this point have not included a term that accounts for respiration, "the cost of doing business," or in the terminology of wastewater engineering, "endogenous decay." Mathematically, the rate of an organism's respiratory losses may be represented using a first-order respiration or decay coefficient. The rate of decay of a population of organisms can be written as:

$$\frac{dX}{dt} = -k_d X \quad \textbf{(5-14)}$$

where k_d is the respiration-rate coefficient (time^{-1}). In some situations (e.g., wastewater-treatment systems), the definition of the term k_d is expanded to include other losses, for example, settling and predation. Integration of Equation 5-14 (as described in Section 3.2.2) yields the analytical solution for a first-order decay:

$$X_t = X_0 \, e^{(-k_d t)} \quad \textbf{(5-15)}$$

Note the similarity of this equation to that developed earlier for exponential growth (Equation 5-5). Later, we will look at simplifications of these growth models where Equation 5-15 will be quite useful.

5.2.2.6 *Biokinetic Coefficients*

The terms μ_{max}, K_s, Y, and k_d are commonly referred to as biokinetic coefficients because they provide information about the manner in which substrate and biomass change over time (kinetically). Values for these coefficients may be derived from thermodynamic calculations or through field and laboratory experimentation; literature compilations of coefficients derived in this fashion are available.

Table 5-1. Ranges and Typical Values for Selected Biokinetic Coefficients for the Activated Sludge Process Applied in Treatment of Wastewater

Coefficient	Range of Values	Typical Value
μ_{max}	0.1–0.5 hr^{-1}	0.12 hr^{-1}
K_s	25–100 mg BOD_5/L	60 mg BOD_5/L
Y	0.4–0.8 mg VSS/mg BOD_5	0.6 mg VSS/mg BOD_5
k_d	0.0020–0.0030 hr^{-1}	0.0025 hr^{-1}

From Metcalf and Eddy, 1989.

Table 5-1 provides some representative values for the biokinetic coefficients as applied in municipal wastewater treatment.

5.2.2.7 *Batch Growth: Putting It All Together*

Respiration and the growth-mediating mechanisms introduced earlier may be integrated into a single expression describing population growth in a batch culture:

$$\frac{dX}{dt} = \left[\mu_{max}\left(1 - \frac{X}{K}\right)\left(\frac{S}{K_s + S}\right) - k_d\right]X \qquad \textbf{(5-16)}$$

Substrate utilization can then be related to Equation 5-16 through the yield coefficient:

$$\frac{dS}{dt} = -\frac{1}{Y}\left[\mu_{max}\left(1 - \frac{X}{K}\right)\left(\frac{S}{K_s + S}\right) - k_d\right]X \qquad \textbf{(5-17)}$$

Although of considerable importance in natural systems, the carrying-capacity term is not typically included in biokinetic models for municipal wastewater engineering because these systems are engineered to operate below their maximum sustainable biomass.

Figure 5-10 illustrates substrate utilization and the attendant phases of population growth in batch culture according to Equations 5-16 and 5-17. Four phases of growth are identified:

1. The *lag* or *acclimation phase* where growth is slow as cells take up nutrients, activate enzyme systems, and equilibrate with their physical and chemical environment;
2. The *exponential* or *log growth phase*, where growth is rapid and well approximated by the exponential model, $dX/dt > 0$;

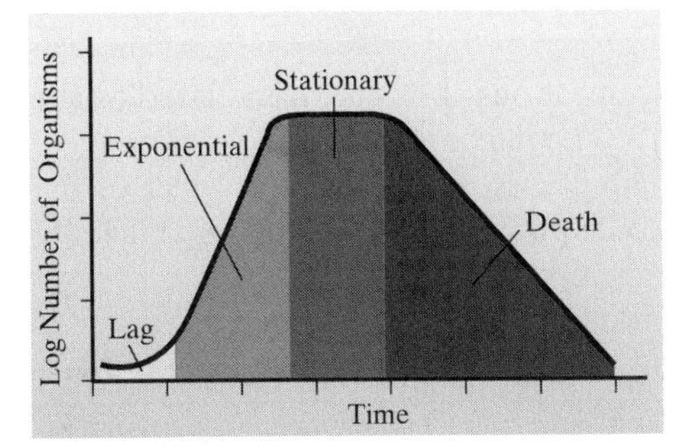

Figure 5-10. Population dynamics in batch culture illustrating four commonly observed regions: a lag phase, an exponential growth phase, a stationary phase, and a death phase. During the lag phase, growth is slow as cells take up nutrients, activate enzyme systems, and equilibrate with their physical and chemical environment. During the exponential or log growth phase, growth is rapid and well approximated by the exponential model, $dX/dt > 0$. During the stationary phase, growth is balanced by death and steady-state conditions are reached, $dX/dt = 0$. The death phase is characterized by resource limitation and/or buildup of toxic materials which inhibit growth; therefore, losses control, $dX/dt < 0$.

3. The *stationary phase*, where growth slows due to resource depletion or crowding and is balanced by losses to respiration (as defined in Section 5.3.1) and predation; therefore, steady state is reached, $dX/dt = 0$;
4. The death phase, where resource limitation and/or the buildup of toxic by-products inhibits growth and losses to respiration and predation control biomass dynamics, $dX/dt < 0$.

Certain simplifying assumptions regarding growth conditions during the exponential and death phases permit the calculation of substrate and biomass changes at those times. Example 5.5 illustrates some of those calculations.

EXAMPLE 5.5. SIMPLIFIED CALCULATIONS OF SUBSTRATE AND BIOMASS

The differential equations that describe biomass and substrate dynamics (Equations 5-16 and 5-17) contain nonlinear terms that require specialized numerical methods for their solution. By applying certain simplifying assumptions, however, quite a bit can be learned about the dynamics of microbial populations and communities. Consider a population of microorganisms with the following characteristics growing in batch culture: initial biomass, $X_0 = 10$ mg DW/L; maximum specific growth rate, $\mu_{max} = 0.3$ day^{-1}; half-saturation constant, $K_s = 1$ mg/L; carrying capacity, $K = 100{,}000$ mg DW/L; respiration-rate coefficient, $k_d = 0.05$ day^{-1}; initial substrate concentration, $S_0 = 2{,}000$ mg S/L; and yield coefficient, $Y = 0.1$ mg DW/mg S.

1. Determine whether this population will ever approach its carrying capacity.

SOLUTION

The change in substrate and biomass concentrations over time are related by the yield coefficient as given by Equation 5-12. The maximum attainable biomass of this population, based on substrate availability, is given as the product of the maximum potential change in substrate concentration and the yield coefficient:

$$dX = dS\ Y = \frac{2{,}000 \text{ mg S}}{\text{L}} \times \frac{0.1 \text{ mg DW}}{\text{mg S}} = 200 \text{ mg DW}$$

This is well below the carrying capacity of 100,000 mg DW/L; therefore, the population will run out of substrate and never approach the carrying capacity.

2. Calculate the population biomass after the first three days of growth, assuming no lag phase.

SOLUTION

Early in the growth phase when substrate concentrations are high (Monod term, $S/(K_s + S), \rightarrow 1$) and biomass concentrations are low (carrying-capacity term, $1 - X/K, \rightarrow 1$), Equation 5-16 reduces to:

$$\frac{dX}{dt} = (\mu_{max} - k_d)\ X$$

integrating yields:

$$X_t = X_0\ e^{(\mu_{max} - k_d)t}$$

$$X_3 = \frac{10 \text{ mg DW}}{\text{L}} \times e^{(0.3 \text{ day}^{-1} - 0.05 \text{ day}^{-1})3 \text{ day}} = \frac{21 \text{ mg DW}}{\text{L}}$$

3. Calculate the substrate concentration after the first three days of growth.

SOLUTION

The change in substrate concentration over the 3-day period is given by Equation 5-12:

$$\frac{dS}{dt} = -\frac{1}{Y}\frac{dX}{dt}$$

dX/dt over 3 days, from the previous calculation is $X_3 - X_0 = 21 - 10 = 11$ mg DW/L, and

$$\frac{dS}{dt} = -\left(\frac{0.1 \text{ mg DW}}{\text{mg S}}\right)^{-1} \times \frac{11 \text{ mg DW}}{\text{L}} = \frac{-110 \text{ mg S}}{\text{L}}$$

and the substrate concentration after three days of growth is given by:

$$S_3 = S_0 - \frac{dS}{dt} = 2000 - 110 = \frac{1890 \text{ mg S}}{\text{L}}$$

4. If the population peaks at 100 mg DW/L when the substrate runs out, calculate the biomass 10 days after the peak.

SOLUTION

When substrate is exhausted, the Monod term = 0 and Equation 5-16 reduces to Equation 5-14 and its analytical solution, Equation 5-15: In this case, the peak population decays according to first-order kinetics:

$$X_t = X_0 \text{ e}^{-k_d t} = 100 \text{ e}^{(-.05 \text{ day}^{-1} \times 10 \text{ day})} = \frac{61 \text{ mg DW}}{\text{L}} \text{ 10 days after the peak}$$

5.2.2.8 *Growth Models and Human Population*

Despite the complexity of their reproduction, populations of humans can be simulated using the types of models described in this chapter. Figure 5-11 shows human population growth since the Stone Age. At present, the growth rate for the more developed regions of the world is ~0.015 year^{-1} (1.5% per year or 15 births per 1,000 people per year); in lesser developed regions it is ~0.025 year^{-1}

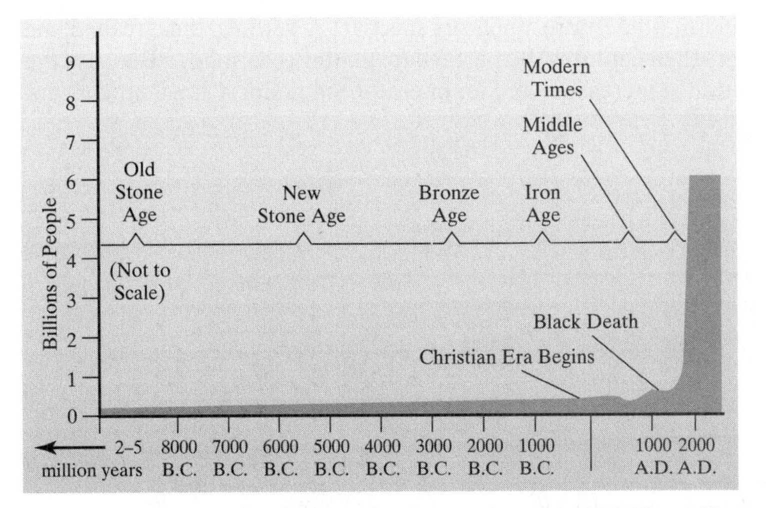

Figure 5-11. Human population growth over time. (Adapted from Enger et al., Environmental Science, 1983, Wm. C. Brown Publishers with permission of the McGraw-Hill Companies.)

(Henry and Heinke, 1996). At this rate, the population of the Earth will double every 25–50 years. Readers should ask themselves several questions: (1) What population model best fits these data? (2) What does this model suggest will happen in the future? (3) Are humans subject to the same limiting factors as other organisms? (4) If so, what are the implications for food, energy, waste disposal, and interactions with other organisms as humans approach the carrying capacity of their environment? We should consider also the following admonition, attributed to Albert Bartlett of the University of Colorado: "The greatest shortcoming of the human race is our ability to understand the exponential function."

ADVANCED TOPIC. COMBINING UNDERSTANDING OF MASS BALANCES AND MICROBIAL GROWTH TO ANALYZE WASTEWATER TREATMENT

The majority of municipal wastewater treatment plants in the United States employ a biological treatment process termed "activated sludge" to remove dissolved organic material from the wastewater. This process results in effluent low in both organic matter and suspended solids.

Figure 5-12 shows a schematic of the activated sludge process with a control volume added for our mass balance. In this process wastewater is passed through a biological reactor (called the aeration basin) where a diverse group of microorganisms are maintained in suspension by pumping air into the reactor. Because the suspension of wastewater and suspended organisms appears mixed, we will model the aeration basin as a completed mixed flow reactor (CMFR). The organisms convert dissolved organic matter into nitrate, water, gaseous CO_2 and particulate organic matter (more microorganisms). A settling tank (called the secondary clarifier) then captures the particulate matter (i.e., solids). Because the organism population is increasing and an operator maintains a constant concen-

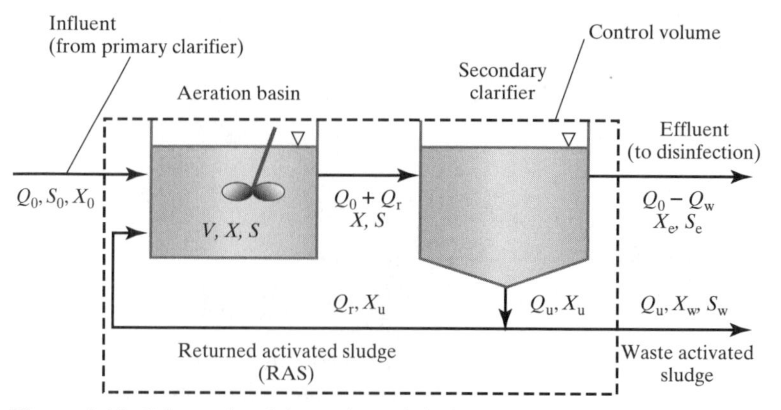

Figure 5-12. Schematic of the activated sludge process.

tration of microorganisms in the aeration basin, some organisms must be removed (or "wasted") from the secondary clarifier. In addition, some microorganisms are recycled back into the aeration basin to "seed" the biological reactor with a metabolically active group of organisms.

Here we will set up and analyze two mass balances, conducted on dissolved organic matter and solids, which when combined with our understanding of microbial growth will allow us to determine the volume of the aeration basin. In all these expressions, Q represents flow, m^3/d; S = substrate concentration (usually measured as mg BOD or COD per L); X is solids (i.e., biomass) concentration, measured as mg SS/L or mg VSS/L; and V is the aeration tank volume, m^3.

MASS BALANCE 1 ON SOLIDS (I.E., BIOMASS)

In English, the mass balance on solids is:

biomass in + biomass produced due to growth = biomass out

Using Figure 5-12 and the stated control volume, the mathematical expression which describes the solids mass balance is:

$$Q_0 X_0 + V \frac{dX}{dt} = (Q_0 - Q_w)\, X_e + Q_w X_w$$

Assuming Monod kinetics (Equation 5-10) and first-order decay (Equation 5-14), the overall biomass growth in the aeration basin which is a result of growth and decay can be written as:

$$\frac{dX}{dt} = \frac{\mu_{max} SX}{K_s + S} - k_d X$$

This growth term can be substituted into the mass balance expression for dX/dt. In addition, we can assume that X_0 and X_e are very small in relation to X (i.e., $X_0 = X_e = 0$). This is a good assumption because the concentration of biomass in the reactor, X, is maintained at approximately 2,000 to 4,000 mg SS/L while the influent solids into the aeration basin (X_0) might be 100 mg SS/L and the effluent solids, X_e, less than 25 mg/L. After performing this substitution and making the assumptions of low solids concentration, the resulting expression can be rearranged to yield:

$$\frac{\mu_{max} S}{K_s + S} = \frac{Q_w X_w}{VX} + k_d$$

The above expression will be used later. Next we will perform a mass balance on the dissolved organic material which is substrate for the organisms.

MASS BALANCE 2 ON SUBSTRATE

In English, the substrate mass balance is written as:

$$\text{Substrate in} - \text{substrate consumed by organisms} = \text{substrate out}$$

Thus, the substrate mass balance can be written as:

$$Q_0S_0 - V\frac{dS}{dt} = (Q_0 - Q_w)\,S + Q_wS$$

Here the effluent substrate, S_e, and the substrate in the waste sludge, S_w, are assumed to equal S (i.e., $S = S_e = S_w$) because the secondary clarifier's purpose is to remove solids, not to biologically transform substrate. Remember from Equations 5-12 and 5-13 that the yield coefficient relates the change in substrate concentration to the change in biomass concentration. Thus, the change in substrate concentration with time, dS/dt, can be written as:

$$\frac{dS}{dt} = \frac{1}{Y}\frac{\mu_{max}S}{K_s + S}X$$

The above expression can be substituted into the substrate mass balance and the overall expression can be rearranged, which results in:

$$\frac{\mu_{max}S}{K_s + S} = \frac{Q_0Y}{VX}(S_0 - S)$$

Note that the left side of the two final rearranged expressions obtained from the solids and substrate mass balances are the same. These two expressions can be set equal to yield our final expression:

$$\frac{Q_wX_w}{VX} = \frac{Q_0Y}{VX}(S_0 - S) - k_d$$

This expression can be used to solve for the volume of the aeration basin. This is because all other terms can be either measured or fixed. Remember, Y and k_d are biokinetic coefficients which are either measured or estimated (see Table 5-1 for example values), S is really the plant effluent concentration which is typically set by a state regulatory permit, S_0 and Q_0 are influent parameters which are a function of things such as population and community wealth and business activity, and Q_w and X_w are items a plant operator can control.

The term on the left side of the above equation is an important expression for design and operation of an activated sludge plant. It is the inverse of a term written mathematically as Θ_c which is referred to as sludge age, solids retention time (SRT), or mean cell retention time (MCRT). These terms equal the following expression:

$$\frac{VX}{Q_wX_w}$$

If you look closely at the units of this expression you will see that it has units of time. Typical sludge ages range from approximately 5 to 15 days. It refers to the average time a microorganism spends in the activated sludge process before it is expelled, or wasted, from the system. Think about the organisms: they feed in the aeration basin and then rest in the secondary clarifier, are recycled back into the aeration basin to feed, then are sent back to the secondary clarifier to rest. This process is repeated many times until the organism is finally removed from the process, or is wasted.

5.3 ENERGY FLOW IN ECOSYSTEMS

The character of the many and varied ecosystems represented on the Earth is determined to a large extent by their physical setting. Consider the changes in flora (plants) and fauna (animals) observed over the course of a long car trip, especially if traveling north–south or through dramatic changes in elevation. The physical setting includes climatic factors such as temperature (extreme values and duration of seasons), sunlight (day length and annual variation), precipitation (extremes and annual distribution), and wind. Other significant features of the physical setting include soil physics (particle size) and chemistry (pH, organic content, nutrients).

Given an appropriate physical setting, organisms require only two things from the environment: (1) *energy* to provide power, and (2) *chemicals* to provide substance. Chemical elements are cycled within an ecosystem, and thus continued function does not require that they be imported. However, the rate of this (biogeochemical) cycling can be an important determinant of ecosystem activity, as will be discussed later in this chapter. Energy flows through and propels ecosystems, that is, it does not cycle, but rather is converted to heat and lost for useful purposes forever. In summary . . . "chemicals circulate—energy dissipates."

5.3.1 Energy Capture and Use: Photosynthesis and Respiration

The Sun is responsible, directly or indirectly, for virtually all of the Earth's energy. Sunlight incident on an aquatic or terrestrial ecosystem is trapped by plant pigments, primarily *chlorophyll*, and that light energy is converted to chemical energy through a process termed *photosynthesis*. The stored chemical energy is subsequently made available for use by organisms through *respiration*. Figure 5-13*a* provides a simplified representation of the photosynthesis process. Photosynthesis can be represented mathematically as follows:

$$CO_2 + H_2O + \Delta \rightarrow C(H_2O) + O_2 \tag{5-18}$$

where Δ is the Sun's energy and $C(H_2O)$ is a general representation of organic carbon (e.g., glucose, $C_6H_{12}O_6$ or $6 \cdot C(H_2O)$). The free-energy change (ΔG) for photosynthesis (Equation 5-18) is positive; thus, the reaction could not proceed

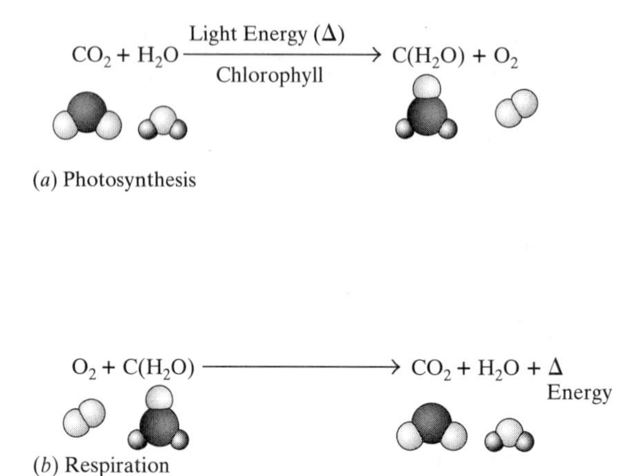

Figure 5-13. (*a*) Simplified version of photosynthesis, the process in which the Sun's energy is captured by pigments such as chlorophyll and converted to chemical energy stored in the bonds of simple carbohydrates, e.g. $C(H_2O)$. More complex molecules (e.g., sugars, starches, cellulose) are then formed from simple carbohydrates. (*b*) Simplified version of respiration, the reverse of the photosynthetic process. The energy stored in chemical bonds (e.g., carbohydrates) is released to support metabolic needs. (Adapted from *Environmental Science* by Kupchella and Hyland, © 1986, by permission of Prentice-Hall, Inc., Upper Saddle River, NJ.)

without the input of energy from the Sun. Chlorophyll acts as an antenna, absorbing the light energy, which is then stored in the chemical bonds of the carbohydrates produced by this reaction. Oxygen is an important byproduct of the process.

Respiration is the process by which the chemical energy stored through photosynthesis is ultimately released to do work in bacteria, plants, and animals:

$$C(H_2O) + O_2 \rightarrow CO_2 + H_2O + \Delta \qquad \textbf{(5-19)}$$

Figure 5-13*b* provides a simplified representation of respiration. The reverse of photosynthesis, this reaction releases stored energy, making it available for cell maintenance, reproduction, and growth. This energy, denoted as (Δ) in Equation 5-19, is equal to the free energy of reaction. Organisms are able to capture and utilize only a fraction (5–50%) of the total free energy of this reaction, and thus all forms of life are, by nature, rather inefficient.

Respiration is what may be described chemically as an oxidation-reduction or redox reaction. In a redox reaction (as explained in Chapter 3), an electron donor (in this case, carbon) loses electrons (is oxidized; its valence state becomes more positive) and an electron acceptor (in this case, oxygen) gains electrons (is reduced; its valence state becomes more negative). This redox reaction can be written in terms of the following two half-reactions. First the oxidation of the organic carbon:

$$C(H_2O) + H_2O \rightarrow CO_2 + 4H^+ + 4e^- \qquad \textbf{(5-20)}$$

where the valence state of carbon goes from (0) in $C(H_2O)$ to (4+) in CO_2, yielding 4 electrons. And second, the reduction of oxygen:

$$O_2 + 4e^- + 4H^+ \rightarrow 2H_2O \quad \textbf{(5-21)}$$

where the valence state of oxygen goes from (0) in O_2 to (2−) in H_2O, gaining 4 electrons. The two half-reactions can be added to yield the overall reaction presented in Equation 5-19. Note that there is no net change in electrons; they are simply redistributed.

Microbial ecologists refer to the respiration described in Equation 5-19 as *aerobic respiration*, because oxygen is utilized as the electron acceptor. Some bacteria, termed *obligate aerobes* or often simply aerobes, rely exclusively on oxygen as an electron acceptor and fail to grow in its absence. At the opposite extreme are microbes that cannot tolerate oxygen, termed *strict anaerobes*. Microbes that can tolerate but not utilize oxygen are termed *obligate anaerobes*. When oxygen is absent, *anaerobic respiration* takes place, utilizing a variety of other compounds as electron acceptors.

The major anaerobic reactions are presented below. There are many bacteria that can utilize oxygen as an electron acceptor, but in its absence may utilize either nitrate or sulfate. Such bacteria are termed *facultative aerobes* and have a distinct ecological advantage over strict or obligate anaerobes or aerobes in environments that may be periodically devoid of oxygen. The term *anoxic* is used in environmental engineering to describe conditions where the supply of oxygen has been exhausted, but where nitrate (an alternate electron acceptor) is present.

Equations 5-22 to 5-26 show the redox reactions for oxidation of organic matter using a variety of alternate electron acceptors (i.e., nitrate, manganese, ferric iron, etc.). Note that these equations are not stoichiometrically balanced, so that participating species in the reactions may be more clearly emphasized:

$$\textit{Nitrate: } C(H_2O) + NO_3^- \rightarrow N_2 + CO_2 + HCO_3^- + H_2O \quad \textbf{(5-22)}$$

$$\textit{Manganese: } C(H_2O) + Mn^{4+} \rightarrow Mn^{2+} + CO_2 + H_2O \quad \textbf{(5-23)}$$

$$\textit{Ferric iron: } C(H_2O) + Fe^{3+} \rightarrow Fe^{2+} + CO_2 + H_2O \quad \textbf{(5-24)}$$

$$\textit{Sulfate: } C(H_2O) + SO_4^{2-} \rightarrow H_2S + CO_2 + H_2O \quad \textbf{(5-25)}$$

$$\textit{Fermentation: } C(H_2O) \rightarrow CH_4 + CO_2 \quad \textbf{(5-26)}$$

In the environment, these reactions are thought to take place in the sequence listed, the order of their favorability from a thermodynamic perspective. Thus reduction of oxygen proceeds first, followed by nitrate, manganese, ferric iron, sulfate, and finally fermentation occurs. This order is termed the *ecological redox sequence*, with each process carried out by different types of bacteria (e.g., nitrate reducers and sulfate reducers). Fermentation (Equation 5-26) differs from the other reactions (Equations 5-19 and 5-22 through 5-25) in that organic matter is oxidized without an external electron acceptor. In this case the organic compound serves as the electron donor. This mode of anaerobic metabolism typically results in two end products, one of which is oxidized with respect to the substrate and one that is reduced. In the production of alcohol, for example, glucose ($C_6H_{12}O_6$, with C in the (0) valence state) is fermented to ethanol (CH_2CH_3OH, with C reduced to the (2−) valence state) and carbon dioxide (CO_2, with C oxidized to the (4+) valence state). Fermentation is a biochemical process mediated by yeasts and certain bacteria. Methanogenesis is a type of fermentation in which methane (CH_4) is an endproduct.

The anaerobic processes summarized above are important in many natural systems, e.g., lake sediments and wetlands. Figure 5-14 illustrates the relative contribution of oxygen and various alternate electron acceptors to the oxidation of organic matter in the bottom waters of Onondaga Lake, New York. Approximately one-third of the organic matter decomposition during the summer period was accomplished aerobically, that is, with oxygen as the terminal electron acceptor, with the balance utilizing the alternate electron acceptors identified in Equations 5-22 through 5-26. Accumulation of the reduced species resulting from these reactions can lead to water quality problems (e.g., Fe and Mn contribute color and taste and odor and H_2S is malodorous and toxic to aquatic life).

In wastewater treatment, systems are sometimes engineered to promote denitrification (Equation 5-22), yielding the removal of both nitrogen in the form of nitrate (as N_2 gas) and organic carbon (as CO_2). Because it is much more soluble in water than oxygen, nitrate can be used in the remediation of groundwater pollution, where it is added to the subsurface in carefully controlled

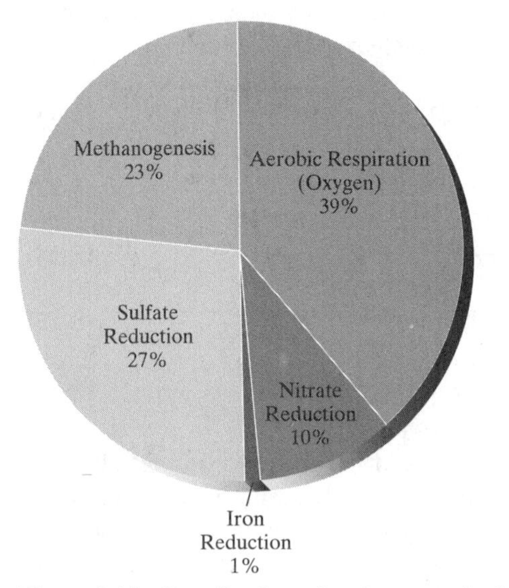

Figure 5-14. Contribution of various terminal electron acceptors to the oxidation of organic matter in the bottom waters of Onondaga Lake, New York. (Effler, 1997.)

amounts to facilitate degradation of certain hazardous organic chemicals (e.g., toluene and naphthalene) through dentrification. Most odor problems in wastewater-conveyance and treatment systems results from chemicals produced through anaerobic processes. Biofilters (depicted in Figure 4-6) are used in some situations to treat malodorous air emissions such as H_2S.

5.3.2 Trophic Structure in Ecosystems

In addition to energy, organisms require a source of carbon. Organisms that obtain their carbon from inorganic compounds (in Equation 5-18, CO_2) are called *autotrophs* (loosely translated as self-feeders). Photosynthetic organisms (green plants, including algae, and some bacteria) which use light as their energy source and nitrifying bacteria, which use ammonia (NH_3) as their energy source, fall into this category. The simple carbohydrates produced through photosynthesis and the more complex organic chemicals synthesized later are collectively termed *organic matter*. Organisms that depend on organic matter produced by others to obtain their carbon are termed *heterotrophs* (loosely translated as other-feeders). This carbon source could be a simple molecule such as methane (CH_4) or a more complex chemical. Animals and most bacteria derive both their carbon and energy from organic matter and fall into this category.

The amount of organic matter present at any point in time is the system's *biomass* (g C or DW/L, gC or DW/m^2) and the rate of production of biomass is the system's *productivity* (g C or DW/L-day, gC or DW/m^2-day). *Primary production* refers to the photosynthetic generation of organic matter by plants and certain bacteria, for example, algae in lakes and field crops on land. *Secondary production* refers to the generation of organic matter by non photosynthetic organisms, that is, those that consume the organic matter originating from primary producers to gain energy and materials and that in turn generate more biomass through growth. Examples of secondary producers include zooplankton in aquatic systems and cattle on land.

ADVANCED TOPIC. TROPHY BY ANY OTHER NAME WOULD TASTE THE SAME

The word trophic, from the Greek trophe or nourishment, refers to nutrition. Biologists have developed a nutritional classification based on an organism's source of carbon and energy. Organisms that utilize preformed (synthesized by others), organic compounds as their carbon source are called *heterotrophs* (literally, "other-feeders"). Organisms that use carbon dioxide as their carbon source are called *autotrophs* (literally, "self-feeders"). Within the categories of autotrophy and heterotrophy, chemotrophs are recognized as those organisms that utilize chemical compounds for energy, and phototrophs are those that depend on the Sun's energy. These four trophic strategies can be used to develop a classification scheme for nutrition:

NUTRITIONAL GROUPS

Chemoautotrophs
- Carbon source: carbon dioxide
- Energy source: inorganic chemicals
- Examples: nitrifying, hydrogen, iron, and sulfur bacteria

Chemoheterotrophs
- Carbon source: organic compounds
- Energy source: organic compounds
- Examples: most bacteria, fungi, protozoa, and animals

Photoautotrophs
- Carbon source: carbon dioxide
- Energy source: light
- Examples: purple- and green-sulfur bacteria, cyanobacteria, algae, and plants

Photoheterotrophs
- Carbon source: organic compounds
- Energy source: light
- Examples: purple- and green-nonsulfur bacteria

The light energy fixed in simple organic compounds (e.g., sugars) by autotrophs is later made available to support life processes. In many microorganisms, plants, and animals, the energy harvest begins with glycolysis, a process in which the energy stored in the chemical bonds of a six-carbon sugar (glucose) are broken down to yield two molecules of the three-carbon compound pyruvate and two molecules of ATP (adenosine triphosphate), a compound important in energy transfer in cells. Energy production then proceeds along one of two pathways. The first is fermentation, an anaerobic process with a low energy yield that produces CO_2, organic acids and/or alcohols, and CH_4 as waste products. The second is respiration, a process with a high energy yield that produces CO_2 and water as waste products. When oxygen is the terminal electron acceptor, it is termed aerobic respiration. In the absence of oxygen (anaerobic respiration) other compounds (nitrate, oxidized Mn and Fe, sulfate, and CO_2) may be used as terminal electron acceptors.

The trophic or feeding structure in ecosystems is composed of the abiotic environment and three biotic components (producers, consumers, and decomposers). *Producers*, most often plants, assimilate simple chemicals and utilize the Sun's energy to produce and store complex, energy-rich compounds that provide an organism substance and stored energy. Those organisms that eat plants, extracting energy and chemical building blocks to make more complex substances, are termed *primary consumers* or *herbivores*. Those that consume herbivores are called *secondary consumers* or *carnivores*. Additional carnivorous trophic levels are possible (tertiary and quaternary consumers). Consumers that eat both plant and animal material are termed *omnivores*. Figure 5-15*a* illustrates the various nutrition or *trophic levels* in a simple aquatic food chain, a linear subset of the more complex relationships and interactions that make up a *food web* (Figure 15-5*b*). The concept of a food web is not limited to natural systems. Example 5.6 illustrates a food web for the activated sludge process in biological-waste treatment, an engineered system.

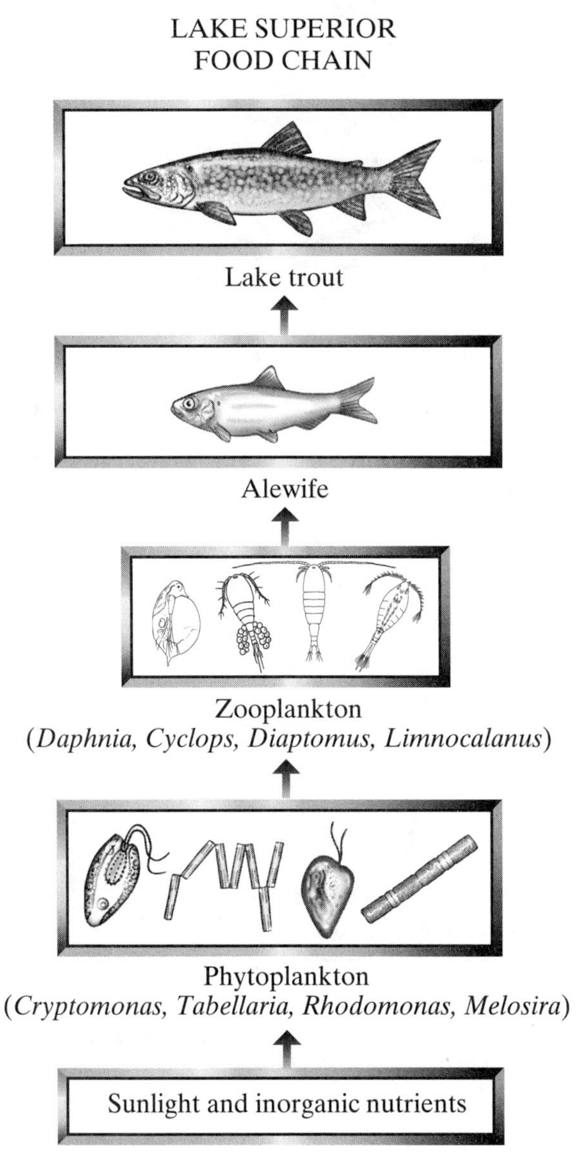

(*a*)

Figure 5-15. (*a*) An aquatic food chain for Lake Superior. Phytoplankton are primary producers, zooplankton are primary consumers or herbivores, and the alewife and lake trout are secondary and tertiary consumers, respectively, both carnivores.

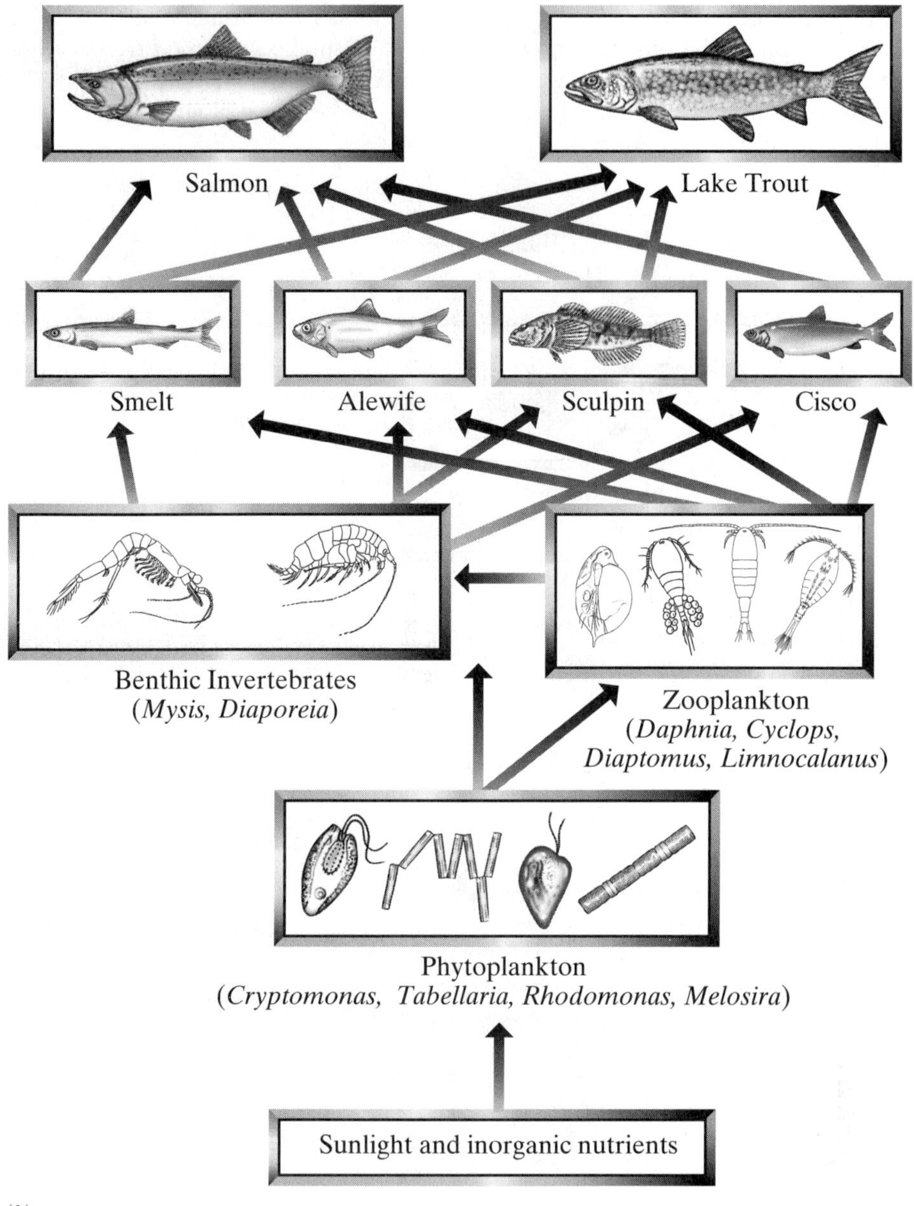

(*b*)

Figure 5-15 (continued). (*b*) A food web for Lake Superior illustrating the more complex interrelationships commonly found in an ecosystem.

The freshly dead or partially decomposed remains and wastes (e.g., fecal matter) of organisms are collectively termed *detritus*. This material serves as a food source for a special group of consumers called *detritivores*, a classification that includes both *detritus feeders* (e.g., macroinvertebrates and microbial grazers) and

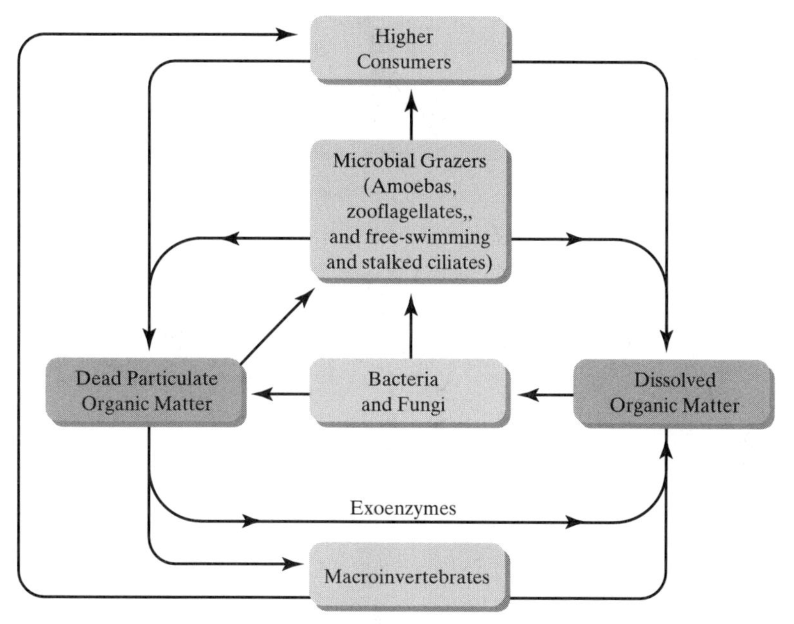

Figure 5-16. The food chain and web illustrated in Figure 5-15 implies that all movement of energy is "upward." Detritus (dead organic matter) generated during material transfer moves "down" and is processed by detritivores that recycle chemicals and energy, making them available once again for upper trophic levels. This microbial loop recovers a significant amount of energy that would be otherwise lost to the ecosystem. However, the decomposer–consumer loop is not self-perpetuating (as is the producer–consumer loop) due to inefficiencies described in the text.

decomposers (e.g., bacteria and fungi). These organisms play a critical role in extracting energy and recycling nutrients that would otherwise be largely unavailable to the biota (Figure 5-16). Without detritivores, dead plant and animal remains would accumulate faster than abiotic decomposition mechanisms (e.g., hydrolysis and photolysis) could remove them. The portion of Figure 5-16 involving microbial grazers, fungi, and bacteria is sometimes termed the *microbial loop*, a reference to its role in nutrient and energy cycling. It is interesting to note that consumers are typically integral members of ecosystems; however, their presence is not essential to producers and decomposers, which have the ability to perpetuate themselves cyclically (Raven et al., 1995).

EXAMPLE 5.6. THE FOOD WEB IN THE ACTIVATED SLUDGE PROCESS

Food chains and food webs exist in engineered systems as well. Figure 5-17 depicts the food web of the activated sludge process, a component of a biological-waste treatment system. The food web is somewhat truncated, both laterally (primary producers are unimportant because the waste provides a source of organic matter) and vertically (higher consumers are absent because the system is

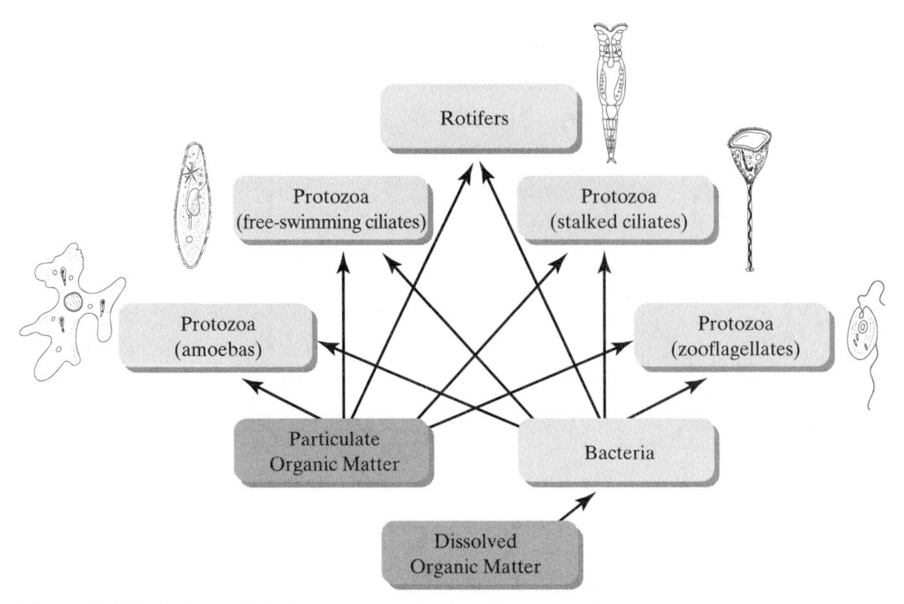

Figure 5-17. Activated sludge process food web.

engineered to top out at a point where the remaining particulate matter is easily removed by sedimentation).

Microbiologists investigating the activated sludge food web have proposed a pattern of community succession associated with the aeration of an organic waste. Different organism groups predominate depending on the degree of stabilization of the waste. At first, amoeboid and zooflagellate protozoans dominate, utilizing the dissolved and particulate organic matter initially present. Next, zooflagellate and free-swimming ciliate protozoans increase in numbers, feeding on developing populations of bacteria. Finally, stalked ciliates and rotifers become most abundant, feeding from the surfaces of activated sludge floc. Plant operating practices (e.g., the rate of sludge wastage, which controls the average time biological solids remain in the system (termed sludge age, SRT, MCRT)) dictate the degree of stabilization and thus the successional position of the microbial community.

5.3.3 Thermodynamics and Energy Transfer

The First Law of Thermodynamics states that energy cannot be created or destroyed, but can be converted from one form to another. Applied to an ecosystem, this law suggests that no organism can create its own energy supply. For example, plants rely on the Sun for energy and grazing animals rely on plants (and thus indirectly on the Sun). Organisms use the food energy that they produce or assimilate to meet metabolic requirements for the performance of work, for example, cell maintenance, growth, and reproduction. Thus, ecosystems must import energy, and the needs of individual organisms must be met by transformations of that energy. The Second Law of Thermodynamics states that in every

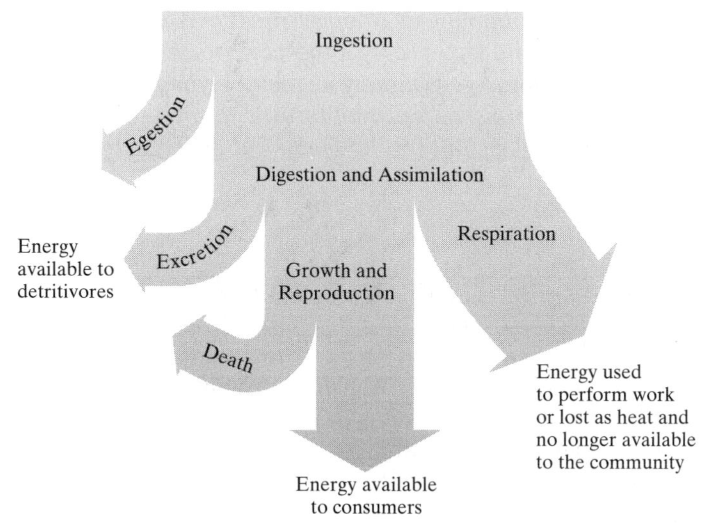

Figure 5-18. A substantial part of the energy ingested by organisms is lost to egestion, excretion, death, and respiration. This inefficiency of energy transfer has a bearing on issues ranging from wastewater treatment microbiology to world population growth.

energy transformation, some energy is lost to heat and becomes unavailable to do work. In the food web, the inefficiency of energy transfer is reflected in losses (Figure 5-18) to decomposition (potentially recycled through the microbial loop) and respiration (heat). Because of this inefficiency, relatively little energy is left after being transferred up several levels of the energy pyramid (Figure 5-19).

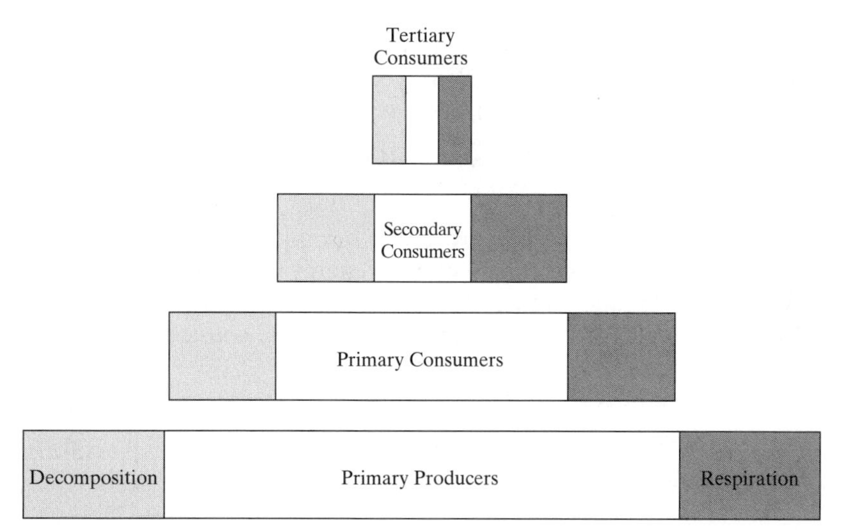

Figure 5-19. An energy pyramid showing the loss of energy to detritivory and respiration with movement up a food chain. A surprisingly small fraction of the energy originally fixed remains available for transfer to higher trophic levels.

This explains why it takes a large amount of primary producers to support a single organism at the top of the food chain (e.g., a top predator may require a very large "range" or territory to support its energy needs). The inefficiencies of energy transfer also have great bearing on our ability to feed a hungry world. Example 5.7 demonstrates the social significance of feeding at different levels on the food chain.

EXAMPLE 5.7. COWS AND POTATOES: TRANSFER OF ENERGY UP THE HUMAN FOOD CHAIN

The principles of thermodynamics and energy transfer introduced earlier demonstrate the tremendous inefficiency of consuming at the top of the food pyramid. For example each American annually consumes an average of 190 pounds of meat. However, more than 60 million people could be fed on the grain saved if Americans reduced their meat intake by just 10%. As human population increases worldwide, so will the demand for land, the food produced there, and the attendant pollution problems as natural ecosystems are converted to agricultural ecosystems. Enger et al. (1983) suggest that well-developed countries such as the United States will have to choose between helping less-developed countries and maintaining good relations or isolating themselves to live in relative splendor. An alternative to this is to invest in changes of lifestyle, eating lower on the food chain, and reducing the energy demand placed on our environment.

5.4 BIOCHEMICAL OXYGEN DEMAND, (BOD), THEORETICAL OXYGEN DEMAND, (ThOD), AND CHEMICAL OXYGEN DEMAND (COD)

5.4.1 Definition of BOD, CBOD, and NBOD

Organisms derive the energy required for maintenance of metabolic function, growth, and reproduction through the process of respiration. Both organic and inorganic matter may serve as sources of that energy. As explained earlier, *chemoheterotrophs* are organisms that utilize organic matter (i.e., $C(H_2O)$) as a carbon and energy source and, under aerobic conditions, consume oxygen in obtaining that energy:

$$C(H_2O) + O_2 \rightarrow CO_2 + H_2O + \Delta \quad \textbf{(5-27)}$$

Chemoautotrophs are organisms that utilize CO_2 as a carbon source and inorganic matter as an energy source, and usually consume oxygen in obtaining that energy.

An example of chemoautotrophy is *nitrification*, the microbial conversion of ammonia to nitrate:

$$NH_3 + 2O_2 \rightarrow NO_3^- + H_2O + H^+ + \Delta \qquad \textbf{(5-28)}$$

In these microbially mediated redox reactions (Section 5.3.1) the electron donors are $C(H_2O)$ and NH_3 and the electron acceptor is O_2. Noting that oxygen is consumed in both reactions, *biochemical oxygen demand* (BOD) can be defined as the amount of oxygen utilized by microorganisms in performing the oxidation. The reactions described by Equations 5-27 and 5-28 are differentiated based on the source compound for the electron donor: *carbonaceous* or CBOD and *nitrogenous* or NBOD.

BOD is a measure of the "strength" of a water or wastewater: the greater the concentration of ammonia–nitrogen or degradable organic carbon, the higher the BOD. Note that chemical strength (mg $C(H_2O)$/L or mg NH_3-N/L) is expressed here in terms of its impact on the environment (oxygen consumed, mg BOD/L).

Dissolved oxygen is a critical requirement of the organism assemblage associated with a diverse and balanced aquatic ecosystem. Water-quality managers seek to maintain levels of dissolved oxygen in lakes and rivers that will support that assemblage. Domestic and industrial wastes often contain high levels of BOD, which if discharged untreated, would seriously deplete oxygen reserves and reduce the diversity of aquatic life. To prevent degradation of receiving waters, engineered systems (i.e., treatment facilities) are constructed where the supply of BOD and oxygen, the availability of the microbial populations that mediate the process, and the rate at which the oxidations themselves (Equations 5-27 and 5-28) proceed may be carefully controlled. The efficiency of BOD removal is a common performance characteristic of wastewater-treatment plants, and BOD is a major feature of treatment-plant discharge permits.

5.4.2 Sources of BOD

The simple carbohydrates produced through photosynthesis are used by plants and animals to synthesize more complex carbon-based chemicals such as sugars and fats. These compounds are utilized by organisms as an energy source, exerting a carbonaceous oxygen demand (Equation 5-27). In addition, plants utilize ammonia to produce proteins—complex, carbon-based chemicals with amino groups ($—NH_2$) as part of their structure. Proteins are ultimately broken down (*proteolysis*) to peptides and then amino acids. The process of *deamination* then further breaks down the amino acids, yielding a carbon skeleton (CBOD) and an amino group. Conversion of the amino group to ammonia (ammonification) completes the degradation process. The ammonia is then available to exert a nitrogenous oxygen demand (Equation 5-28) when utilized by microorganisms. Figure 5-20 illustrates the chemical structure of some representative carbonaceous and nitrogenous compounds.

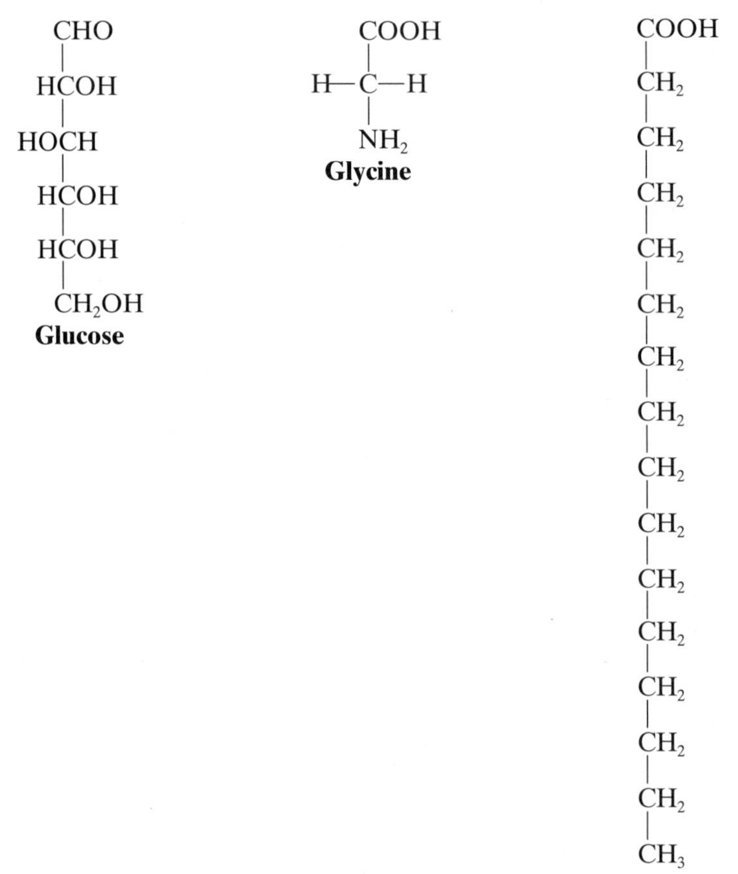

Figure 5-20. The chemical structure of some representative carbonaceous and nitrogenous compounds. Wastewater contains a vast number of different organic chemicals including sugars (20–25%), amino acids (40–60%) and fatty acids (10%). (From Metcalf and Eddy, 1989.)

Even unpolluted natural waters contain some BOD, associated with the carbonaceous and nitrogenous organic matter derived from the watershed and from the waters themselves (e.g., decaying algae and macrophytes, leaf litter, fecal matter from aquatic organisms). Dissolved oxygen levels in surface waters (excluding those with excessive algal photosynthesis and attendant O_2 production) are often below the saturation level (see Example 3-14 and Section 5.5.1) due to this "natural" BOD. Domestic wastewater and many industrial wastes are highly enriched in organic matter compared with natural waters. Proteins and carbohydrates constitute 90% of the organic matter in sewage. Sources include feces and urine from humans; food waste from sinks; soil and dirt from bathing, washing, and laundering; plus various soaps, detergents, and other cleaning products. Wastes from certain industries, for example, breweries, canneries, meat-

Table 5-2. The BOD of Selected Waste Streams

Origin	5-day BOD (mgO_2/L)	Origin	5-day BOD (mgO_2/L)
River	2	Beet sugar factory	10,000
Domestic wastewater	200	Tannery	15,000
Pulp and paper mill	400	Brewery	25,000
Commercial laundry	2,000	Cherry canning factory	55,000

From Nemerow, 1971.

processing plants, and pulp and paper and textile producers, also have elevated levels of organic matter. Table 5-2 presents some representative 5-day BOD values for different waste streams (5-day BOD is described in detail in Sec. 5.4.6).

5.4.3 Theoretical Oxygen Demand

Given the significance of BOD to environmental quality and its ubiquity in waste streams, it is important for engineers to have an approach for determining the oxygen demand of water and wastewater. Theoretical oxygen demand (ThOD, mg O_2/L), calculated from the stoichiometry of the oxidation reactions involved, is a useful parameter in this regard. For the oxidation of a simple carbohydrate ($C(H_2O)$ in Equation 5-27) to carbon dioxide and water, 1 mole (or 32 g) of oxygen is consumed for every mole (or 30 g) of carbohydrate oxidized. The stoichiometric coefficient for determining the ThOD of this carbonaceous waste is thus 32/30 or 1.07. A waste stream containing 300 mg/L of simple carbohydrate (written as $C(H_2O)$) would have a carbonaceous theoretical oxygen demand of 320 mg/L. The calculation is similar for the oxidation of ammonia (NH_3 in Equation 5-28) to nitrate: 2 moles (or 64 g) of oxygen are consumed for every mole (or 14 g) of ammonia–nitrogen (NH_3-N) oxidized; note that ammonia is reported as mg N/L at 14 g/mole, not as mg NH_3/L at 17 g/mole). The stoichiometric coefficient for oxidation of nitrogenous wastes is thus 64/14 or 4.57. A waste containing 50 mg/L of NH_3-N would have a nitrogenous theoretical oxygen demand of 229 mg/L. The total ThOD of a waste stream containing multiple chemicals (e.g., ammonia and organic matter) is determined by adding the contributions of the component compounds, here 549 mg/L.

The stoichiometry for oxidation of ammonia is invariant (and Equation 5-28 holds in all cases) because ammonia–nitrogen occurs in only one form with nitrogen in a single valence state. This is not the case for carbon-based compounds (Equation 5-27 does not hold in all cases), as they exist in a wide variety of forms with carbon in several valence states. For this reason, the reaction stoichiometry for each carbon-based compound must be inspected and the oxidation equations balanced individually.

A general approach for calculation of carbonaceous ThOD is offered by the following 3-step process.

Step 1. Write the equation describing the reaction for oxidation of the carbon-based chemical of interest to carbon dioxide and water (e.g., for benzene, C_6H_6):

$$C_6H_6 + O_2 \rightarrow CO_2 + H_2O$$

Step 2. Balance the equation in the following sequence: (i) balance the number of carbon atoms; (ii) balance the number of hydrogen atoms; and (iii) balance the number of oxygen atoms. For benzene one would (i) place a 6 in front of CO_2 to balance the carbon; (ii) place a 3 in front of H_2O to balance the hydrogen; and (iii) place a 7.5 in front of the oxygen to balance the oxygen. The balanced equation for this example is then:

$$C_6H_6 + 7.5O_2 \rightarrow 6CO_2 + 3H_2O$$

Step 3. Use the stoichiometry of the balanced chemical reaction, applying unit conversions, to determine the carbonaceous ThOD.

Example 5.8 provides some examples for steps 1 and 2 (writing and balancing the equations). Example 5.9 carries the calculation through all three steps for propanol, an organic chemical that contains carbon, hydrogen, and oxygen, and Example 5.10 demonstrates the 3-step process applied to a waste that contains both carbonaceous and nitrogenous oxygen demand. Some chemicals are more complex yet and may contain a variety of inorganic components. If the chemical also contains sulfur, it is assumed to be converted to SO_4^{2-}; if the chemical contains phosphorus, it is assumed to be converted to PO_4^{3-}; if the chemical contains nitrogen, the nitrogen is assumed to be converted to NH_3 and then oxidized to nitrate; if the chemical contains any halogens (e.g., chlorine, bromine), they are assumed to be converted to the respective halogen ion (e.g., Cl^-, Br^-). The Advanced Topic box which follows Example 5.10 explains how to determine the ThOD of these types of chemicals.

EXAMPLE 5.8. BALANCING THE REACTIONS USED TO DETERMINE ThOD

Balance the reactions for converting the following chemicals to CO_2 and H_2O. The chemicals are: ethylene glycol ($C_2H_6O_2$), phenol (C_6H_6O), acetic acid ($C_2H_4O_2$), and oleic acid ($C_{18}H_{34}O_2$).

SOLUTION

In order to balance equations for the conversion of simple organic chemicals (those that contain C, H, and O only) follow the following two steps. Step 1 is to write out the reaction of the organic chemical plus oxygen going to carbon dioxide and water. Step 2 is to (i) balance the number of carbons by placing the appropriate integer in front of the carbon dioxide; (ii) balance the hydrogens by placing the appropriate integer in front of the water; and (iii) balance the oxygen by placing the appropriate integer in front of the oxygen. The following chemical reactions are stoichiometrically balanced.

Ethylene glycol, $C_2H_6O_2$

$$C_2H_6O_2 + 2.5O_2 \rightarrow 2CO_2 + 3H_2O$$

Phenol, C_6H_6O

$$C_6H_6O + 7O_2 \rightarrow 6CO_2 + 3H_2O$$

Acetic acid, $C_2H_4O_2$

$$C_2H_4O_2 + 2O_2 \rightarrow 2CO_2 + 2H_2O$$

Oleic acid, $C_{18}H_{34}O_2$

$$C_{18}H_{34}O_2 + 25.5O_2 \rightarrow 18CO_2 + 17H_2O$$

EXAMPLE 5.9. DETERMINING ThOD FOR A CHEMICAL CONSISTING OF C, H, AND O

Determine the carbonaceous ThOD of a waste that contains 100 mg/L of propanol (C_3H_7OH).

SOLUTION

First, write out the equation describing the oxidation of propanol to carbon dioxide and water:

$$C_3H_7OH + O_2 \rightarrow CO_2 + H_2O$$

Second, balance the equation as described earlier and illustrated in Example 5.8:

$$C_3H_7OH + 4.5O_2 \rightarrow 3CO_2 + 4H_2O$$

This reaction shows that for each mole of propanol oxidized, 4.5 moles of oxygen are required. The ThOD is then calculated from the reaction's stoichiometry (4.5 moles of oxygen per mole of propanol oxidized), applying unit conversions, as follows:

$$\frac{300 \text{ mg propanol}}{\text{L}} \times \frac{\text{g}}{1{,}000 \text{ mg}} \times \frac{1 \text{ mole propanol}}{60 \text{ g propanol}} \times \frac{4.5 \text{ mole } O_2}{\text{mole propanol}}$$

$$\times \frac{32 \text{ g } O_2}{\text{mole } O_2} \times \frac{1{,}000 \text{ mg}}{\text{g}} = \frac{720 \text{ mg}}{\text{L}} \text{ (the carbonaceous ThOD)}$$

EXAMPLE 5.10. DETERMINATION OF CARBONACEOUS, NITROGENOUS, AND TOTAL ThOD

A waste contains 300 mg/L of simple carbohydrate $C(H_2O)$ and 50 mg/L of NH_3-N. Calculate the theoretical carbonaceous oxygen demand, the theoretical nitrogenous oxygen demand, and total ThOD of the waste.

SOLUTION

First, write out the balanced equation describing the oxidation of simple carbohydrate to carbon dioxide and water:

$$C(H_2O) + O_2 \rightarrow CO_2 + H_2O$$

This reaction shows that for each mole of simple carbohydrate oxidized, 1 mole of oxygen is required. The carbonaceous theoretical oxygen demand is found from the reaction's stoichiometry, applying unit conversions, as follows:

$$\frac{300 \text{ mg carbohydrate}}{\text{L}} \times \frac{\text{g}}{1{,}000 \text{ mg}} \times \frac{1 \text{ mole carbohydrate}}{30 \text{ g carbohydrate}} \times \frac{1 \text{ mole } O_2}{\text{mole carbohydrate}} \times \frac{32 \text{ g } O_2}{\text{mole } O_2} \times \frac{1{,}000 \text{ mg}}{\text{g}} = \frac{320 \text{ mg}}{\text{L}} \text{ (the carbonaceous ThOD)}$$

Next, write out the balanced equation describing the oxidation of ammonia–nitrogen to nitrate–nitrogen.

$$NH_3 + 2O_2 \rightarrow NO_3^- + H^+ + H_2O$$

This reaction shows that for each mole of ammonia oxidized, 2 moles of oxygen are required. Note that the ammonia concentration is reported as mg N/L, not mg NH_3/L. The nitrogenous theoretical oxygen demand can be found from the reaction's stoichiometry, applying unit conversions, as follows:

$$\frac{50 \text{ mg } NH_3\text{-N}}{\text{L}} \times \frac{\text{g}}{1{,}000 \text{ mg}} \times \frac{1 \text{ mole } NH_3\text{-N}}{14 \text{ g } NH_3\text{-N}} \times \frac{2 \text{ mole } O_2}{\text{mole } NH_3\text{-N}} \times \frac{32 \text{ g } O_2}{\text{mole } O_2} \times \frac{1{,}000 \text{ mg}}{\text{g}} = 229 \text{ mg/L (the nitrogenous ThOD)}$$

The sum of the carbonaceous and nitrogenous components, 320 mg/L + 229 mg/L, yields the total ThOD, 549 mg/L. The solubility of oxygen in water equilibrated with the atmosphere only ranges from 7.6 mg/L to 14.6 mg/L, depending on the temperature. Note how the presence of organic matter and ammonia can quickly deplete available oxygen in a lake or river.

ADVANCED TOPIC. DETERMINING THE ThOD FOR CHEMICALS THAT ALSO CONTAIN S, P, Cl, AND/OR N

The ThOD can also be determined for chemicals consisting of more than C, H, and O. To calculate the carbonaceous theoretical oxygen demand for complicated organic chemical reactions, a generalized stoichiometric equation can be written as:

$$C_nH_mO_eX_kN_jS_iP_h + aO_2 \rightarrow nCO_2 + \frac{(m - k - 3j - 2i - 3h)}{2} H_2O$$
$$+ kHX + jNH_3 + iH_2SO_4 + hH_3PO_4$$

where

$$a = n + \frac{m - k - 3j - 2i - 3h}{4} - \frac{e}{2} + 2i + 2h$$

and X represents the sum of all halogens. For the combined carbonaceous and nitrogenous theoretical oxygen demand of complicated organic chemicals, the generalized stoichiometric equation can be written as:

$$C_nH_mO_eX_kN_jS_iP_n + bO_2 \rightarrow nCO_2 + \frac{m - k - j - 2i - 3h}{2} H_2O$$
$$+ kHX + jHNO_3 + iH_2SO_4 + hH_3PO_4$$

where

$$b = n + \frac{(m - k - j - 2i - 3h)}{4} - \frac{e}{2} + \frac{3j}{2} + 2i + 2h$$

and X represents the sum of all halogen species.

After the balanced equation is obtained, the stoichiometric amount of oxygen required to oxidize a specified concentration of organic chemical to CO_2, H_2O

and the appropriate inorganic end products is calculated. The values for a and b from the preceding equations provide the stoichiometric amount of oxygen required (mole O_2/mole organic) for carbonaceous and combined carbonaceous and nitrogenous ThOD, respectively. This value can be multiplied by the ratio of the molecular weights of oxygen and the organic chemical to get a mass-based stoichiometric conversion factor for that organic chemical (mass O_2/mass chemical).

5.4.4 BOD Kinetics

The ThOD calculation defines the oxygen requirement for complete oxidation of ammonia–nitrogen to nitrate–nitrogen or a carbon-based compound to carbon dioxide and water. ThOD does not, however, offer any information regarding the likelihood that the reaction will proceed to completion. For nitrogenous BOD this is not a serious issue, as ammonia–nitrogen is readily oxidized and the theoretical and actual nitrogen oxygen demand are identical.

Carbonaceous compounds, on the other hand, are not all easily or completely oxidized (biodegraded) and the rate of that oxidation may vary widely among different sources of organic matter. For example, the carbon contained in a styrofoam cup is not as biodegradable as the carbon contained in tree leaves, and both are less biodegradable than the carbon present in sugar. Thus, while a mass of carbon present as sugar or tree leaves or styrofoam may have the same ThOD, their actual oxygen demand may be substantially different. Further, most wastes are a complex mixture of chemicals (e.g., styrofoam + tree leaves + sugar), present in differing amounts, each with a different level of biodegradability. This characteristic of oxygen-demanding wastes is accommodated through BOD kinetics.

As shown graphically in Figure 5-21, consider the oxidation of organic matter in a water sample as a function of time, where y_t is the CBOD exerted (oxygen consumed, mg O_2/L) and L_t is the CBOD remaining (potential to consume oxygen, mg O_2/L) at any time, t. At $t = 0$, no CBOD has been exerted ($y_{t=0} = 0$) and all of the potential for oxygen consumption remains ($L_{t=0} = L_0$, the *ultimate* CBOD). As the oxidation process begins, oxygen is consumed (CBOD is exerted and y_t increases) and the potential to consume oxygen consumption is reduced (CBOD remaining, L_t, decreases). The rate at which CBOD is exerted is rapid at first, but later slows and eventually reaches zero when all of the biodegradable organic matter has been oxidized. The total amount of oxygen consumed in oxidizing the waste defines the ultimate CBOD (L_0).

The exponential decline (shown in Figure 5-21b) is the potential for oxygen consumption of CBOD remaining (L). It can be modeled as a first-order decay:

$$\frac{dL}{dt} = -k_L\, L \qquad \textbf{(5-29)}$$

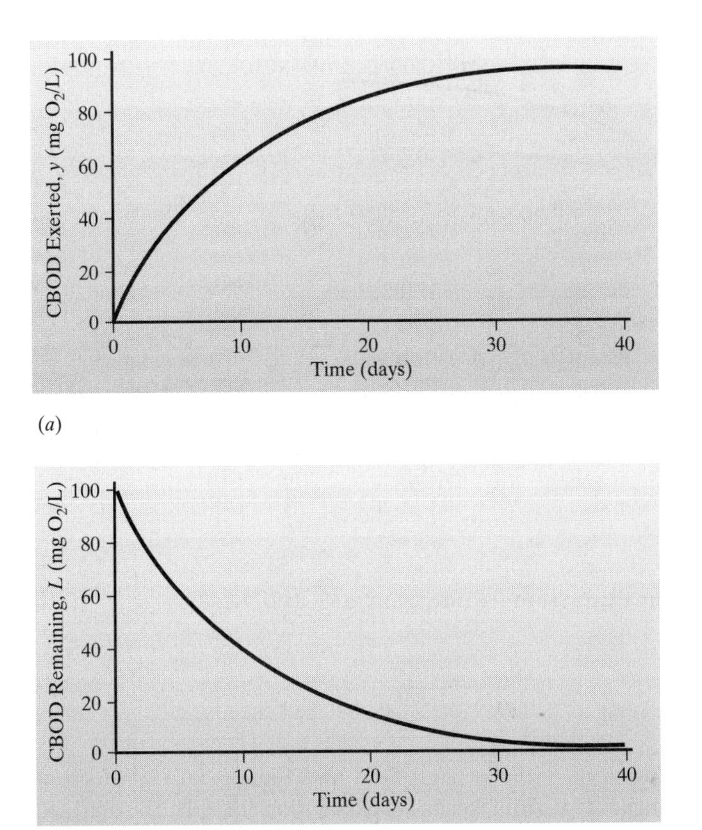

Figure 5-21. (*a*) CBOD exerted, y and (*b*) CBOD remaining, L, as functions of time.

where k_L is the CBOD reaction rate coefficient (day^{-1}). Equation 5-29 can be integrated to yield the analytical expression that is depicted in Figure 5-21*b*:

$$L_t = L_0 \, e^{(-k_L t)} \qquad \textbf{(5-30)}$$

Note that the CBOD exerted at any time t (y_t) (see Figure 5-21*a*) is given by the difference between the ultimate CBOD and the CBOD remaining:

$$y_t = L_0 - L_t \qquad \textbf{(5-31)}$$

Substituting Equation 5-30 into Equation 5-31 yields:

$$y_t = L_0(1 - e^{-k_L t}) \qquad \textbf{(5-32)}$$

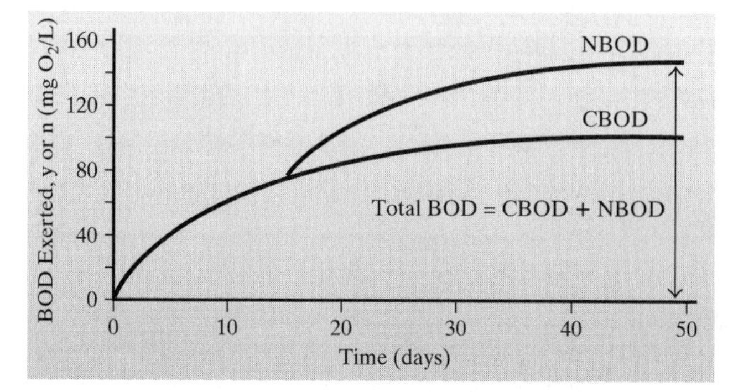

Figure 5-22. First- (CBOD) and second- (NBOD) stage biochemical oxygen demand. Exertion of nitrogenous demand lags that of carbonaceous demand because nitrifiers (Sections 5.3.1 and 5.4.1) grow more slowly that those that derive their energy from organic carbon.

and rearranging, yields an expression for the ultimate CBOD:

$$L_0 = \frac{y_t}{(1 - e^{-k_L t})} \tag{5-33}$$

An understanding of Equations 5-29 through 5-33 and Figure 5-21 is of great value in converting among the various forms of expression for CBOD that are widely used in engineering applications.

NBOD behaves in an almost identical fashion. The exertion of NBOD follows first-order kinetics, and Equations 5-29 through 5-33 apply, substituting n, N, and k_N for y, L, and k_L. The ultimate NBOD (N_0) is calculated from the ammonia–nitrogen content of the sample, based on the stoichiometry of Equation 5-28 (4.57 mg O_2 consumed per mg NH_3-N oxidized). Figure 5-22 shows how the NBOD is initiated well after the CBOD has begun to be exerted.

5.4.5 The CBOD Rate Constant

The reaction rate coefficient (k_L) utilized in CBOD calculations, is a measure of the biodegradability of a waste. Figure 5-23 illustrates the relationship between k_L and the rate of CBOD exertion. The magnitude of k_L is related to the nature of the waste (its biodegradability), but k_L values vary little for a given water or waste. Typical ranges for this coefficient are presented in Table 5-3. As with other microbially mediated processes, values for the CBOD and NBOD reaction rate coefficients also vary with temperature.

Values for the CBOD reaction-rate coefficient may be determined experimentally in the laboratory according to the Thomas Slope Method. Measurements of CBOD exerted (y_t) are made daily for 7–10 days. Values for the parameter

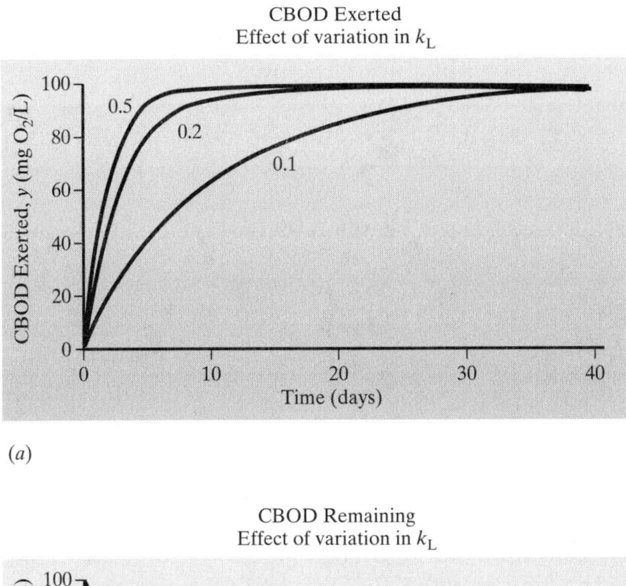

(*a*)

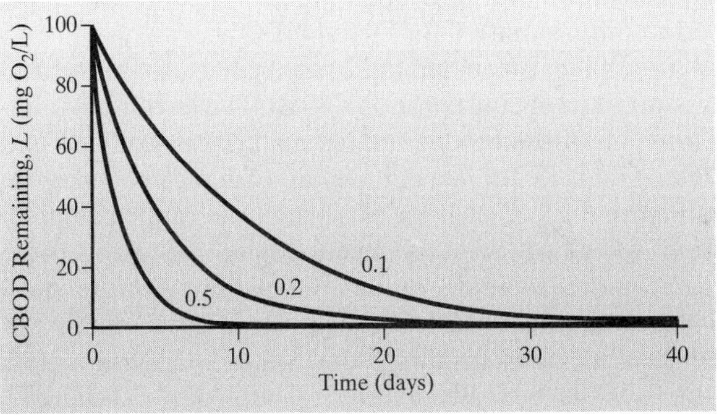

(*b*)

Figure 5-23. Variations in the rate at which organic matter is stabilized is reflected in the reaction-rate constant, k_L. (*a*) CBOD exerted (y, mg/L); (*b*) CBOD remaining (L, mg/L).

Table 5-3. Ranges of Values for the CBOD Reaction Rate Coefficient

Environment	k_L (day^{-1})
Untreated sewage	0.35–0.70
Treated sewage	0.10–0.35
Unpolluted river water	<0.05

From Davis and Cornwell, 1991; Chapra, 1997.

$(t/y)^{1/3}$ are then calculated and k_L is determined from the slope and intercept of a plot of $(t/y)^{1/3}$ versus t according to Equation 5-34:

$$k_L = 6.01 \frac{\text{slope}}{\text{intercept}} \tag{5-34}$$

Once determined for a particular waste source, the rate constant usually displays little variation.

5.4.6 Measurement of BOD

The kinetics described in the previous section can be applied within the context of laboratory measurements to quantify the BOD of a water or wastewater sample. The technique is termed a bioassay: a test in which organisms ("bio") are used to assess ("assay") the amount of a target substance present in a sample. Here, microorganisms are used to degrade (oxidize) the many different oxygen-demanding compounds that constitute the BOD of a water or wastewater sample. The assay may be designed to differentiate CBOD and NBOD.

In the BOD assay, a water sample is incubated under controlled environmental conditions for several days and oxygen consumption (y_t, BOD exertion) is monitored. After a period of time, all of the biodegradable material is oxidized and the ultimate CBOD is determined as the asymptote, as is depicted in Figure 5-21*a*. It is apparent from Figure 5-21*a* that it takes a long time (approximately 20 days for 95% completion) for all of the carbonaceous oxygen demand to be exerted (i.e., all of the organic matter to be degraded). Such a wait is impractical, and thus a shortened, 5-day measurement of BOD exertion is utilized (the details of the test are provided below). During this 5-day period, approximately two-thirds of the ultimate BOD is typically exerted. The 5-day CBOD (y_5, mg O_2/L) is calculated as the difference between the initial and final dissolved oxygen concentration during a 5-day incubation. The ultimate CBOD (L_0) is calculated from the y_5 value using Equation 5-33, with $t = 5$.

The BOD determination is truly a fundamental procedure in environmental engineering, and thus an expanded presentation of the technique is merited. The following description of the BOD test methodology summarizes the detailed method outlined in Standard Methods. The requirements for proper determination of BOD are:

1. ***Existence of appropriate microorganisms.*** It is fundamental to the BOD test that microorganisms capable of breaking down (oxidizing) the biodegradable organic matter be present in the sample. Such organisms are typically present in sufficient numbers as a community or mixed culture (e.g., numerous bacteria species, protozoa, etc.) in untreated domestic wastewater, treated (but not disinfected) wastewater, and most surface waters. Disinfected wastes, wastes subject to high temperatures or extreme pH, and some surface waters may have insufficient populations of the required or-

ganisms. In such cases, the microbes are added to the samples as "seed" to ensure the potential for oxidation of organic matter. Microbes for use as seed can be purchased or obtained from aerobic units in a biological waste-treatment facility.

2. ***Constant and appropriate temperature.*** Nearly all biochemical reactions are temperature dependent. In the BOD test, samples are incubated at 20 ± 1°C to facilitate comparison of results for various sites and among different laboratories. Microbially-mediated reactions typically increase with temperature over the range of temperatures common in temperate regions (0 to 30°C). It would be inappropriate to measure the BOD of a waste at, for example, 4°C, and then extrapolate that result in estimating receiving water impacts at 20°C. The temperature dependence of BOD must always be considered in engineering applications. The mathematical treatment of temperature dependence of the BOD rate constant was illustrated in Example 3.7.
3. ***Time required.*** As described earlier, the DO of an incubated BOD sample decreases over time; however, because it follows first-order kinetics (exponential decay), "complete" oxidation of the degradable organic matter may take several weeks. The standard incubation time for the BOD test is 5 days, leading to the notation y_5. This period permits acclimation of the microbes to their physical and chemical environment (lag phase growth) and significant (although typically incomplete) oxidation during the exponential growth phase. The time of incubation is not critical to the test, but must be clearly noted in expressing the result (e.g., y_2 or y_{10}). The 5-day BOD (y_5 or BOD_5) is the standard procedure; however, longer incubations can be used to gain information regarding the kinetics of the process.
4. ***Differentiation between CBOD and NBOD.*** A chemical (2-chloro-6-(trichloro methyl) pyridine, or TCMP) may be added to inhibit the oxidation of ammonia by nitrifying bacteria. The NBOD is then calculated as the difference between inhibited and uninhibited samples. If an inhibitor is added, results are reported as CBOD; without inhibitor addition, results are reported as BOD.
5. ***Adequate DO.*** The dissolved oxygen of the sample must not be completely depleted before the end of the test. Oxygen depletion leads to underestimation of the 5-day BOD, because the potential oxygen-demand of the organic matter present is not fully realized. This problem is addressed by mixing the sample with dilution water (see Example 5.13 below).
6. ***Adequate nutrients.*** Some water or wastewater samples may not contain inorganic nutrient levels sufficient to support unlimited microbial growth. To prevent this possibility nutrient elements such as P, Mg, Fe, Ca, and S are added.
7. ***Preventing DO gains during incubation.*** Samples are incubated in the dark to prevent photosynthetic production of oxygen. In addition, the bottle utilized in the BOD test is designed to eliminate the entrapment of air bubbles during sealing.

8. ***Appropriate pH.*** Microbial activity is often pH specific. In the standard BOD test, samples are neutralized to a pH of 6.5 to 7.5 by the addition of a phosphate buffer.
9. ***Toxic effects.*** Samples suspected to contain a potentially toxic material (e.g., heavy metals or organic compounds in industrial wastes) are typically diluted. The presence and nature of toxicity generally requires separate and additional studies.

EXAMPLE 5.11. LABORATORY DETERMINATION OF CBOD

A 15-mL wastewater sample is placed in a standard 300-mL BOD bottle and the bottle is filled with dilution water. The bottle had an initial dissolved-oxygen concentration of 8 mg/L and a final dissolved-oxygen concentration of 2 mg/L. A blank (a BOD bottle filled with dilution water) run in parallel showed no change in dissolved oxygen over the 5-day incubation period. The CBOD reaction-rate coefficient for the waste is 0.4 day^{-1}. Calculate the 5-day (y_5) and ultimate (L_0) CBOD of the wastewater.

SOLUTION

The y_5 is the amount of oxygen consumed over the 5-day period corrected for the dilution of the original sample. This can be written as:

$$y_5 = \frac{[\mathrm{DO_{initial}} - \mathrm{DO_{final}}]}{\left[\frac{\text{mL sample}}{\text{total test volume}}\right]} = \frac{\left[\frac{8\text{ mg O}_2}{\text{L}} - \frac{2\text{ mg O}_2}{\text{L}}\right]}{\left[\frac{15\text{ mL}}{300\text{ mL}}\right]} = \frac{120\text{ mg O}_2}{\text{L}}$$

Equation 5-33, with $t = 5$ and $k_L = 0.4$ day^{-1}, is then applied:

$$L_0 = \frac{y_5}{(1 - e^{(-k_L\,t)})} = \frac{120}{(1 - e^{(-0.4\text{ day}^{-1}\times 5\text{ day})})} = \frac{138\text{ mg O}_2}{\text{L}}$$

which is the ultimate BOD.

EXAMPLE 5.12. LABORATORY DETERMINATION OF BOD

Due to poor scheduling, students set up a CBOD test on the Monday before a holiday (initial dissolved oxygen = 9 mg/L). On the third day, the students measure a dissolved oxygen of 6 mg/L and mistakenly reaerate the sample so that the dissolved oxygen goes up to 8 mg/L. They then go home for the holiday. After an 18-day interval, they return to school and measure the dissolved oxygen to be 3 mg/L. The original sample size was 6 mL and the total sample volume was 300 mL (the balance being dilution water). Assume that the ultimate CBOD was reached in 14 days and that $k_L = 0.1\ \text{day}^{-1}$. Calculate the 5-day ($y_5$) and ultimate ($L_0$) CBOD of the sample.

SOLUTION

The first part can be answered by understanding how the CBOD test is performed. To solve the problem, determine y, the CBOD exerted, for the period of day 1–3 and then the period of day 4–18:

$$y_{(\text{day } 1\text{–}3)} = \frac{\left[\dfrac{9\ \text{mg}}{\text{L}} - \dfrac{6\ \text{mg}}{\text{L}}\right]}{\dfrac{6\ \text{mL}}{300\ \text{mL}}} = \frac{150\ \text{mg}}{\text{L}}$$

$$y_{(\text{day } 4\text{–}18)} = \frac{\left[\dfrac{8\ \text{mg}}{\text{L}} - \dfrac{3\ \text{mg}}{\text{L}}\right]}{\dfrac{6\ \text{mL}}{300\ \text{mL}}} = \frac{250\ \text{mg}}{\text{L}}$$

The problem states that the ultimate demand is exerted in 14 days; therefore, no additional oxygen demand was exerted after day 14. Thus the ultimate CBOD, L_0, is $150 + 250 = 400$ mg O_2/L.

The second part can be answered by applying the appropriate kinetic expression (Equation 5-32) and the ultimate CBOD determined in the first part):

$$y_5 = 400\ \text{mg/L}(1 - e^{-[0.1\ \text{day}^{-1} \times 5\ \text{day}]}) = \frac{157\ \text{mg } O_2}{\text{L}}$$

which is termed the 5-day BOD or BOD_5.

EXAMPLE 5.13. DILUTION WATER AND THE BOD TEST

As noted in Table 5-2, the 5-day BOD of domestic wastewater influent is on the order of 200 mg/L. Determine whether or not there would be enough oxygen present in an undiluted sample of wastewater influent to support the 5-day BOD test. Assume that dissolved-oxygen concentration in the waste sample is 6 mg O_2/L. A standard "BOD bottle" has a sample volume of 300 mL (0.3 L).

SOLUTION

A 0.3-L volume exerting an oxygen demand of 200 mg/L over 5 days would require 60 mg of dissolved oxygen ($200 \times 0.3 = 60$). However, a 0.3-L volume containing 6 mg O_2/L has only 1.8 mg of dissolved oxygen available for consumption ($6 \times 0.3 = 1.8$). This is much less than the 60 mg of oxygen required to oxidize the waste; thus, the oxygen resources would be depleted well before the end of the 5-day test, violating the procedural guidelines introduced earlier. The results of a test performed incorrectly in this fashion would suggest a 5-day BOD of 6 mg O_2/L (1.8 mg O_2 consumed/0.3 L)—a gross underestimate of the true value.

To prevent this error, "strong" wastes (samples with a high BOD) are mixed with aerated dilution water so that the oxygen demand of the sample does not exceed the oxygen supply available. Typical dilution factors provided by Standard Methods are: 0.0 to 1.0% for strong industrial wastes; 1 to 5% for untreated domestic wastewater; 5 to 25% for biologically treated wastewater; and 25 to 100% for polluted river water. Dilution water also contains nutrients and a pH buffer to provide a suitable environment for the microorganisms that carry out the oxidation.

5.4.7 The BOD Test: Limitations and Alternatives Such as Chemical Oxygen Demand

While the 5-day BOD test remains a fundamental tool for assessing water quality, concerns regarding its logistics and accuracy have led to proposals for replacement by other measures. While relatively simple to perform, the test has three major shortcomings: (1) the time required to obtain results—5 days is almost unthinkable in today's world of real-time data acquisition; (2) the fact that it may not accurately assay waste streams that degrade over a time period longer than 5 days; and (3) the inherent inaccuracy of the procedure, largely due to variability in seed. Other analytical measures, such as total organic carbon (TOC), provide greater accuracy and precision, but fail to readily distinguish biodegradable (*labile*) and nonbiodegradable (*refractory*) organic carbon—precisely the objective of the BOD test in the first place.

Some organic chemicals may resist biodegradation. Others may be toxic to microorganisms. The BOD test thus cannot be applied to these wastes to provide, for example, a measure of the efficiency of treatment. Here a different test, one for *chemical oxygen demand* (COD; see Section 3.5 for a description of the test), is applied. In this case, rather than using microorganisms, a strong chemical agent is added to completely oxidize the waste. Results are expressed in oxygen equivalents (mg O_2/L), that is, the amount of oxygen required to completely oxidize the waste. The test is relatively quick (3 h) and correlation to BOD_5 is easy to establish on a particular wastestream. For example, domestic wastewaters typically have a BOD_5/COD ratio of 0.4–0.5 (Metcalf and Eddy, 1989). Comparison of CBOD and COD results can help identify the occurrence of toxic conditions in a waste stream or point to the presence of biologically resistant (refractory) wastes. For example, a BOD_5/COD ratio approaching one may indicate a highly biodegradable waste, while a ratio approaching zero suggests a poorly biodegradable material.

5.5 DISSOLVED OXYGEN AND BOD IN RIVERS

Dissolved oxygen is required to maintain a balanced community of organisms in lakes and rivers. When an oxygen-demanding waste is added to a river or stream, the rate at which oxygen is consumed in oxidizing that waste (*deoxygenation*) may exceed the rate at which oxygen is resupplied from the atmosphere (*reaeration*). This can lead to depletion of oxygen resources, with concentrations falling far below saturation levels as shown in Figure 5-24. When oxygen levels drop

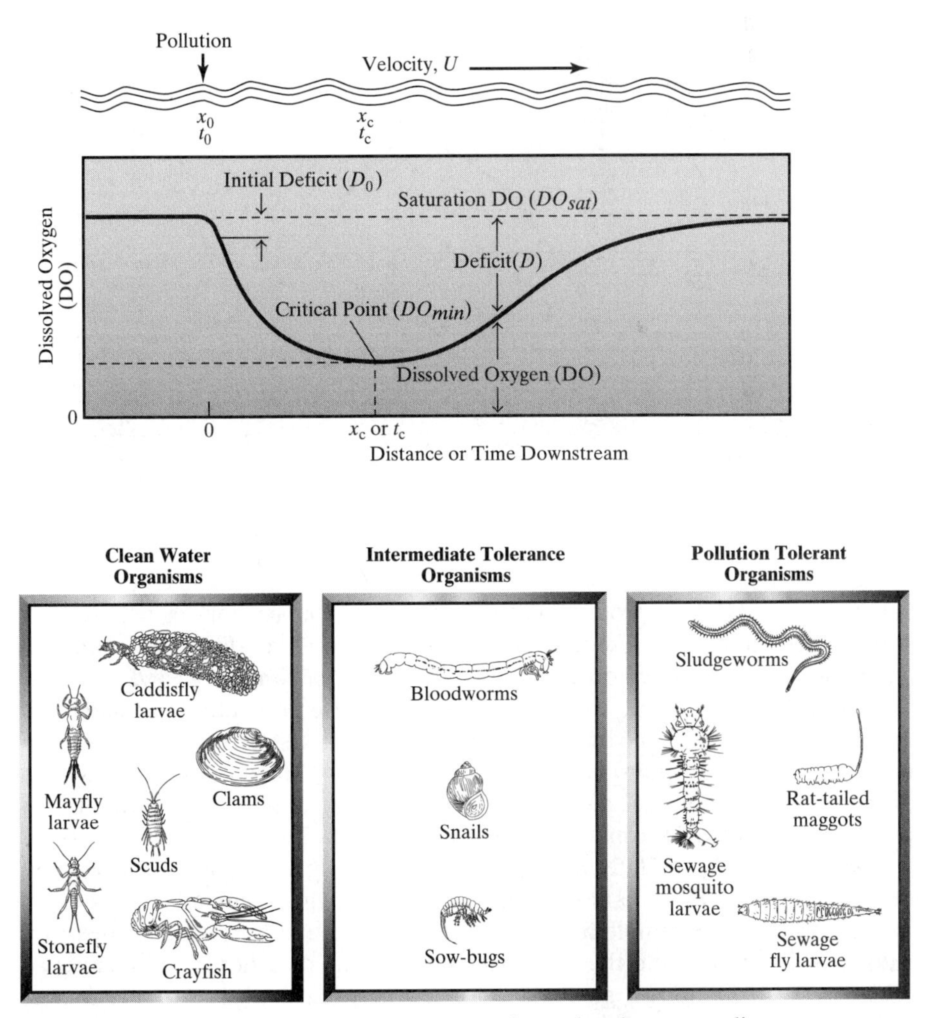

Figure 5-24. The dissolved oxygen sag curve and associated water quality zones reflecting impacts on physical conditions and the diversity and abundance of organisms.

	Stream Zones				
	Clean Water	Degradation	Damage	Recovery	Clean Water
Physical conditions	Clear water; no bottom sludge	Floating solids; bottom sludge	Turbid water; malodorous gases; bottom sludge	Turbid water; bottom sludge	Clear water; no bottom sludge
Fish species	Cold or warm water game and forage fish; trout, bass,	Pollution-tolerant fish; carp, gar, buffalo	None	Pollution-tolerant fish; carp, gar, buffalo	Cold or warm water game and forage fish; trout, bass,
Benthic invertebrate	clean water	intermediate tolerance	pollution tolerant	intermediate tolerance	clean water

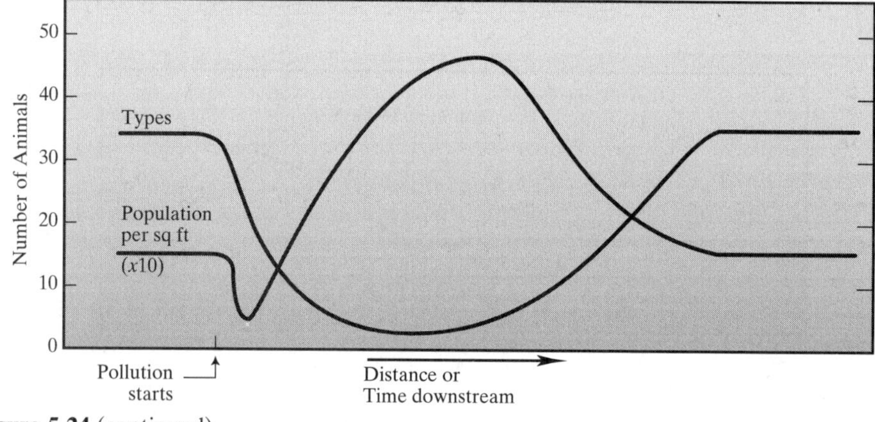

Figure 5-24 (continued).

below 4–5 mg O_2/L, reproduction by fish and macroinvertebrates is impacted. Oxygen depletion is often severe enough that anaerobic conditions develop with an attendant loss of biotic diversity (many species are eliminated) and poor aesthetics (turbidity and odor problems). Figure 5-24 also illustrates the response of stream biota to pollution from oxygen-demanding wastes. Civil and environmental engineers are responsible for seeing that rivers are not overloaded with oxygen-demanding wastes and for designing, building, and operating treatment plants to remove BOD from the waste stream before discharge to surface waters.

Consideration of the fate of CBOD following discharge to a river is a useful starting point for examination of the impact of oxygen-demanding wastes on oxygen resources. Example 5.14 applies mixing-basin concepts developed previously in Chapter 4, and the BOD kinetics developed earlier in this chapter to examine the oxidation of an organic waste following discharge to and mixing with a river.

EXAMPLE 5.14. MIXING BASIN CALCULATION FOR BOD

A waste with a 5-day CBOD (y_5) of 200 mg/L and a k_L of 0.1 day^{-1} is discharged to a river at a rate of 1 m^3/s. Calculate the ultimate CBOD (L_0) of the waste before discharge to the river. Assuming instantaneous mixing after discharge, calculate the ultimate CBOD of the river water after it has received the waste. The river has a flow rate of 9 m^3/s and a background ultimate CBOD of 2 mg/L upstream of the waste discharge. Finally, calculate the ultimate CBOD (L_0) and CBOD$_5$ (y_5) in the river 50 km downstream of the point of discharge. The river has a width of 20 m and a depth of 5 m.

SOLUTION

This problem has several steps. First, calculate the ultimate CBOD of the waste prior to discharge:

$$L_{0,\text{waste}} = \frac{y_5}{(1 - e^{-k_L \times 5\ \text{day}})} = \frac{\dfrac{200\ \text{mg}}{L}}{(1 - e^{-0.1\ \text{day}^{-1} \times 5\ \text{day}})} = \frac{508\ \text{mg}}{L}$$

Second, perform a mass-balance mixing-basin calculation to determine the ultimate CBOD after the waste has been discharged and mixed with the river. The general relationship for calculation of the concentration of any chemical in a mixing basin (C_{mb}) is:

$$C_{mb} = \frac{C_{\text{up}}\ Q_{\text{up}} + C_{\text{in}}\ Q_{\text{in}}}{Q_{mb}}$$

Here the total flow, Q_{mb}, equals $Q_{\text{up}} + Q_{\text{in}}$ and the ultimate CBOD equals:

$$L_{0,mb} = \frac{\dfrac{2\ \text{mg}}{\text{L}} \times \dfrac{9\ \text{m}^3}{\text{s}} + \dfrac{508\ \text{mg}}{\text{L}} \times \dfrac{1\ \text{m}^3}{\text{s}}}{\dfrac{10\ \text{m}^3}{\text{s}}} = \frac{52.6\ \text{mg}}{\text{L}}$$

This is the value of the ultimate CBOD of the river water after it has received the waste. In order to answer the last two questions concerning the ultimate CBOD and 5-day CBOD 50 km downstream of discharge, first calculate the 5-day CBOD of the river water after it has received the waste.

$$y_t = L_0(1 - e^{-k_L t})$$

$$y_{5,mb} = 52.6\ \frac{\text{mg}}{\text{L}} \times (1 - e^{-0.1\ \text{day}^{-1} \times 5\ \text{day}})$$

$$y_{5,mb} = 20.7\ \frac{\text{mg}}{\text{L}}$$

Next, calculate the ultimate CBOD 50 km downstream of the point of discharge. As the waste travels downstream, it will decay and deplete oxygen according to first-order kinetics. The river downstream of the mixing zone can be modeled as a plug flow reactor (PFR). Therefore,

$$L_t = L_0\, e^{-k_L t}$$

However, the time of travel needs to be calculated. The river velocity is given by:

$$U = \frac{Q}{A} = \frac{Q}{W\,H} = \frac{10\,\frac{\text{m}^3}{s}}{20\text{ m} \times 5\text{ m}} = 0.1\,\frac{\text{m}}{\text{s}} \times 86{,}400\,\frac{\text{s}}{\text{day}} \times \frac{\text{km}}{1{,}000\text{ m}} = 8.64\,\frac{\text{km}}{\text{day}}$$

The time of travel is then determined by dividing the distance by the stream velocity:

$$t = \frac{x}{U} = \frac{50\text{ km}}{8.64\,\frac{\text{km}}{\text{day}}} = 5.87\text{ day}$$

This value can then be used to determine the ultimate CBOD 5.78 days downstream:

$$L_{0,50\text{ km}} = L_{0,mb}\, e^{-k_L t} = 52.6 \times e^{-0.1\text{ day}^{-1} \times 5.78\text{ day}} = \frac{29.5\text{ mg}}{\text{L}}$$

and a 5-day CBOD of:

$$y_t = L_0(1 - e^{-k_L t})$$

$$y_{5,50\text{ km}} = 29.5 \times (1 - e^{-0.1\text{ day}^{-1} \times 5\text{ day}}) = \frac{11.6\text{ mg}}{\text{L}}$$

In Example 5.14, more than 23 mg O_2/L of ultimate CBOD is exerted over the 50-km stretch downstream of the discharge. In order to appreciate the effect of this demand on oxygen resources, it is necessary to understand the capacity of water to hold oxygen (saturation) and the rate at which oxygen can be resupplied from the atmosphere (reaeration).

5.5.1 Oxygen Saturation

As discussed in Section 3.4.2, the amount of oxygen that can be dissolved in water at a given temperature—its equilibrium or *saturation concentration*—may be determined through a knowledge of the Henry's constant (assuming no ionic strength effects):

$$DO_{sat} = K_H \, P_{O_2} \tag{5-35}$$

where DO_{sat} is the saturation dissolved oxygen concentration (moles O_2/L), K_H is the Henry's constant (1.356×10^{-3} moles/L-atm at 20°C), and P_{O_2} is the partial pressure of oxygen in the atmosphere (~21% or 0.21 atm).

The value of the Henry's constant varies with temperature (see Section 3.4.2.1), and thus the saturation concentration of dissolved oxygen will vary as well. Example 5.15 illustrates the calculation of the saturation dissolved-oxygen concentration.

EXAMPLE 5.15. DETERMINATION OF DISSOLVED-OXYGEN SATURATION

Determine the saturation dissolved-oxygen concentration, DO_{sat}, at 20°C.

SOLUTION

Determine the DO_{sat} from the appropriate temperature-dependent Henry's constant and the oxygen partial pressure:

$$DO_{sat} = \frac{1.356 \times 10^{-3} \text{ moles}}{\text{L-atm}} \times 0.21 \text{ atm} = \frac{2.85 \times 10^{-4} \text{ moles } O_2}{\text{L}}$$

and converting to mg O_2/L:

$$DO_{sat} = \frac{2.85 \times 10^{-4} \text{ moles } O_2}{\text{L}} \times \frac{32 \text{ g } O_2}{\text{mole } O_2} \times \frac{1{,}000 \text{ mg } O_2}{\text{g } O_2} = \frac{9.1 \text{ mg } O_2}{\text{L}}$$

Note that the words "dissolved-oxygen saturation concentration" and the "solubility of oxygen" can be used interchangeably.

The value for DO_{sat} ranges from about 14.6 mg O_2/L at 0°C to 7.6 mg O_2/L at 30°C, typical temperature extremes for natural and engineered treatment systems. This is why fish with high oxygen requirements seek cold waters and why the impacts of oxygen-demanding wastes on water quality may be greatest in the summer when water temperatures are high. Also, streamflows are typically lower in warmer months, offering less dilution of the waste. In addition, the concentration of oxygen in water decreases as the salinity or salt concentration increases (because of the salting-out effect discussed in Section 3.1.3), although this is typically only of importance in marine or estuarine conditions.

5.5.2 The Oxygen Deficit

The *oxygen deficit* (D, mg O_2/L) is defined as the departure of the ambient dissolved-oxygen concentration from saturation, and is defined as:

$$D = DO_{sat} - DO_{act} \tag{5-36}$$

where DO_{act} is the ambient or measured dissolved-oxygen concentration (mg O_2/L). The calculation is illustrated in Example 5.16.

EXAMPLE 5.16. OXYGEN DEFICIT

Determine the dissolved-oxygen deficit, D, at 20°C for a stream with an ambient dissolved-oxygen concentration of 5 mg O_2/L.

SOLUTION

From Example 5.15, the DO_{sat} at 20°C was determined to be 9.1 mg O_2/L. Applying knowledge of the oxygen deficit (shown mathematically in Equation 5-36) yields:

$$D = 9.1 - 5 = 4.1 \text{ mg } O_2/L$$

See that the actual DO in this case is 5 mg/L, which is below the saturation level. There may be organic matter or ammonia nitrogen in this stream that is causing a depletion of oxygen.

Note that negative deficits may also occur when ambient oxygen concentrations exceed the saturation value. This happens in lakes and rivers under quiescent, nonturbulent conditions when algae and macrophytes are actively photosynthesizing and thus producing dissolved oxygen.

5.5.3 The Oxygen Mass Balance

It is now apparent that the BOD exertion expected following discharge of a waste to a stream (e.g., 29.5 mg O_2/L over 50 km in Example 5.14) may exceed the oxygen resources of a system, even at saturation (e.g., 7.6 to 14.6 mg O_2/L). The shortfall (oxygen present minus oxygen required) must be made up through atmospheric exchange, that is, reaeration. Sometimes the demand for deoxygenation exceeds the potential for reaeration, and anaerobic conditions develop. The dynamic interplay between the oxygen source (reaeration) and sink (deoxygenation) terms can be examined through a mass balance on oxygen in the river. Deoxygenation occurs as BOD is exerted and is well described by Equation 5-29. The rate of reaeration is proportional to the deficit and is also described using first-order kinetics:

$$\frac{dO_2}{dt} = k_2\,D - k_1\,L \tag{5-37}$$

Here the in-stream deoxygenation rate coefficient (k_1, day^{-1}) is comparable to (and for the purposes of this book the same as) the laboratory or "bottle" CBOD reaction rate coefficient (k_L), but also includes in-stream phenomena, for example, sorption and turbulence and roughness effects. The reaeration rate coefficient (k_2, day^{-1}) varies with temperature and the turbulence (velocity and depth) of the stream and ranges from approximately 0.1 to 1.2 day^{-1}.

In practice, the mass balance is written in terms of deficit:

$$\frac{dD}{dt} = k_1\,L - k_2\,D \tag{5-38}$$

which simply reverses the order of the source–sink terms. Equation 5-38 can be integrated, yielding an expression that describes the oxygen deficit at any location downstream of an arbitrarily established starting point, for example, the point where a waste is discharged to a river:

$$D_t = \frac{k_1\,L_0}{(k_2 - k_1)}\left(e^{-k_1 t} - e^{-k_2 t}\right) + D_0\,e^{-k_2 t} \tag{5-39}$$

where L_0 and D_0 are the ultimate CBOD and oxygen deficit at the starting point (initial, $t = 0$), and D_t is the oxygen deficit at some downstream location. The notation t refers to time of travel, defined here as the time (in days) required for a parcel of water to travel a distance x downstream. Therefore, $t = x/U$, where x is distance downstream (km) and U is the river velocity (km/d). The time–distance relationship permits expression of the analytical solution for the oxygen deficit in terms of x, the distance downstream of the starting point:

$$D_x = \frac{k_1\,L_0}{(k_2 - k_1)}\left(e^{-k_1 x/U} - e^{-k_2 x/U}\right) + D_0\,e^{-k_2 x/U} \tag{5-40}$$

Equation 5-40 is called the *Streeter-Phelps model* and was developed in the 1920s for studies of pollution in the Ohio River.

5.5.4 The Dissolved-oxygen Sag Curve and the Critical Distance

The discharge of oxygen-demanding wastes to a river yields a characteristic response in oxygen levels that is termed the *dissolved-oxygen sag curve*, depicted in Figure 5-24. It can be noted from that figure that a typical dissolved-oxygen sag curve has three phases of response: (1) an interval where dissolved-oxygen levels fall because the rate of deoxygenation is greater than the rate of reaeration ($k_1 \times L > k_2 \times D$); (2) a minimum (termed the *critical point*) where the rates of deoxygenation and reaeration are equal ($k_1 \times L = k_2 \times D$); and (3) an interval where dissolved-oxygen levels increase (eventually reaching saturation) because BOD levels are being reduced and the rate of deoxygenation is less than the rate of reaeration ($k_1 \times L < k_2 \times D$).

The location of the critical point and the oxygen concentration at that point are of greatest interest because this is where conditions are the worst. Design calculations for remediation of water quality are based on this location because if standards are met for the critical point, they will be met elsewhere. The location of the critical point can be determined by first using Equation 5-41 to determine the critical time, and then multiplying the critical time by the river velocity to determine the critical distance:

$$t_{crit} = \frac{1}{k_2 - k_1} \ln\left[\frac{k_2}{k_1}\left(1 - \frac{D_0(k_2 - k_1)}{k_1 L_0}\right)\right] \quad \textbf{(5-41)}$$

The oxygen deficit at the critical distance can then be found by substituting the critical time into Equation 5-39 to determine the oxygen deficit at the critical distance. Knowledge of DO_{sat} then allows determination of the actual DO concentration at the critical distance. Example 5.17 illustrates this approach and suggests opportunities for its application in river management.

EXAMPLE 5.17. DETERMINING FEATURES OF THE DO SAG CURVE

After receiving the discharge from a waste treatment plant, a river has a dissolved-oxygen concentration of 8 mg O_2/L and an ultimate CBOD of 20 mg O_2/L. The saturation dissolved-oxygen concentration is 10 mg O_2/L, the deoxygenation rate coefficient, k_1, is 0.2 day^{-1}, and the reaeration rate coefficient, k_2, is 0.6 day^{-1}. The river travels at a velocity of 10 km/day. Calculate the location of the critical point (time and distance) and the oxygen deficit and concentration at the critical point.

SOLUTION

First determine the initial DO deficit at the point of discharge using Equation 5-36:

$$D_0 = DO_{sat} - DO_{act} \qquad D_0 = 10 - 8 = 2 \text{ mg } O_2/\text{L}$$

Then, use Equation 5-41 to determine the critical time and knowledge of the river's velocity to determine the critical distance:

$$t_{crit} = \frac{1}{k_2 - k_1} \ln\left[\frac{k_2}{k_1}\left(1 - \frac{D_0(k_2 - k_1)}{k_1\, L_0}\right)\right]$$

$$t_{crit} = \frac{1}{0.6 \text{ day}^{-1} - 0.2 \text{ day}^{-1}} \times$$

$$\ln\left[\frac{0.6 \text{ day}^{-1}}{0.2 \text{ day}^{-1}} \times \left(1 - \frac{2 \text{ mg/L} \times (0.6 \text{ day}^{-1} - 0.2 \text{ day}^{-1})}{0.2 \text{ day}^{-1} \times 20 \text{ mg/L}}\right)\right]$$

$$t_{crit} = 2.2 \text{ day} \qquad x_{crit} = 2.2 \text{ day} \times 10 \frac{\text{km}}{\text{day}} = 22 \text{ km}$$

Finally, use Equation 5-39 to determine the oxygen deficit and Equation 5-36 to determine the actual dissolved-oxygen concentration for the critical time as just calculated:

$$D_t = \frac{k_1\, L_0}{(k_2 - k_1)}\left(e^{-k_1 t} - e^{-k_2 t}\right) + D_0\, e^{-k_2 t}$$

$$D_t = \frac{0.2 \text{ day}^{-1} \times 20 \text{ mg/L}}{(0.6 \text{ day}^{-1} - 0.2 \text{ day}^{-1})} \times \left(e^{-0.2 \text{ day}^{-1} \times 2.2 \text{ day}} - e^{-0.6 \text{ day}^{-1} \times 2.2 \text{ day}}\right)$$

$$+ \, 2 \text{ mg/L}\; e^{-0.6 \text{ day}^{-1} \times 2.2 \text{ day}}$$

$$D_t = 4.3 \frac{\text{mgO}_2}{L} \qquad DO = 10.0 - 4.3 = 5.7 \ \text{mgO}_2/L$$

Note that this deficit occurs 22 km downstream from the point of initial discharge.

Engineers will use calculations such as those presented above in concert with waste treatment design protocols to safeguard a receiving water. If the dissolved oxygen at the critical point violates a water quality standard, additional design modifications are necessary. Engineers will seek to identify a BOD for the waste discharge which will result in a mixing basin BOD (see Example 5.14) which does not lead to a violation.

5.6 MATERIAL FLOW IN ECOSYSTEMS

Earlier it was pointed out that, given an appropriate physical setting (e.g., light, temperature), organisms require only two things from the environment: (1) *energy* to provide power, and (2) *chemicals* to provide substance. The role of energy in driving ecosystem function was discussed in Section 5.3, where it was noted that energy does not cycle, but rather flows through ecosystems, becoming dissipated as it is converted to heat and lost as useful energy forever. In contrast, chemicals are used over and over again within an ecosystem. That is, they cycle and thus need not be imported for the ecosystem to continue to function. However, the rate of cycling can be an important determinant of ecosystem activity.

Table 5-4. The Chemical Elements that Make Up Living Things*

Major Macronutrients (>1% dry organic wt)	Relatively Minor Macronutrients (0.2–1.0% dry organic wt)	Micronutrients (<0.2% dry organic wt)
Carbon, hydrogen, nitrogen, oxygen, phosphorus	Calcium, chlorine, copper, iron, magnesium, potassium, sodium, sulfur	Aluminum, boron, bromine, chromium, cobalt, fluorine, gallium, iodine, manganese, molybdenum, selenium, silicon, strontium, tin, titanium, vanadium, zinc

From Kupchella and Hyland, 1986.

*Note that many of these elements are included among the major ions represented in surface water, groundwater, and the ocean (Chapter 2).

The organisms that make up the biosphere need a variety of chemicals to sustain growth and life function. However, of the more than 100 chemical elements in the Periodic Table, only 30 are constituents of living things. Table 5-4 shows that some elements are required in relatively large amounts and are termed *macronutrients*; others, needed in only trace quantities, are called *micronutrients*. Each of these, in various chemical forms (e.g., carbon, hydrogen, and oxygen in organic matter, nitrogen and sulfur in amino acids, phospholipids in cell membranes, and in enzyme systems), is synthesized into biomass.

The stoichiometric ratio of the elements in living things is relatively constant across the spectrum of organisms that make up the biosphere. Consider, for example, the contribution of selected chemicals (C, H, O, N, P, and S) to the composition of the seed-bearing land plants. Based on molar ratios presented in Table 5-5, plant material could be represented stoichiometrically as: $C_{388}H_{567}O_{277}N_{23}S_1P_1$ (note that the unconventional subscript of "1" for S and P is added for emphasis).

Despite the fact that some chemicals are required only in trace quantities (e.g., S and P), their availability may regulate the productivity of an entire ecosystem.

Table 5-5. Major Chemical Constituents of Seed-Bearing Land Plants

Element	% Dry Weight	Mole Ratio
Carbon	45	388
Oxygen	43	277
Hydrogen	5.5	567
Nitrogen	3.1	23
Sulfur	0.2	1
Phosphorus	0.3	1

From Lerman, 1988.

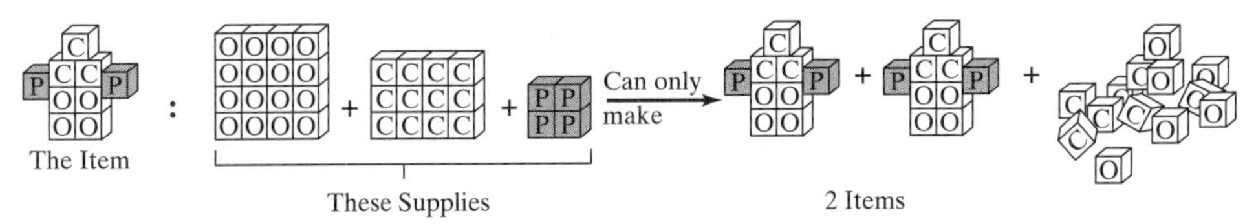

Figure 5-25. An illustration of the Limiting Nutrient Concept or Liebig's Law of the Minimum. Here, the production of an "item" such as an algal cell is limited by the availability of phosphorus. Although carbon and oxygen are present in excess, only two "items" can be produced. Addition of phosphorus, e.g. through cultural activity, would permit the production of additional "items," a potential problem in some environments as discussed in the next section.

The "Limiting Nutrient Concept" or "Liebig's Law of the Minimum" states that "the yield of any organism will be determined by the abundance of the substance that, in relation to the needs of the organism, is least abundant in the environment." Figure 5-25 illustrates the principle of nutrient limitation. This concept explains why the Monod model (Section 5.2.2.3) need only focus on a single substrate in simulating resource-limited population growth.

Therefore, it is the availability of these chemical elements ("least abundant in the environment in relation to the needs of the organism") that governs ecosystem production. In theory, chemical elements should never be in short supply in ecosystems because they are not "used up," but rather are continuously cycled. In practice, however, the availability of an element can be limited (1) by its chemical form, (2) by its physical location, and (3) by the rate at which it is cycled. Chemical elements exist in the environment in a variety of chemical forms. For example, a common organic form of carbon is sugar ($C_6H_{12}O_6$), and common inorganic forms of carbon include carbon monoxide (CO), bicarbonate (HCO_3^-) and carbon dioxide (CO_2). Animals can utilize sugar as a carbon source, but cannot make use of the others listed here. Plants, on the other hand, require CO_2 as their carbon source and cannot make use of external supplies of sugar. Thus *chemical form* is important.

Most organisms occupy a "thin" film of air, water, and soil near the Earth's surface. Elements may be lost to the biosphere for short or (geologically) long periods when they reside in the upper atmosphere, the deep ocean, or are incorporated into rocks. For example, nitrogen can be present in the atmospheric reservoir as N_2 gas, in organic forms that constitute detritus or living organisms, or in inorganic forms in rocks, soil, and water. Thus, *physical location* is important.

As organisms die and become subject to decomposition, the complex compounds that constitute biomass are broken down into simpler units, yielding hydrocarbons (primarily containing C, H, and O) and "releasing" other nutrients (N, P, S, etc.) in soluble forms (e.g., NH_4^+, $H_2PO_4^-$, HS^-). These simpler compounds are then subject to further microbial and chemical transformation and may ultimately become available once again for uptake by living organisms.

The process of cycling (uptake, assimilation, decay, transformation, and release) is common to all living things and extends to most components of the

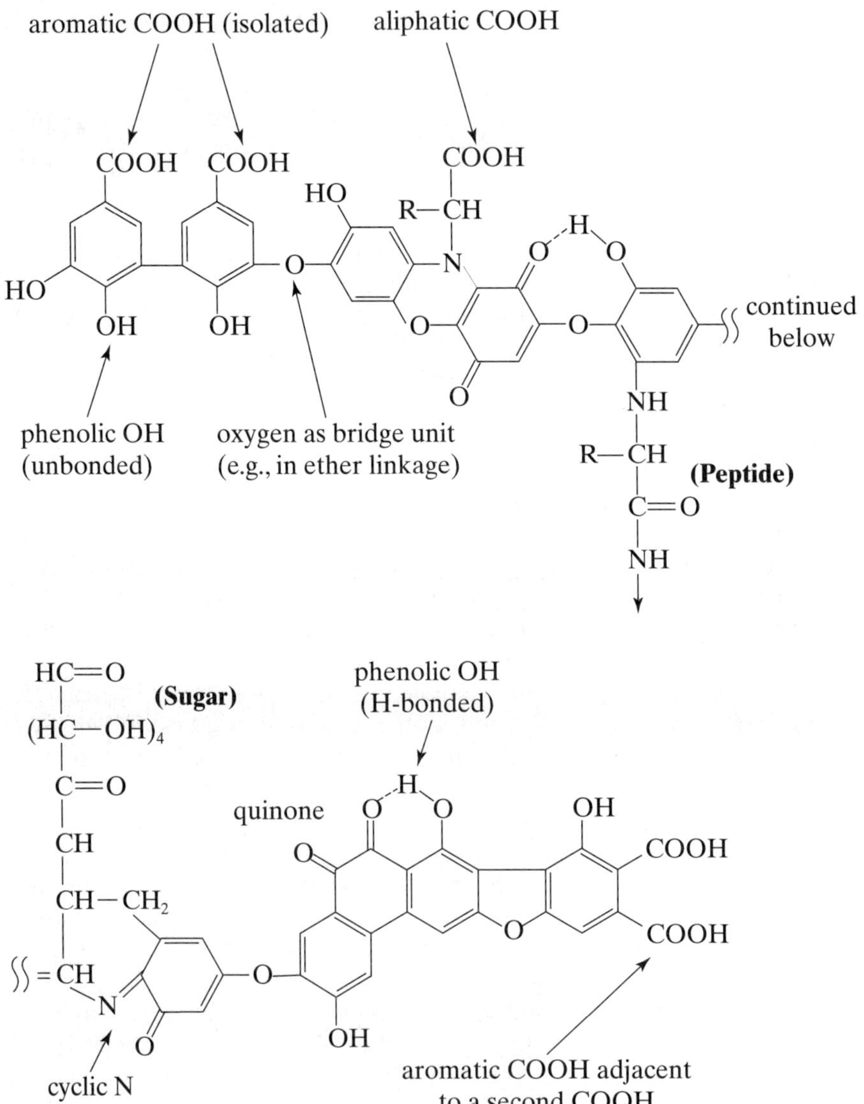

Figure 5-26. Model of a "simple" humic acid with a molecular weight of approximately 1,500. Humic acids range in size from molecular weight of 100 to 1,000,000! These natural organic compounds are found in many environments, including surface waters, soils, and activated sludge. Note the complex chemical structure and functional groups such as carboxyl (—COOH) and phenolic (—OH). These functional groups can deprotonate, releasing the H^+, which results in a negatively charged functional group. These groups have the ability to complex commonly occurring cations such as Ca^{2+} and Mg^{2+} as well as heavy metals such as Pb^{2+}, Cu^{2+}, and Cd^{2+}. This can result in increases or decreases in organism growth as well as reductions in free ion metal toxicity, depending on the strength of the cation/humic acid complex formed. In addition, note the many hydrophobic aromatic groups on the humic acid which a hydrophobic organic chemical can partition into. (Figure obtained from *Humus Chemistry: Genesis, Composition, Reactions*, Stevenson, 1994. Reprinted by permission of John Wiley & Sons, Inc.)

abiotic environment (e.g., mineral reserves). Elements move from one reservoir to another in cycles. Cycling may involve long, complex loops or short loops and thus *rate of recycle* is important.

Not all of the organic material in plants (and to a slight degree in animals) is readily decomposed. The relatively stable (refractory) organic matter is termed "humus" (perhaps thought of as the organic "skeleton") which accumulates in soils (e.g., the dark, rich soil of a forest floor). Humus is also the end product of composting yard wastes, paper, and food. This leads to the commonly noted "earthy" smell of compost. While the material sources and microbial communities differ, the end result is comparable.

Figure 5-26 shows a diagram of the chemical structure of a model soil humic acid which illustrates the complex structure of natural organic matter. Humus is important to soils for many reasons including: a) its large surface area which aids in water retention, b) its numerous carboxyl (—COOH) and phenolic (—OH) groups which can dissociate H^+ and result in negative sites which can complex nutrient cations (e.g., K^+, Mg^{2+}, Ca^{2+}) and heavy metals, c) its ability, in some cases, to provide a "slow-release" or long-term source of nutrients such as N and P (however the potential to use humus as a fertilizer is typically very limited).

Humus is found in any degraded or partially degraded source of organic material. In addition, in municipal wastewater, humics may contribute to the difference between the measured COD and BOD, because they are not readily degraded by microorganisms (i.e., $COD > BOD$).

Living things are important participants in cycles, mediating the movement of chemicals from one reservoir to another, from one location to another, and from one chemical form to another. Because of the involvement of organisms, these are termed *biogeochemical cycles*. Of the 30 chemical elements listed in Table 5-4 as being constituents of living things, five are of particular importance in environmental engineering (C, O, N, P, and S). In addition, the hydrologic cycle is of interest because it plays an important role in moving chemical elements through the ecosphere. Attention will now be directed to a consideration of these key cycles.

5.6.1 Hydrologic Cycle

Figure 5-27 depicts the hydrologic cycle. Water moves among various environmental reservoirs through *precipitation*, *runoff*, and *evapotranspiration*—component processes in the *hydrologic cycle*. Evapotranspiration includes evaporation (conversion of the Sun's energy to latent heat as it removes water from wet surfaces converting it to water vapor) and transpiration (passage of water vapor from the leaves of plants to the atmosphere). Quantification of these processes permits the development of a water balance (a special case of the mass balance, Section 4.1), illustrated here in microcosm for leachate generation in a landfill (Example 5.18). The concept of the hydrologic cycle could be presented elsewhere in this book. We have chosen to present it here because we feel it is a major reason for determining how materials flow through aquatic, terrestrial, and subsurface ecosystems.

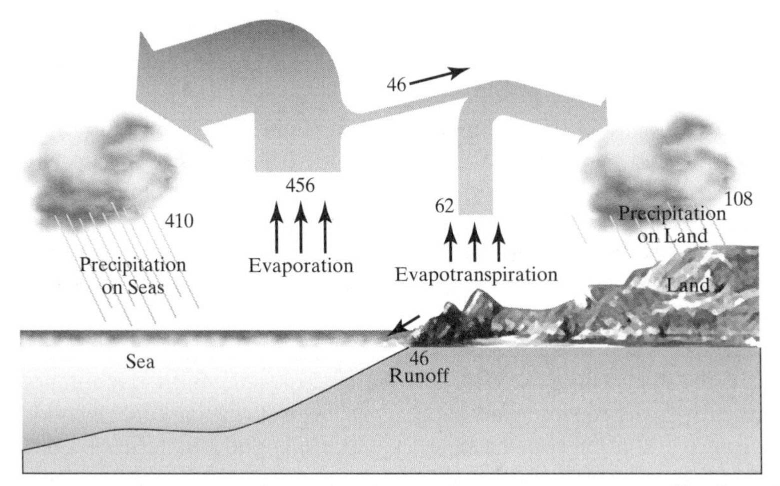

Figure 5-27. The hydrologic cycle (units of water transfer are 10^{12} m^3/year). (Data from Budyko, 1974.)

EXAMPLE 5.18. WATER BALANCE ON A LANDFILL TO DETERMINE LEACHATE GENERATION

The amount of water that leaches through and out of a sanitary landfill is an important determinant of the potential for contamination of adjacent surface water and groundwater. This "leachate" is usually very high in BOD, COD, N, P, and may also contain some metals and toxic organic chemicals. A simplified water balance on a landfill can be constructed to determine some of the parameters that influence leachate production. The water balance can be written as:

$$\frac{dS}{dt} = P - R - E - L$$

where S is storage, the volume of water present as moisture in the landfill; P is the annual precipitation volume (cm rain/year times the landfill surface area); R is runoff, the fraction of the annual precipitation that leaves via the sloped landfill surface (often expressed as a dimensionless "runoff coefficient" times P); E is evapotranspiration, the volume of water lost due to evaporation and transpiration by plants (cm/year times the landfill surface area); and L is the leachate volume produced (cubic cm per year). Initially the only moisture in the landfill is that added in the refuse and any cover soils. Over time, however, some precipitation percolates into the landfill. The landfill can then be envisioned as a sponge, which over a long time span will reach its water-holding capacity. At this time no additional water can be absorbed and a steady state is reached where dS/dt equals 0 and the volume of leachate generated is equal to:

$$L = P - R - E$$

Engineers working on regional problems have little control over precipitation amounts, but can influence runoff in designing the landfill cover and roughness. Evapotranspiration is influenced by the local climate and the type of vegetation that develops on the landfill cover.

Calculate the volume of leachate generated annually (in liters) at a 10-acre landfill where the annual precipitation is 30 cm/yr, the landfill's surface is a sandy soil with a 7% slope (corresponding runoff coefficient of 0.17), and E is 10 cm/yr.

SOLUTION

Assume steady-state conditions have been met and use the preceding equation to solve for L:

$$L = \left[\frac{30\text{ cm}}{\text{y}} - 0.17\left(\frac{30\text{ cm}}{\text{y}}\right) - \frac{10\text{ cm}}{\text{y}}\right] \times 10\text{ acres} \times \frac{\text{m}}{100\text{ cm}}$$

$$\times \frac{4{,}046.8\text{ m}^2}{\text{acre}} \times \frac{1{,}000\text{ L}}{\text{m}^3}$$

$$= 6.0 \times 10^6\text{ L per year}$$

Note that this large amount of leachate must be contained or treated. Table 2-7 listed the concentration of typical leachate constituents. Remember, leachate has relatively high concentrations of many compounds. Therefore, it can potentially contaminate groundwater if not properly managed. Many local wastewater-treatment plants accept landfill leachate, but operators carefully meter its delivery into the plant in order to not upset the delicate biological treatment processes.

Almost all of the world's water (>95%) is located in the oceans, where high concentrations of salts make it virtually of no use for drinking or irrigation. A relatively small fraction (10%) of the water leaving the ocean through evaporation reaches land, falling there together with water originating as evapotranspiration from land surfaces and evaporation from inland waters. Nearly 60% of the water falling on land is lost through evapotranspiration, with the balance moving back to the oceans as surface runoff or groundwater. Table 5-6 describes the various stocks of water on Earth. Although only a small ($\ll$1%) percentage of the Earth's water is *held in* the atmosphere, extremely large volumes are *cycled through* the atmosphere, $\sim 1.6 \times 10^{12}$ m^3/day. The residence time of water in the atmosphere is short, on the order of 8 days, whereas the residence time of the ocean is about 2,700 years. Thus it should be apparent that the role of the atmosphere in the hydrologic cycle is one of circulation, not storage.

It is estimated that only 0.3% of the water stock of the Earth is freshwater available for human use. This number drops by a factor of 100 if inaccessible or

Table 5-6. Stocks of Water on Earth

Source	Volume (km^3)	% of Total Water	% of Freshwater
Oceans	1,338,000,000	96.5	
Groundwater			
Fresh	10,530,000	0.76	30.1
Saline	12,870,000	0.93	
Soil Moisture	16,500	0.0012	0.05
Polar Ice	24,023,500	1.7	68.6
Other Ice and Snow	340,600	0.025	1.0
Lakes			
Fresh	91,000	0.007	0.26
Saline	85,400	0.006	
Marshes	11,470	0.0008	0.03
Rivers	2,120	0.0002	0.006
Biological Water	1,120	0.0001	0.003
Atmospheric Water	12,900	0.001	0.04
Total Freshwater	35,029,210	2.5	100.0
Total Water	1,385,984,610	100.0	

From UNESCO, 1978.

highly polluted stocks are omitted. The task of providing a safe and palatable drinking water supply falls to individuals such as civil and environmental engineers and hydrogeologists. There are several items to consider when selecting a particular water source as a supply for drinking, industrial, or agricultural applications. Surface waters such as rivers have relatively large flows, but are often turbid (suspended solids concentrations of 10 to >100 mg/L) and may be contaminated with domestic, industrial, and agricultural wastes. Thus, they require extensive treatment before they can be used for drinking. Groundwater often has a higher quality (i.e., less contamination) than surface water because of the filtering capacity of soil; however, it may have high concentrations of dissolved ions (Ca^{2+}, Mg^{2+}, Fe^{2+}, HS^-) that can lead to scale formation or taste and odor problems in drinking water. Groundwater formations (aquifers) also may be less capable of meeting flow demands than are surface waters. This is evidenced in many parts of the world (including the United States) where the rate at which groundwater is pumped exceeds the rate of natural recharge. This lowers groundwater tables, resulting in the depletion of a valuable resource as well as creating land subsidence problems. Furthermore, if contaminated, an aquifer may be very difficult and costly to clean up. The oceans offer an abundant and relatively clean source of water; however, the desalination process is prohibitively expensive and is applied primarily in the Middle East where oil revenues support treatment costs. Water reclamation is the processing of wastewater for reuse in municipal, industrial, or agricultural applications. As a potential drinking water source, reclaimed water is relatively expensive and suffers from aesthetic problems. However, reclamation is becoming more important as cities expand and population in dry climates such as the American West increases.

There is a fixed amount (supply) of freshwater on Earth; however, the demand continues to grow as population and per capita use increase. In the future, arid regions will likely be faced with water supply issues of increased complexity. All these issues require new engineering solutions, including the most obvious, water conservation. Even with water supplies of sufficient quantity, poor sanitation in much of the developing world prevents access to water supplies free of pathogens. Thus, the need for engineering services in these nations is rapidly expanding.

5.6.2 Carbon Cycle

Key features of the carbon cycle include photosynthesis and respiration, presented previously as Equations 5-18 and 5-19, respectively. These processes are important to ecosystem energy balances, primary production, and oxygen levels in lakes and rivers. Remember that photosynthesis is carried out by plants and some bacteria, and that respiration is carried out by all organisms, including those that photosynthesize. The release of carbon dioxide through fossil-fuel combustion (a kind of "cultural" respiration) is of concern because of its potential for contributing to changes in the global climate (greenhouse effect). Basically, the "fixed carbon reservoir" present as long-buried plants and animals (fossil fuels) is being extracted and released to the atmosphere as CO_2, free to enter the dynamic global carbon cycle once again.

The carbon cycle is a major feature of the biochemistry of wastewater treatment. The process of anaerobic digestion of sludge provides an excellent demonstration of carbon cycling as mediated by bacterial action/interaction. Residuals (sludge) from wastewater-treatment plants are often subjected to anaerobic digestion prior to ultimate disposal (e.g., land application or incineration). Sludge treatment through anaerobic digestion further degrades the organic portion of the sludge solids, stabilizing the residuals. The end products of anaerobic digestion are (1) gas (approximately two-thirds as methane (CH_4) and one-third as carbon dioxide (CO_2) with trace amounts of hydrogen sulfide (H_2S); (2) stabilized (refractory) organics; and (3) new microbial biomass.

There are two major classes of bacteria involved in the anaerobic digestion process: acid-formers and methanogens. Acid-formers are responsible for converting complex organic compounds into simpler organic acids. Methanogens convert the organic acids produced by acid-formers to methane and carbon dioxide. The methanogens are unable to use carbohydrates, proteins, and other complex hydrocarbons; thus, they are dependent upon the acid-formers for substrate (carbon source). However, the methanogens are quite sensitive to pH levels. If the continued production of organic acids (largely acetic acid ~ vinegar) exceeds their consumption (and the buffering capacity of the system), the system becomes acidic and the growth of methanogens is inhibited, a condition known as "pickled" sludge. Methanogens are strict anaerobes that produce methane by fermentation. Similar reactions take place more slowly in landfills, which are, in effect, solid-waste digesters. The capture and utilization of methane from landfills as a fuel has received increased attention in recent years and is examined in Example 5.19.

EXAMPLE 5.19. METHANE PRODUCTION AT A LANDFILL

Significant quantities of methane can be produced from solid-waste landfills. The production of methane occurs by the anaerobic decomposition of organic matter yielding methane and carbon dioxide. If the solid waste is assumed to have the chemical composition of $C(H_2O)$, the reaction can be written as:

$$2C(H_2O) \rightarrow CH_{4(g)} + CO_{2(g)}$$

Gases generated from landfills are typically either vented to the atmosphere or collected for energy production. Because methane's specific gravity is less than air, it can accumulate below buildings or in other enclosed spaces (e.g., basements) where it can be a safety hazard because of its explosiveness.

The U.S. Environmental Protection Agency has reported that in 1995, solid-waste production in the United States was 4.3 lb/person-day. Calculate the potential methane production in one year for a landfill that serves a community of 50,000 people. Assume that none of the solid waste is recycled, that all the solid waste can decompose rapidly, and that the density of methane equals 0.717 kg/m^3.

SOLUTION

Calculate the amount of solid waste produced annually, then use the reaction stoichiometry for methane production to determine the volume of methane produced:

$$\frac{4.3 \text{ lb}}{\text{person-day}} \times 50{,}000 \text{ people} \times \frac{365 \text{ days}}{\text{yr}} \times \frac{0.454 \text{ kg}}{\text{lb}} = 3.56 \times 10^7 \text{ kg solid waste}$$

Assume that all this solid waste is in the form of $C(H_2O)$ and use the stoichiometry of the assumed reaction to determine the mass of methane produced:

$$\frac{3.56 \times 10^7 \text{ kg } CH_2O}{\text{yr}} \times \frac{\text{mole } CH_2O}{30 \text{ g}} \times \frac{\text{mole } CH_4}{2 \text{ mole } CH_2O} \times \frac{16 \text{ g } CH_4}{\text{mole } CH_4} = \frac{9.5 \times 10^6 \text{ kg } CH_4}{\text{yr}}$$

Use the density to determine the volume of CH_4 produced annually:

$$\frac{9.5 \times 10^6 \text{ kg CH}_4}{\text{yr}} \times \frac{\text{m}^3 \text{ CH}_4}{0.717 \text{ kg}} = \frac{1.3 \times 10^7 \text{ m}^3 \text{ CH}_4}{\text{yr}}$$

This estimate is on the high side. We assumed that all of the solid waste was biodegradable under anaerobic conditions and that it was degraded rapidly. In fact, many biodegradable components of municipal solid waste may be composted, recycled, or sewered. Also, the rate of anaerobic decomposition of organic matter (and methane production) is slower than under aerobic conditions and the typical design of sanitary landfills actually minimizes biological degradation by minimizing infiltration of water and dissolved oxygen. Finally, the rate of methane production in municipal landfills is relatively rapid the first several years as the readily degradable material decomposes, and then tapers off over a period that can exceed 20 years.

5.6.3 Oxygen Cycle

The oxygen and carbon cycles are closely linked, as discussed previously in this chapter. The most important oxygen source is photosynthesis and the most important oxygen sink is respiration. Billions of years ago, photosynthetic production introduced oxygen to the atmosphere, permitting the evolution of aerobic organisms. Subsequent photoconversion of some of the oxygen to ozone led to the development of a high-altitude (stratospheric) ozone layer that screens out harmful ultraviolet radiation, permitting organisms to live at the surface of the Earth. Today, scientists and engineers are interested in ozone, a part of the oxygen cycle, because of its harmful health effects in the lower atmosphere (smog) and its beneficial effects in destroying UV radiation in the upper atmosphere. These two concepts were discussed in Section 3.6.

5.6.4 Nitrogen Cycle

The biogeochemical transformations embodied in the nitrogen cycle (Figure 5-28) are important in a variety of applications in both natural and engineered systems. As a result of their association with bacteria and plants, many features of the nitrogen cycle are linked with the oxygen and carbon cycles as well. Plants take up and utilize nitrogen in the form of ammonia or nitrate, chemicals typically in short supply in agricultural soils, thus leading to requirements for fertilization. Certain bacteria and plant (e.g., legumes and clover) species can also utilize atmospheric nitrogen (N_2), converting it to ammonia through a process termed *nitrogen fixation*. Plants incorporate ammonia and nitrate into a variety of organic compounds, such as proteins and nucleic acids, important to the energy budget of food webs. Organic nitrogen compounds released to the environment as de-

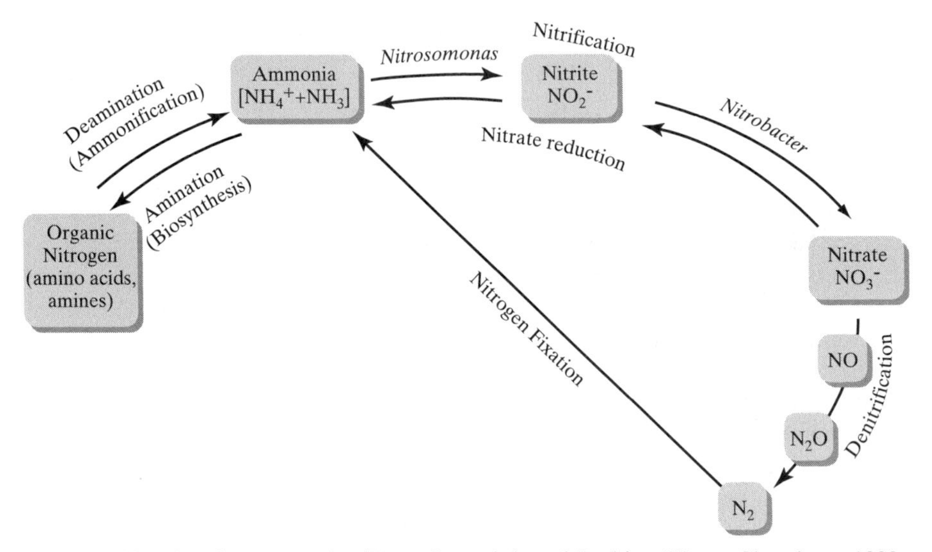

Figure 5-28. The nitrogen cycle. (From Snoeyink and Jenkins, Water Chemistry, 1980. Reprinted by permission of John Wiley & Sons, Inc.)

tritus (Section 5.3.2) support a complex biochemical cycling process (Equation 5-42) that has nitrate as its stable end product. The process is of environmental importance due to the potential production of toxic levels of ammonia and due to oxygen consumption in supporting the oxidation.

$$Organic - N \rightarrow NH_3 \rightarrow NO_2^- \rightarrow NO_3^- \qquad \textbf{(5-42)}$$

The conversion of organic-N (e.g., proteins) to ammonia, a process termed *ammonification*, is the first step in this process which occurs in lakes and rivers as well as in wastewater-collection, -conveyance, and -treatment systems. The subsequent bacterial conversion of ammonia to nitrite and nitrate, introduced previously as NBOD in Equation 5-28, is termed *nitrification*. Ammonia is toxic to aquatic life at high concentrations, and thus conversion to nitrate in waste treatment is desirable. The transformation of ammonia to nitrate requires oxygen (exertion of NBOD) and can lead to oxygen depletion in rivers and streams. Nitrification is an integral part of the waste-treatment process (where the required oxygen can be added), reducing the NBOD of the waste stream prior to discharge. Excessive nitrate in groundwater is of concern if the aquifer is to be used as a drinking water supply. Nitrate can pose a health threat to infants by interfering with oxygen transfer in the bloodstream. This problem is of increasing importance in agricultural regions where excessive fertilizer applications lead to the formation and accumulation of nitrate.

Nitrate is also readily utilized by plants and algae, and discharges to surface waters (especially in marine environments) can create nuisance algal growth and attendant water-quality problems, as discussed later in this chapter. Thus, it may

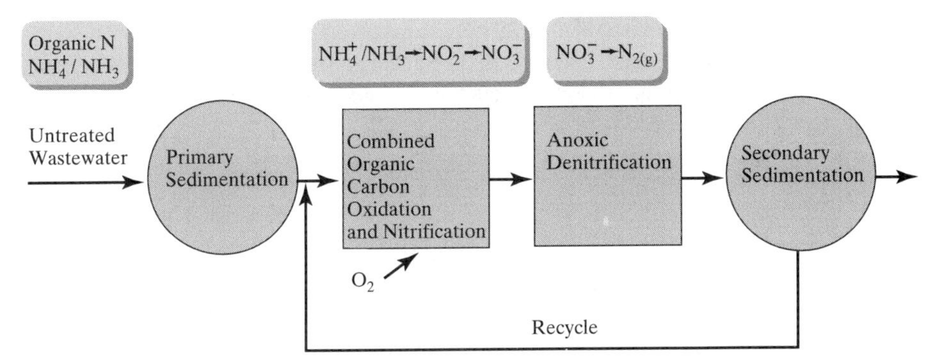

Figure 5-29. Schematic showing how a conventional municipal wastewater-treatment plant can be configured for removal of nitrogen. Also note the change in the chemical form of nitrogen as it passes through the treatment plant.

be desirable in wastewater treatment to convert nitrate to nitrogen gas with subsequent release to the atmosphere (which is already 79% N_2 by volume). The conversion process, termed *denitrification*, is microbially mediated and proceeds under anoxic conditions as follows (assuming $C(H_2O)$ as a representative source of organic carbon):

$$5C(H_2O) + 4NO_3^- + 4H^+ \rightarrow 2N_{2(g)} + 5CO_{2(g)} + 7H_2O \qquad \textbf{(5-43)}$$

In denitrification nitrate replaces oxygen as the electron acceptor in the oxidation of organic carbon. One of many configurations for achieving combined nitrification/denitrification in wastewater treatment for removal of nitrogen is presented as Figure 5-29. Provisions are sometimes made for adding a carbon source ($C(H_2O)$ in Equation 5-43; often methanol in practice) prior to the denitrification step, as organic carbon (BOD) levels may have been significantly reduced at this point in the process stream. Example 5.20 expands on this concept.

EXAMPLE 5.20. REMOVAL OF NITROGEN DURING WASTEWATER TREATMENT

Nitrogen can be removed from municipal wastewater by operating a portion of the plant to facilitate the denitrification process (see Figure 5-29 for a schematic). In this microbially mediated reaction, nitrate (NO_3^-) serves as the electron acceptor, organic carbon serves as the electron donor, and nitrogen gas (N_2) is a product. Organic carbon is supplied directly through the waste stream (that not degraded during aerobic treatment) or it is supplied externally (e.g., methanol, CH_3OH, is added).

Assume that a wastewater-treatment plant with a flow of 2 MGD (million gallons per day) must remove 22 mg N/L from its influent stream. In the early stages of the process stream, most of the nitrogen exists as organic nitrogen and ammonia. The organic nitrogen is largely converted to ammonia and the ammonia to nitrate as the waste proceeds through the plant, leaving essentially all of the nitrogen present as nitrate. Calculate the methanol addition (kg) required to remove all of the remaining nitrogen.

SOLUTION

The concentration of nitrate–nitrogen is reported as 22 mg N/L, which is the method of reporting concentration as a common constituent (see Section 2.5.2). Convert this value to mg NO_3^-/L, consistent with the form of the denitrification reaction presented as Equation 5-43:

$$\frac{22 \text{ mg N}}{\text{L}} \times \frac{\text{g}}{1{,}000 \text{ mg}} \times \frac{\text{mole N}}{14 \text{ g}} \times \frac{\text{mole } NO_3^-}{\text{mole N}} \times \frac{62 \text{ g}}{\text{mole } NO_3^-} \times \frac{1{,}000 \text{ mg}}{\text{g}} = \frac{97 \text{ mg } NO_3^-}{\text{L}}$$

The next step is to write out the balanced chemical equation for the denitrification reaction, substituting methanol for the generic carbon source presented in Equation 5-43. Here methanol serves as the electron donor and nitrate as the electron acceptor (the following equation assumes no biomass production). Then use the stoichiometry of this reaction to determine the required concentration of methanol:

$$5CH_3OH + 6NO_3^- \rightarrow 3N_{2(g)} + 5CO_{2(g)} + 7H_2O + 6OH^-$$

The stoichiometry of this equation can be used to determine the required methanol addition:

$$\frac{97 \text{ mg } NO_3^-}{\text{L}} \times \frac{\text{mole } NO_3^-}{62 \text{ g}} \times \frac{5 \text{ mole } CH_3OH}{6 \text{ mole } NO_3^-} \times \frac{32 \text{ g } CH_3OH}{\text{mole } CH_3OH} = \frac{42 \text{ mg } CH_3OH}{\text{L}}$$

Convert this concentration to a mass basis by multiplying by the plant flow:

$$\frac{42 \text{ mg } CH_3OH}{\text{L}} \times \frac{2 \times 10^6 \text{ gal}}{\text{day}} \times \frac{3.78 \text{ L}}{\text{gal}} \times \frac{\text{kg}}{10^6 \text{ mg}} = 320 \text{ kg methanol per day}$$

As will also be seen in Example 5.21, it can be costly to remove nutrients such as phosphorus and nitrogen from domestic sewage. Unfortunately, discharging nutrients to surface waters can adversely influence water quality.

In addition to the processes described in this section, oxides of nitrogen (termed NO_x) are released to the atmosphere through high-temperature combustion (auto exhaust and emissions from industry and power production). These chemicals are converted to nitric acid (HNO_3) in the atmosphere and return to the Earth's surface as acid deposition (see Chapter 1). Oxides of nitrogen also serve as precursors in the production of urban smog (see Section 3.6).

5.6.5 Phosphorus Cycle

Phosphorus is of special interest because it is the nutrient most commonly limiting plant growth in lakes. Excessive discharges of phosphorus can lead to nuisance algal growth, making lakes unpleasant and unavailable for a variety of uses. Phosphorus-bearing minerals are poorly soluble, and thus most surface waters naturally contain very little phosphorus. When phosphorus is mined and incorporated in cleaning agents, with subsequent discharge to surface waters, the biogeochemical cycle of this element is upset, vastly accelerating the rate at which phosphorus is routed through the environment.

Increased use of phosphorus in detergents in the 1950s and 1960s, passing largely untreated through primary (solids removal) and secondary (organic carbon removal) treatment plants, resulted in excessive loading to many surface waters and attendant nuisance algal growth. For example, massive nuisance algal growth in Lake Erie led to loss of oxygen (see Section 5.7.2 for how this influences oxygen levels), fish kills, and deterioration of beach aesthetics as mats of rotting algae washed on shore. Untreated domestic wastewater contains 5 to 15 mg P/L (in both organic and inorganic forms), concentrations more than two orders of magnitude greater than those desired for healthy surface waters (<0.02 mgP/L). Thus, significant P removal has become required as part of the wastewater-treatment process. This is accomplished by tertiary (advanced) treatment through chemical precipitation (with iron or aluminum salts) and/or enhanced biological uptake by microorganisms. Both of these methods convert soluble P into a particulate P, which is collected as sludge solids. Example 5.21 explains P removal by chemical precipitation in greater detail.

EXAMPLE 5.21. DETERMINING CHEMICAL DOSAGE TO REMOVE PHOSPHORUS AT A WASTEWATER-TREATMENT PLANT

A wastewater-treatment plant has an influent phosphorus concentration of 8 mg P/L and a flow of 50 MGD. Most of the influent phosphorus is in the dissolved form (see Section 2.5.2 for a discussion of different chemical forms of phosphorus). Modifications of the activated sludge process, utilizing alternating aerobic and anaerobic environments, can lead to enhanced microbial P uptake, doubling the P content of a typical cell and yielding substantial P removal. However, many wastewater-treatment plants require additional chemical treatment (e.g., precipitation) to achieve effluent standards. Phosphorus removal thus consists of the conversion of influent-dissolved P to a particulate form (as newly formed biomass or a chemical precipitate), which is then removed by sedimentation.

Assume that all the P entering a treatment plant is in the dissolved form and is to be removed by chemical precipitation with ferric chloride ($FeCl_3$) according to the following simplified chemical reaction:

$$Fe^{3+} + PO_4^{3-} \rightarrow FePO_{4(s)}$$

Each mole of $FeCl_3$ added yields one mole of ferric ion (Fe^{3+}), which removes one mole of PO_4^{3-} according to the stoichiometry of this reaction. Calculate the daily $FeCl_3$ addition (kg) required to meet an effluent standard of 1 mg P/L.

SOLUTION

The required phosphorus removal is 8 − 1 = 7 mg P/L. The corresponding $FeCl_3$ concentration is:

$$\frac{7 \text{ mg P}}{\text{L}} \times \frac{\text{g}}{1{,}000 \text{ mg}} \times \frac{\text{mole P}}{31 \text{ g P}} \times \frac{\text{mole } PO_4^{3-}}{\text{mole P}} \times \frac{\text{mole Fe}}{\text{mole } PO_4^{3-}}$$

$$\times \frac{\text{mole } FeCl_3}{\text{mole Fe}} \times \frac{162 \text{ g}}{\text{mole } FeCl_3} = \frac{0.037 \text{ g } FeCl_3}{\text{L}}$$

Convert the required concentration of ferric chloride to a daily mass by multiplying by the flow:

$$\frac{0.037 \text{ g } FeCl_3}{\text{L}} \times \frac{50 \times 10^6 \text{ gal}}{\text{day}} \times \frac{3.78 \text{ L}}{\text{gal}} \times \frac{\text{kg}}{1{,}000 \text{ g}} = \frac{7{,}000 \text{ kg}}{\text{day}}$$

The large daily requirement for ferric chloride carries with it substantial operating costs (both for raw chemical and the additional sludge generated), and thus encourages plant operators to optimize biological removal (as described earlier). The economics of treatment also support reductions in phosphorus inputs at the source, for example, in domestic wastewaters. This was accomplished in some locations through legislative mandates that required that soaps and detergents be phosphate-free. However, phosphorus removal at wastewater-treatment plants is still a major concern and a focus of attention in plant design, operation, and financing.

5.6.6 Sulfur Cycle

Like the oxygen and nitrogen cycles, the sulfur cycle (Figure 5-30) is to a large extent microbially mediated and is thus linked to the carbon cycle. Sulfur reaches lakes and rivers as organic-S, incorporated into materials such as proteins and as

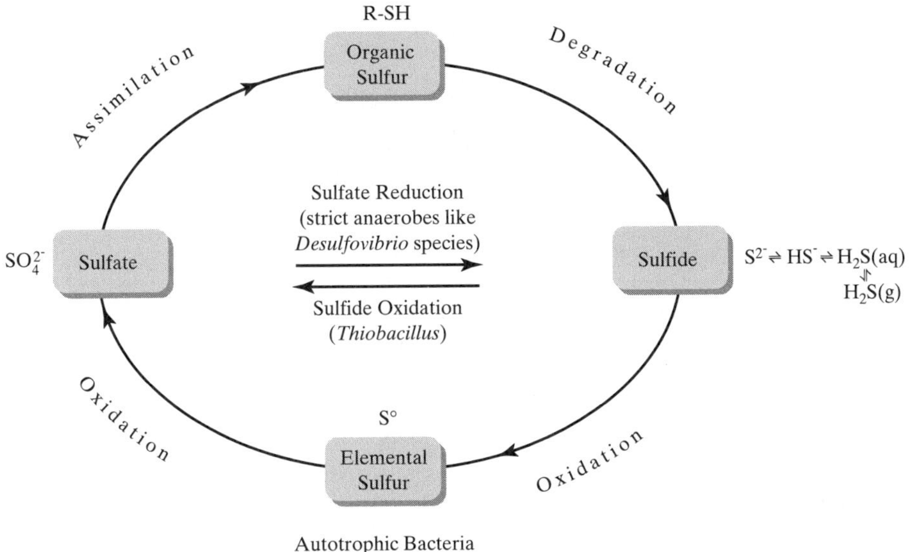

Figure 5-30. The sulfur cycle. Reduction of sulfate (SO_4^{2-}) to hydrogen sulfide (H_2S) can lead to odor problems at wastewater-treatment plants. Oxidation of reduced sulfur can lead to acidification and discoloration of surface waters.

inorganic-S, primarily in the form of sulfates (SO_4^{2-}). Under anaerobic conditions, hydrogen sulfide is produced through the decomposition of amino acids and through *sulfate reduction*, a redox reaction where sulfate replaces oxygen as the electron acceptor in the oxidation of organic carbon:

$$2C(H_2O) + SO_4^{-2} + 2H^+ \rightarrow H_2S + 2CO_2 + 2H_2O \qquad \textbf{(5-44)}$$

Hydrogen sulfide is malodorous and is toxic to aquatic life at very low concentrations. Under aerobic conditions, hydrogen sulfide is oxidized to sulfate, a process that consumes oxygen:

$$H_2S + 2O_2 \rightarrow 2H^+ + SO_4^{2-} \qquad \textbf{(5-45)}$$

Sulfur dioxide (SO_2) is released to the atmosphere through the combustion of fossil fuels, largely by industry and for power production. Sulfur dioxide is converted to sulfuric acid in the atmosphere and returns to the Earth's surface as acid deposition (see Chapters 1 and 3 for further discussion).

Pyrite (FeS_2) is often found in and around geologic formations of materials such as coal and heavy metals (silver, zinc, etc.) that are mined commercially. Exposure of pyrite to the atmosphere results in a three-step oxidation process.

This process is catalyzed by a variety of microorganisms such as *Thiobacillus thiooxidans*, *Thiobacillus ferrooxidans*, and *Ferrobacillus ferrooxidans*:

$$4FeS_2 + 14O_2 + 4H_2O \rightarrow 4Fe^{+2} + 8SO_4^{-2} + 8H^+ \quad \textbf{(5-46)}$$

$$4Fe^{2+} + 8H^+ + O_2 \rightarrow 4Fe^{3+} + 2H_2O \quad \textbf{(5-47)}$$

$$4Fe^{+3} + 12H_2O \rightarrow 4Fe(OH)_{3(s)} + 12H^+ \quad \textbf{(5-48)}$$

which yields mine drainage rich in sulfate, acidity, and ferric hydroxides (a yellowish orange precipitate or floc termed "yellow-boy"). While the sulfate is rather innocuous, acidity lowers the pH of the surface waters (often to levels severely impairing water quality) and the floc covers stream beds, eliminating macroinvertebrate habitat. Figure 5-31 depicts a stream impacted by acid mine drainage. In addition, the low pH of the water dissolves rocks and minerals, releasing hardness and total dissolved solids.

Figure 5-31. Pyrite (FeS_2) is often found in and around geologic formations of desirable materials such as coal, and heavy metals (silver, zinc, etc.), which are mined. The exposure of pyrite to the open atmosphere results in an oxidation process where the products are sulfate (rather innocuous), acidity, and ferric iron hydroxides (a yellowish orange precipitate or floc termed "yellow-boy"). Acidity lowers the pH of surface waters receiving mine runoff, often to levels that severely impair water quality. The floc covers the stream bed creating a habitat unfit for macroinvertebrates. (Thomas Braise/Tony Stone Images/New York, Inc.)

Table 5-7. Chemical Composition of Fuels (% by weight)

Element	Coal (bituminous)*	Crude Oil†	Processed Natural Gas†,‡
C	80	83	75
H	5	13	23
O	5	1	1
N	2	1	1
S	1	2	trace-0.2

*From Lerman after H. J. M. Bowen. 1966. *Trace Elements in Biochemistry*, Academic Press, London. (The remaining 7% is ash.)

†From Lerman after A. I. Levorsen. 1967. *Geology of Petroleum*, 2nd ed., Freeman, San Francisco.

‡Information provided by Wisconsin Power and Light Company.

Natural gas is often referred to as a "clean" energy source. Table 5-7 presents the chemical composition of several fuels, permitting comparison of sulfur content. Note that the carbon content is similar ($\pm 10\%$), but that the sulfur content is an order of magnitude greater in coal and oil than in natural gas. The burning of natural gas therefore produces fewer oxides of sulfur and may thus be considered a cleaner fuel.

5.7 LAKES: NUTRIENTS AND EUTROPHICATION

The biogeochemical cycles described in the preceding section play a major role in governing water quality in lakes. Here, the relationship between the physical environment of lakes, cultural perturbations of biogeochemical cycles, and water quality are considered. Mathematical models supporting lake-management decisions are introduced, and engineered works supporting the protection and restoration of lakes are described.

5.7.1 Thermal Stratification of Lakes and Reservoirs

A major difference between lakes and rivers lies in their mass-transport characteristics. In temperate latitudes, lakes undergo *thermal stratification* twice annually, dividing the system into layers and restricting mass transport. Periods of stratification alternate with periods of complete mixing where mass transport is at a maximum. The restriction of mass transport during stratification influences the cycling of many chemical species (e.g., iron, oxygen, and phosphorus) and can have profound effects on water quality.

The process of thermal stratification is driven by the relationship between water temperature and density (Figure 5-32). Of particular importance here is the fact that the maximum density of water occurs at 3.94°C rather than at 0°C. Thus ice floats and lakes freeze from the top down, instead of from the bottom up, as

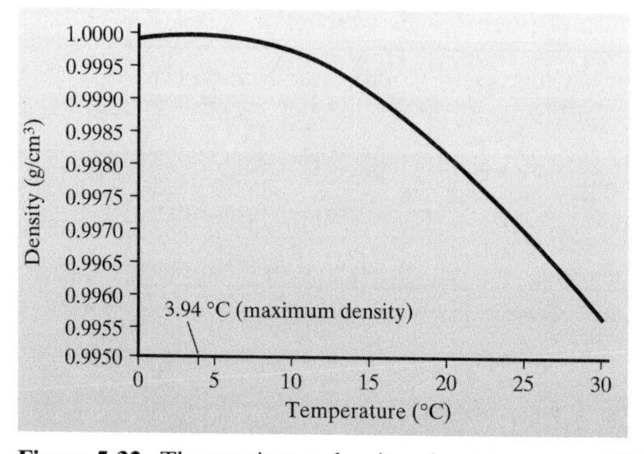

Figure 5-32. The maximum density of water occurs at 3.94°C. Thus water at ~4°C will be found below both colder waters (e.g., ice at 0°C) in winter and warmer waters (e.g., 20°C) in summer.

they would if the maximum density were at 0°C (consider the consequences of the reverse situation). During summer stratification, an upper layer of warm, less dense water floats on a lower layer of cold, more dense water. The layers are assigned special names (Figure 5-33): *epilimnion*, the warm, surface layer that is well-mixed with respect to temperature; *metalimnion*, the region of transition where temperature changes ≥1°C with every meter of depth; and *hypolimnion*, the cold, bottom layer that is also well mixed with respect to temperature. The

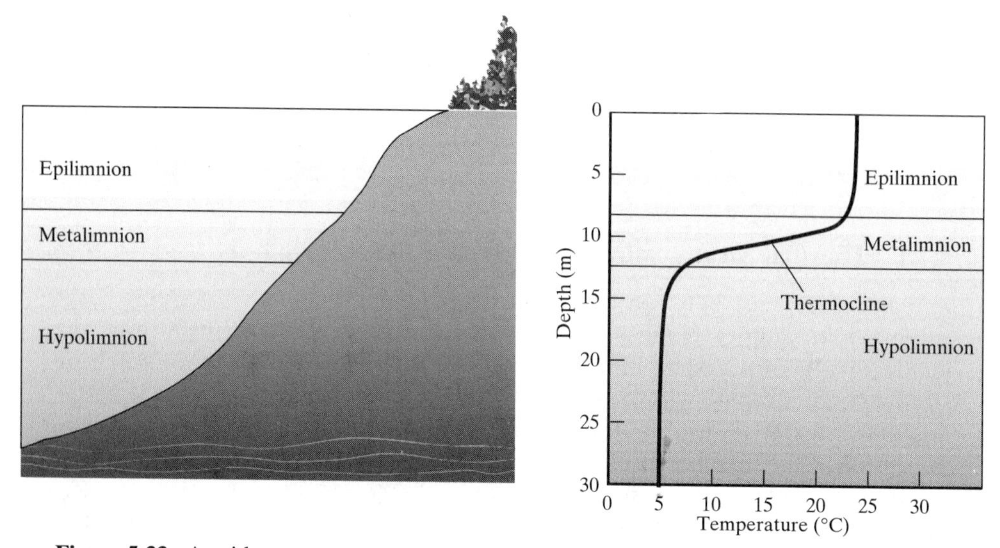

Figure 5-33. A midsummer temperature profile for a thermally stratified lake, identifying the epilimnion, metalimnion (with the thermocline), and the hypolimnion.

plane in the metalimnion where the temperature–depth gradient is steepest is termed the *thermocline*. Figure 5-33 presents a temperature–depth profile for a thermally stratified lake, identifying these three regions.

The stratification process and mixing follows a predictable seasonal pattern, as shown in Figure 5-34. In winter, the lake is thermally stratified with cold (near 0°C) waters near the surface and warmer (2–4°C), more dense waters located near the bottom. As the surface waters warm toward 4°C in spring, they become more dense and sink, bringing colder waters to the surface to warm. The process of mixing by convection, aided by wind energy, serves to circulate the water column, leading to an isothermal condition termed *spring turnover*. As the lake waters continue to warm above 4°C, the lake thermally stratifies. Surface waters become significantly warmer and less dense than the lower waters during this period of *summer stratification*. In the fall, air temperatures decline and heat is lost from the lake more rapidly than it is gained through solar input. As the surface waters cool, they become more dense, sink, and promote circulation through convection, again aided by wind. This phenomenon is called *fall turnover*, and again leads to isothermal conditions. Finally, as the lake cools further, cold, low-density waters gather at the surface and the lake reenters *winter* (or inverse) *stratification*. Lakes with two periods of stratification (winter and summer) and mixing (fall and spring) each year are said to be *dimictic*.

5.7.2 Organic Matter, Thermal Stratification, and Oxygen Depletion

The internal production of organic matter in lakes, resulting from algal and macrophyte growth and stimulated by discharges of growth-limiting nutrients, can dwarf that supplied externally, that is, from wastewater-treatment plants and runoff. This stands in marked contrast to the case for rivers, where poor light conditions often limit internal production and external sources of organic matter dominate. Organic matter produced in the well-lit upper waters of lakes settles to the bottom where it decomposes. Oxygen consumed in the bottom waters through decomposition is not resupplied at a significant rate under stratified conditions (limited mass transport). If the production and decomposition of organic matter exceeds the oxygen resources of the hypolimnion, oxygen depletion will result, as shown in Figure 5-35.

The oxygen resources of productive and unproductive stratified lakes differ significantly, leading to dramatic contrasts. Oxygen depletion leads to acceleration in the cycling of pollutants that reside in lake sediments (especially phosphorus), the generation of several undesired and potentially toxic chemical species commonly associated with anaerobic environments (NH_3, H_2S, CH_4), and the death of fish and bottom-dwelling macroinvertebrates. Oxygen depletion is one of the most important and commonly observed water-quality problems in lakes. It is also important in reservoirs where drinking water intakes are placed at multiple depths, offering operators an opportunity to select the best quality water and avoiding nuisance algal growth near the top and accumulations of noxious chemicals near the bottom.

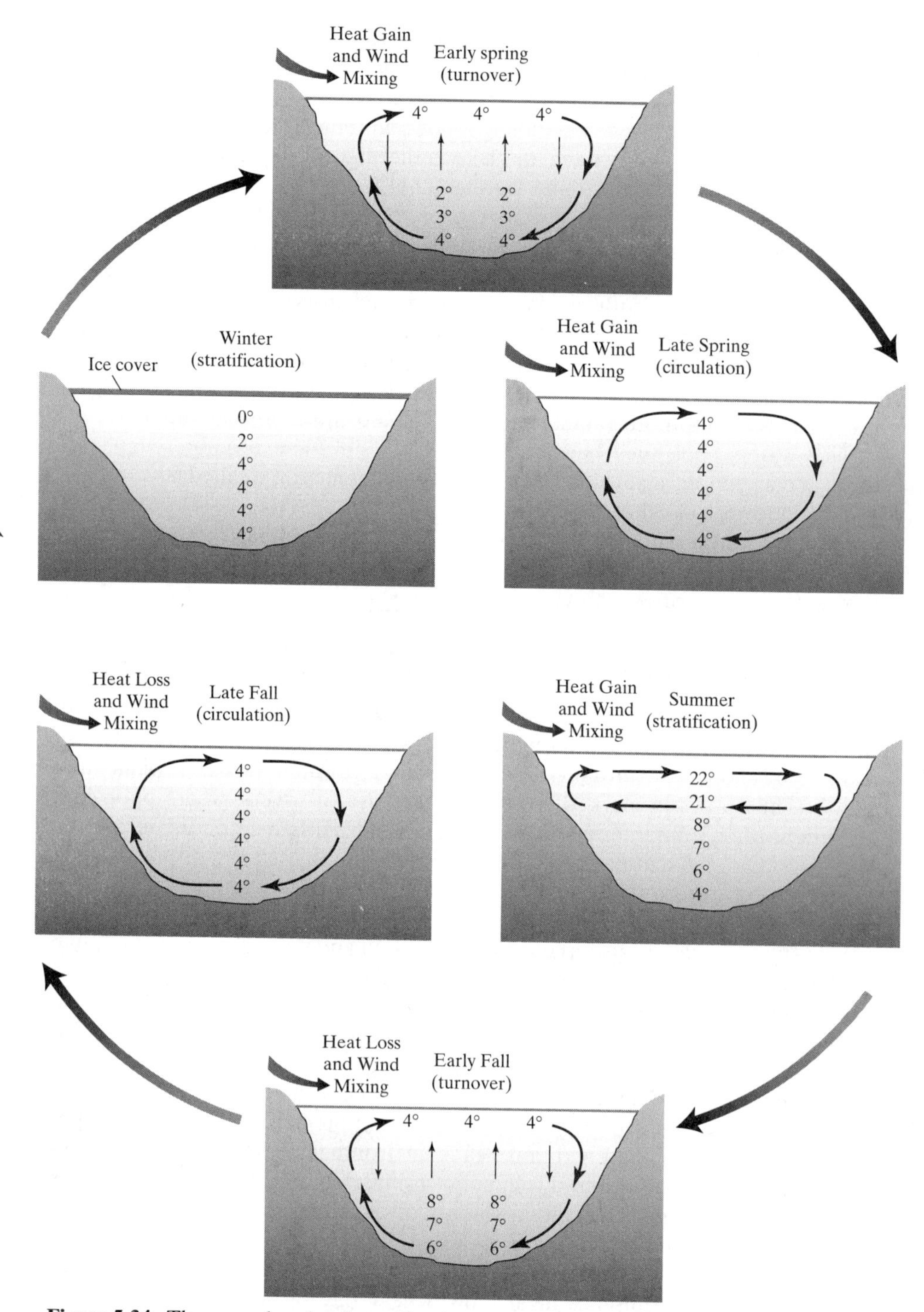

Figure 5-34. The annual cycle of stratification, overturn, and circulation in temperate lakes. Variation in meteorological conditions (temperature, wind speed) may impart significant variation to the timing and extent of these events.

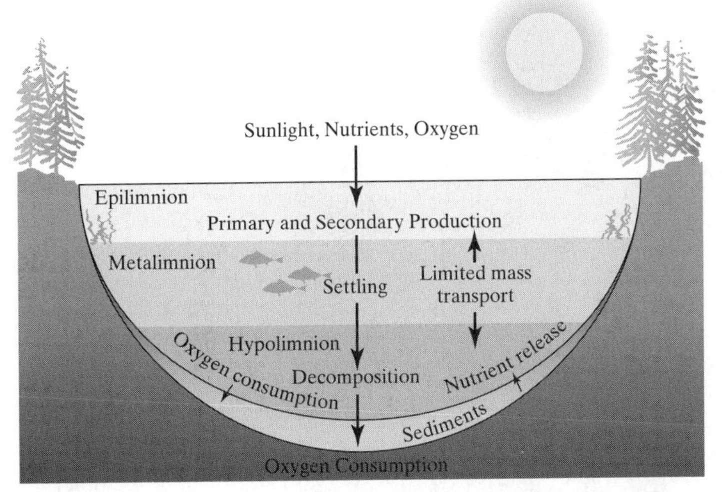

Figure 5-35. The interaction between primary production, thermal stratification, and degradation of water quality in lakes. The oxygen resources of the hypolimnion are largely fixed at stratification. Subsequent production, deposition, and mineralization of organic matter may deplete these reserves leading to poor water quality.

5.7.3 Nutrient Limitations and Trophic State

Trophy is defined as the rate at which organic matter is supplied to lakes, both from the watershed and through internal production. The growth of algae and macrophytes in lakes is influenced by conditions of light and temperature and by the supply of growth-limiting nutrients. Because levels of light and temperature are more or less constant regionally, trophy is determined primarily by the availability of growth-limiting nutrients (recall the Limiting Nutrient Concept in Section 5.6). As mentioned previously, phosphorus is generally found to be the nutrient limiting plant growth in freshwater environments. Because naturally occurring phosphorus minerals are sparingly soluble, *anthropogenic* inputs (those originating through human activities) can have dramatic effects on the rate of algal and macrophyte growth and attendant production of organic matter. Figure 5-36 illustrates the "divided lake," a dramatic whole-lake experiment that demonstrated the importance of phosphorus as a limiting nutrient.

Lakes can be classified according to their trophy or degree of enrichment with nutrients and organic matter, that is, they are classified according to *trophic state*. Three classes are generally recognized: oligotrophic, mesotrophic, and eutrophic. *Oligotrophic* lakes are nutrient poor, have low levels of algae, macrophytes, and organic matter, good transparency, and abundant oxygen. *Eutrophic* lakes are nutrient rich, have high levels of algae, macrophytes, and organic matter, poor transparency, and are often oxygen-depleted in the hypolimnion. *Mesotrophic* lakes are intermediate, often with abundant fish life because they have both elevated levels of organic-matter production and adequate supplies of oxygen. The process of nutrient enrichment of a water body, with attendant increases in

Figure 5-36. Lake 226, the divided lake. D. W. Schindler, a noted Canadian limnologist, demonstrated the role of phosphorus in eutrophication by separating the basins of a lake in northwestern Ontario and adding carbon, nitrogen, and phosphorus to one side and only carbon and nitrogen to the other. The side of the lake receiving phosphorus developed a severe algal bloom (white in this photo), while the side that did not receive phosphorus remained clear (black in this photo). (Courtesy David Schindler.)

organic matter, is termed *eutrophication*. This is considered to be a natural aging process in lakes, part of the succession of newly formed water bodies to dry land as depicted in Figure 5-37. Addition of phosphorus through anthropogenic activities and the resultant aging of the lake is termed *cultural eutrophication*. Variation in land use and population density can lead to a range of trophic states within a given region, for example, from oligotrophic Lake Superior to eutrophic Lake Erie. The restoration of lakes to a trophic state consistent with identified water quality uses is a focus of activity for lake managers and environmental engineers.

5.7.4 Modeling and Managing Eutrophication

Engineers and scientists have established water-quality standards for trophic-state management based on total phosphorus because phosphorus is the growth-limiting nutrient in fresh waters and because phosphorus has been clearly related to other water-quality variables, such as algal biomass, transparency, and oxygen resources. Total phosphorus concentrations of 0.01 and 0.02 g P/m^3 (10 and 20 μg P/L) are generally accepted as the boundaries between oligotrophy and mesotrophy and mesotrophy and eutrophy, respectively. These guidelines have been applied within the context of a steady-state model for total phosphorus in lakes

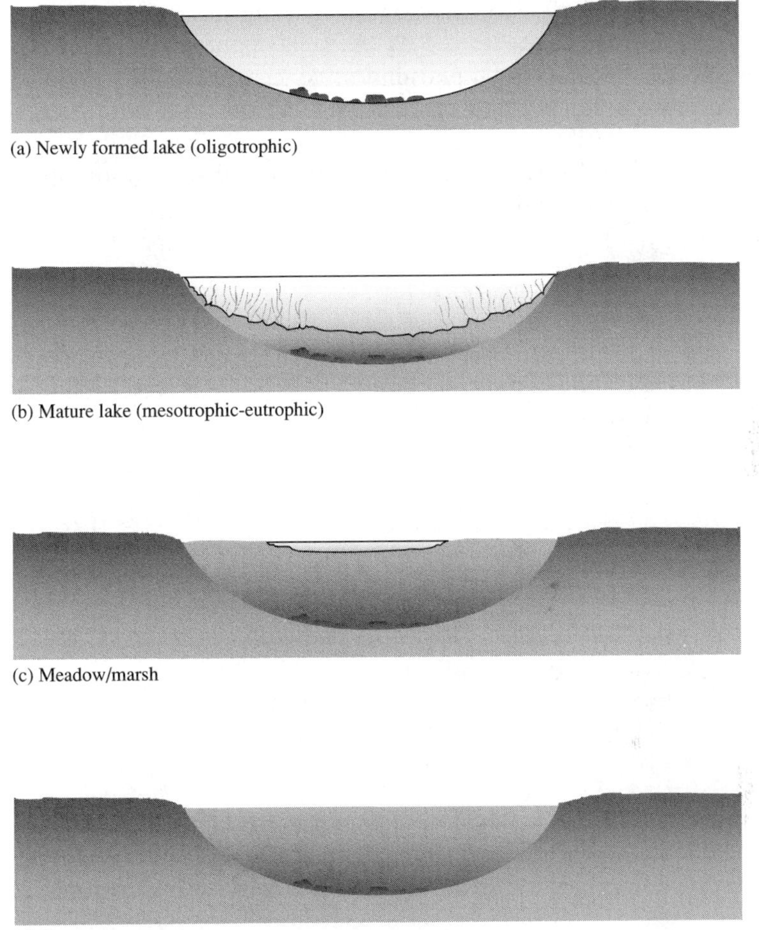

Figure 5-37. One concept of natural succession in lakes suggests that these systems pass through a series of stages as they become enriched with nutrients and organic matter, eventually being transformed to dry land. The rate of lake aging is importantly influenced by local meteorological conditions, the depth of the lake, and the size and fertility of the drainage basin (cf. Horne and Goldman, 1994).

to support water-quality management decisions. The mass balance treats the lake as a completely mixed flow reactor (CMFR, Chapter 4) and is developed from a consideration of phosphorus sources (loading from the watershed) and sinks (outflow and losses to settling):

$$V\frac{dC}{dt} = W - QC - vAC \tag{5-49}$$

where A is the lake surface area (m^2), C is the total phosphorus concentration (g/m^3), Q is the lake outflow (m^3/yr), v is the phosphorus settling velocity (m/yr), V is the lake volume (m^3), and W is the total phosphorus loading (g P/yr). The loading may be further characterized as the sum of the products of the tributary inflow and the tributary total phosphorus concentration ($W = Q\,C_{in}$) and the atmospheric phosphorus flux (J, g P/m^2-yr) and the lake surface area ($W = J\,A$).

At steady state ($dC/dt = 0$), Equation 5-49 can be solved for C_{ss}, the total steady-state phosphorus concentration:

$$C_{ss} = \frac{W}{Q + vA} \tag{5-50}$$

Expressing the load and outflow per unit surface area facilitates comparison among different lakes. Dividing the top and bottom of the right side of Equation 5-50 by the lake area, A, yields:

$$C_{ss} = \frac{\frac{W}{A}}{\frac{Q}{A} + v} \tag{5-51}$$

Defining W' (W/A, g P/m^2-yr) and Q' (Q/A, m^3/m^2-yr) as the area-specific phosphorus and hydraulic loading rates yields:

$$C_{ss} = \frac{W'}{Q' + v} \tag{5-52}$$

Figure 5-38 is a plot of W' as a function of Q' for a typical settling velocity of 10 m/yr and steady-state total phosphorus concentrations of 0.01 and 0.02 g P/m^3, corresponding to the oligotrophic–mesotrophic and mesotrophic–

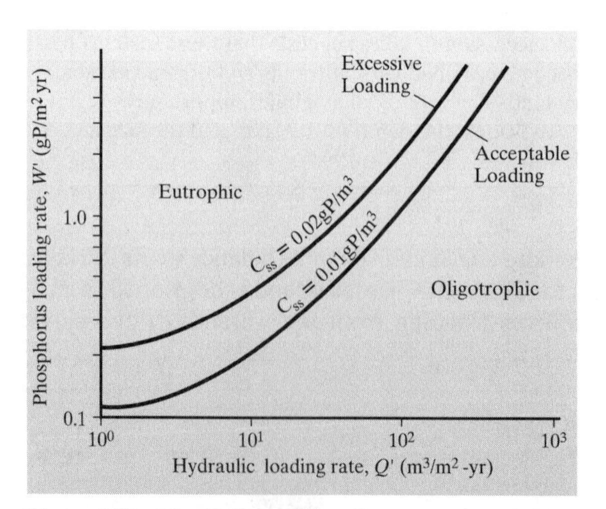

Figure 5-38. The Vollenweider plot, an approach for managing trophic conditions in lakes based on hydraulic and phosphorus loading rates.

eutrophic boundaries, respectively. Loads positioned above the 0.02 g P/m^3 curve are termed "dangerous," and those below the 0.01 g P/m^3 curve, "permissible"; this graph is referred to as a *Vollenweider Plot*, acknowledging its developer, Richard Vollenweider. Environmental engineers and water-quality managers identify acceptable phosphorus loads (g P/m^2-yr) for lakes by calculating the system's area-specific hydraulic loading rate (m^3/m^2-yr) and then consulting the Vollenweider Plot; Example 5.21 illustrates the process.

EXAMPLE 5.22. PHOSPHORUS MANAGEMENT

A lake with a surface area of 2.5 km^2 receives a total phosphorus input of 5 metric tons (MT) per year. The lake receives a water inflow of 1×10^8 m^3/yr. Determine the present trophic state of the lake and calculate the percent reduction in total phosphorus loading required to achieve conditions of oligotrophy.

SOLUTION

First, calculate the areal phosphorus and hydraulic loading rates:

$$W' = \frac{W}{A} = \frac{\dfrac{5 \times 10^6 \text{ g P}}{\text{yr}}}{2.5 \times 10^6 \text{ m}^2} = \frac{2 \text{ g P}}{\text{m}^2\text{-yr}}$$

$$Q' = \frac{Q}{A} = \frac{\dfrac{1 \times 10^8 \text{ m}^3}{\text{yr}}}{2.5 \times 10^6 \text{ m}^2} = \frac{40 \text{ m}^3}{\text{m}^2\text{-yr}}$$

Locating the coordinates W' (2 gP/m^2-yr) and Q' (40 m^3/m^2-yr) on the Vollenweider plot (Figure 5-38) shows that the lake is presently above the "excessive loading curve" and would be classified as eutrophic. The phosphorus loading rate required for oligotrophy is identified as the point on the "acceptable loading curve" for $Q' = 40$ m^3/m^2-yr, i.e. $W' = 0.45$ gP/m^2-yr. Thus, the required loading reduction is:

$$\frac{2 - 0.45}{2} \times 100 = 78\%$$

5.7.5 Lake and Reservoir Restoration and Management

"The most obvious, persistent, and pervasive water quality problem is that of eutrophication. Lakes and reservoirs have deteriorated through excessive additions of plant nutrients, organic matter, and silt, which combine to produce algae and rooted plant biomass, reduced water clarity, and usually decreased lake or reservoir volumes." This statement is from the introduction of the treatise on lake and reservoir management by G. Dennis Cooke and colleagues (Cooke et al., 1993) sets the stage for the evolving role for environmental engineers in the protection and restoration of surface water quality.

With the growing interest in pollution prevention, it should be no surprise that the first choice in lake protection is to reduce or eliminate discharges to the system. Great strides have been made in advanced waste treatment, such as phosphorus removal by chemical or biological means, and in land management practices that reduce phosphorus loads from the watershed. Artificial wetlands and stormwater detention basins have been employed to trap nutrients and heavy metals washed from the land surface in urban environments. Effluent diversion, rerouting of treated effluents to bypass sensitive lakes or reservoirs, has been successfully applied in Europe and North America where treatment or watershed protection have proved impractical.

Remediation or restoration efforts are required in cases where surface waters have been significantly degraded. The application of algicides to reduce phytoplankton blooms and the use of mechanical harvesting to eliminate nuisance growth of macrophytes is common in productive lakes heavily used for recreation and water supply. Even after the reduction or elimination of external nutrient loads, contaminated sediments can recycle enough phosphorus to sustain plant populations and degrade water quality. Here, hypolimnetic aeration, artificial circulation, or chemical inactivation can be applied to trap phosphorus in the sediment and prevent its release and recycle. In extreme cases, lake and reservoir sediments may be removed by dredging; however, difficulties in arranging the disposal of contaminated sediments may limit this approach. Biomanipulation, management of the food web from the top down, is an emerging technology in lake management. Predatory fish are introduced to the system in an effort to remove forage fish and reduce predation on the microcrustacean community that controls algal populations by grazing. In each of these cases, the restoration and management process is appropriately guided by water-quality models, similar to the Vollenweider Plot introduced previously, but of a wide range of sophistication and complexity (see Chapra, 1997).

5.8 ECOSYSTEM HEALTH AND THE PUBLIC WELFARE

All engineering projects should be designed, constructed, and operated in an environmentally safe manner that will ultimately protect human health and the environment. This section introduces three topics relating to chemicals and microorganisms that are important from this perspective: bioconcentration/bioaccumulation, toxicity, and indicator organisms.

5.8.1 Bioconcentration and Bioaccumulation

Bioconcentration is the direct absorption of a chemical into an individual organism (e.g., from water into a fish by the gills). *Bioaccumulation* refers to the accumulation of chemical species both by exposure to contaminated water (bioconcentration) and ingestion of contaminated food, for example, bioconcentration of a contaminant by plankton results in bioaccumulation in fish. Although there are significant inefficiencies in transferring energy and biomass up the food chain, some chemicals (e.g., mercury, DDT) are retained by organisms with little loss. This retention, coupled with the loss of biomass (i.e., oxidation and excretion of organic matter), generates a concentrating effect in each successive level up the food chain. Figure 5-39 depicts this concentrating effect, using mercury as an example. Bioaccumulation of toxic substances has led to severe impacts in many species of wildlife (e.g., bald eagle populations in the Great Lakes region) and, as will be seen in Example 5.22, can contribute significantly to the total human exposure for a particular chemical. Because of these effects and the threat posed to human populations, there is a pressing need to better understand the dynamics of bioaccumulation and its potential impact on the environment. This is one reason why engineers study food webs and material and energy flow in ecosystems.

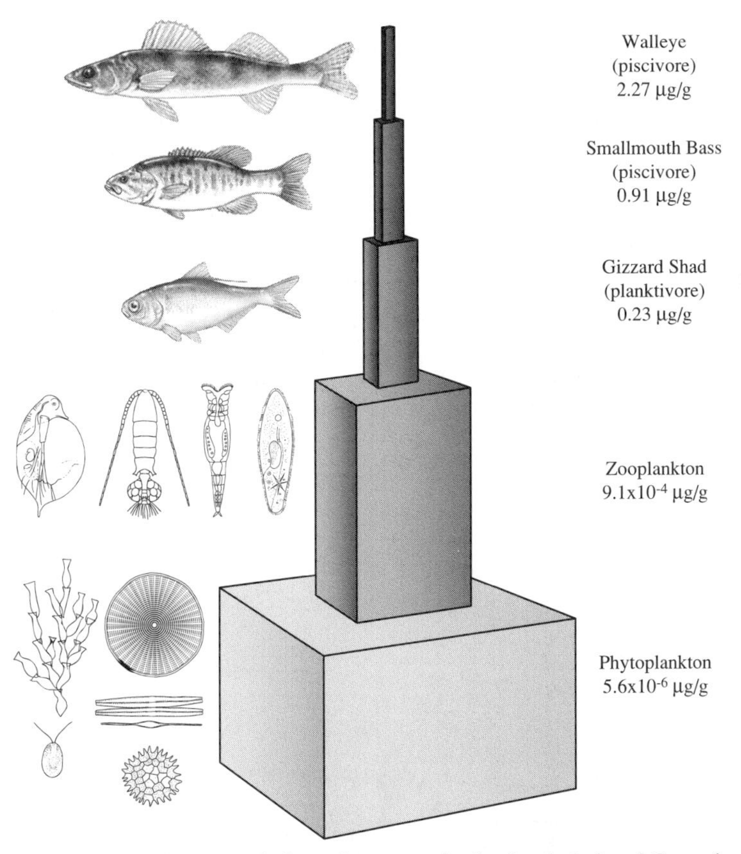

Figure 5-39. Bioaccumulation of mercury in the food chain of Onondaga Lake. Box size represents biomass (decreasing up the food chain through inefficiency of energy transfer) and shading represents the mercury concentration of the biomass (increasing up the food chain because it is retained as biomass is reduced). The concentration of Hg in the water column is ~0.01 μg/L. (Data from Becker and Bigham, 1995).

All chemical compounds would tend to bioconcentrate and bioaccumulate in organisms unless metabolized or excreted. Differences in physiology among organisms creates situations where a chemical passes through one species and is retained by another. However, there are some general rules that can be applied, such as correlations between the physicochemical character of a compound and a property termed the bioconcentration factor (BCF).

The BCF is defined as the ratio at equilibrium of the concentration of a chemical in an organism divided by the concentration of the chemical in the environmental medium (generally air or water). BCFs have been correlated with the hydrophobicity of a chemical (poor affinity for water; good affinity for fats or lipids) as measured by its octanol–water partition coefficient (K_{ow}; see Section 3.4.5.2). Chemicals with a tendency to partition into an organic phase (high K_{ow}) will also partition into the fatty (lipid) portion of a fish or human. Figure 5-40 illustrates the relationship between BCFs and K_{ow}, providing a predictive capacity for BCFs as a function of a compound's K_{ow} as demonstrated in Example 5.22.

Organisms with a high lipid content tend to exhibit greater BCFs, for example, PCB concentrations are typically higher in trout, a fatty fish, than in bass, which are a leaner species. The phenomenon of contaminant accumulation is the basis for fish-consumption advisories in many states. Unfortunately, wildlife do not read. Specific recommendations are made for removal of fatty tissue when cleaning and preparing fish to minimize human consumption (and bioaccumulation) of contaminants. However, exposure to some contaminants such as mercury is not decreased by selective trimming of fatty tissue because the mercury is uniformly distributed in the fish tissues. In this case one can only limit exposure to mercury by controlling the amount of fish consumed per week. Hydrophobic chemicals (such as PCBs and DDT), which are distinguished by a high K_{ow} (typically log $K_{ow} > 5\text{-}6$), also concentrate in the fats of humans, where they may cause adverse effects, such as birth defects and cancer.

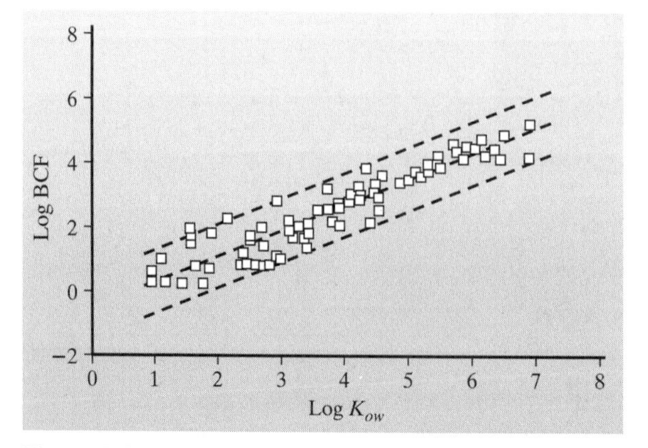

Figure 5-40. Relationship between log K_{ow} and log BCF. The log BCF = 0.85 log K_{ow} − 0.70 (r^2 = 0.897). (From Veith and Kosian, 1983.)

EXAMPLE 5.23. USE OF BIOCONCENTRATION FACTORS

The Saginaw River receives water from a large drainage basin encompassing the east-central portion of Michigan's lower peninsula. The concentration of total dissolved PCBs in the Saginaw River has been determined to range from 1.9 ng/L to 16 ng/L (Verbrugge et al., 1995). Remember that one ng/L equals one part per trillion (ppt).

Concentrations of PCBs in sport fish caught in Saginaw Bay and the Saginaw River have been found to exceed human consumption guidelines and have also been implicated in adverse effects in fish-eating water birds that feed around the bay and river. If the log of the octanol–water coefficient (log K_{ow}) for PCBs is assumed to equal 6.0, what is the expected concentration of PCBs in fish in the Saginaw Bay and Saginaw River? (Assume that the aqueous phase PCB concentration is 16 ng/L.) If it is assumed that an average person drinks 2 L of untreated water daily and consumes 30 g of contaminated fish every day, what route of exposure (drinking water or eating fish) results in the greatest risk from PCBs in one year?

SOLUTION

The BCF relates the aqueous concentration to the concentration in fish:

$$\text{BCF} = \frac{[\text{PCB}_{\text{fish}}]}{[\text{PCB}_{(\text{aq})}]}$$

The problem did not provide a BCF; therefore, we need to either look one up in a reliable data base or estimate it. Assuming the correlation provided in Figure 5-40 is applicable to PCBs and our fish, use the log K_{ow} to estimate the BCF:

$$\log \text{BCF} = 0.85 \log K_{ow} - 0.70$$

$$\log \text{BCF} = 0.85\ (6.0) - 0.70 = 4.4$$

Therefore, the BCF $= 10^{4.4}$

$$10^{4.4} = \frac{[\text{PCB}_{\text{fish}}]}{[\text{PCB}_{(\text{aq})}]} = \frac{[\text{PCB}_{\text{fish}}]}{16\ \text{ppt}}$$

Solve for $[\text{PCB}_{\text{fish}}] = 4.0 \times 10^5$ ppt, which equals 0.40 ppm (or 0.40 mg of PCB/kg of fish). Note that this concentration may be underpredicted because the problem used a BCF and not a BAF. BAFs are more realistic of what occurs in the environment; however, there are fewer BAFs available.

For the second question we need to determine the mass of PCBs that an individual is exposed to over a one-year period. A person drinking 2 L of untreated water per day is exposed in one year to:

$$\frac{2\ \text{L}}{\text{day}} \times \frac{16\ \text{ng}}{\text{L}} \times \frac{\text{g}}{10^9\ \text{ng}} \times \frac{365\ \text{day}}{\text{yr}} = \frac{1.2 \times 10^{-5}\ \text{g PCB}}{\text{yr}}$$

A person consuming 30 g of contaminated fish per day is exposed in one year to:

$$\frac{30 \text{ g fish}}{\text{day}} \times \frac{0.40 \text{ mg of PCB}}{\text{kg of fish}} \times \frac{\text{g}}{1{,}000 \text{ mg}} \times \frac{\text{kg}}{1{,}000 \text{ g}} \times \frac{365 \text{ day}}{\text{yr}} = \frac{4.4 \times 10^{-3} \text{ g PCB}}{\text{yr}}$$

Note that because PCBs biomagnify strongly in the food web, an individual is exposed to a much greater mass of PCBs by ingesting contaminated food than by drinking contaminated water. This amount could be much greater for segments of our population that consume more fish than the average person as well as for wildlife that depend upon fish for food. Also note that for chemicals that persist (do not degrade by natural mechanisms) and also biomagnify, low water concentrations may have a great environmental significance. This effect can be further magnified for chemicals with a high BCF.

5.8.2 Toxicity

Environmental toxicology is a rapidly growing interdisciplinary field dealing with the effects of chemicals on living organisms. Because energy and material are distributed through food webs, it is likely that an impact on one level will be reflected in other levels as well. For example, there is evidence that elevated PCB levels in Lake Michigan fish resulted in adverse health effects to children born from mothers who included such fish in their diet (Jacobson et al., 1990). While bioaccumulation of PCBs may have had no direct adverse effect on the fish, the next trophic level (humans) was impacted.

Toxic effects can be divided into two types: carcinogenic and noncarcinogenic. A carcinogen promotes or induces tumors (cancer), that is, the uncontrolled or abnormal growth and division of cells. Carcinogens act by "attacking" or altering the structure and function of DNA within a cell. Many carcinogens seem to be site-specific, that is, a particular chemical tends to attack a specific organ. In addition, carcinogens may be categorized based on whether they cause direct or indirect effects: *primary carcinogens* directly initiate cancer; *procarcinogens* are not carcinogens but are metabolized to form carcinogens and thus indirectly initiate cancer; *cocarcinogens* are not carcinogens, but enhance the carcinogenicity of other chemicals; and *promoters* enhance the growth of cancer cells.

Classification of a chemical as a "known" human carcinogen requires sufficient evidence that human exposure leads to a significantly higher incidence of cancer. Such evidence (epidemiological data) is often collected from workers in job environments where there is prolonged contact with a chemical. While there are few known human carcinogens (e.g., benzene, vinyl chloride, asbestos, chromium), there are many *probable* carcinogens (e.g., benzo(*a*)pyrene, carbon tetrachloride, cadmium, Mirex, Saccharin, PCBs) and hundreds of *possible* carcinogens. Chemicals are listed as probable human carcinogens when experimental evidence indicates increased cancer risk in test animals and insufficient information is available to show a direct cause–effect relationship for humans.

Noncarcinogenic effects include all other toxicological responses, of which there are countless examples: organ damage (e.g., kidney, liver), neurological damage, suppressed immunity, and birth and developmental (adversely effecting an organism's reproductive ability or intelligence) effects. For example, elevated lead levels in children have been shown to cause learning disorders and lower IQs. The toxic effects that are manifested following exposure to a chemical often result from interference with enzyme (catalyst) systems that mediate the biochemical reactions critical for organ function. In other cases, the endocrine system is impacted. Chemicals collectively known as *endocrine disruptors* exert their effects by mimicking or interfering with the actions of hormones, biochemical compounds that control basic physiological processes such as growth, metabolism, and reproduction. Endocrine disruptors may also cause breast cancer in women. Chemicals identified as endocrine disruptors include pesticides (such as DDT and its metabolites), industrial chemicals (such as some surfactants and PCBs), some prescription drugs, and other contaminants, such as dioxins (National Science and Technology Council, 1996).

The likelihood of a toxicological response is determined by the *exposure* to a chemical: a product of the chemical *dose* and the *duration* over which that dose is experienced. In humans, there are three major exposure pathways: ingestion (eating and/or drinking), inhalation (breathing), and dermal (skin) contact. It is very important to note that basically every compound, if present in large enough quantities, is toxic in some way. It is the dose and duration that determine exposure and thus toxicity.

Some chemicals (e.g., dioxin, TCDD) can be lethal to test animals in very small doses, whereas others create problems only at much higher levels. Table 5-8 lists chemical compounds with widely varying toxicities, with toxicity defined here as causing death, an experimental endpoint that (for test animals) is more readily

Table 5-8. Oral LD_{50} Values for Various Organisms and Chemicals

Chemical	Organism	LD_{50} (mg/kg)
Methyl ethyl ketone	Rat	5,500
Fluoranthene	Rat	2,000
Pyrene	Rat	800
Pentachlorophenol	Mouse	117
Lindane	Mouse	86
Dieldrin	Mouse	38
Sarin (nerve gas)	Rat	0.5

Values from Patnaik, 1992.

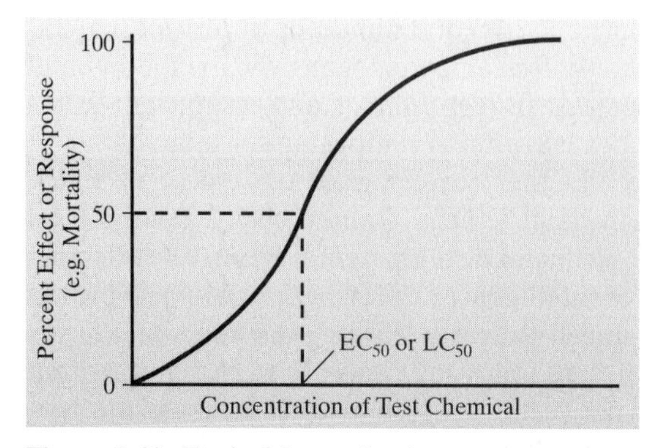

Figure 5-41. Typical form of a dose-response curve used in identifying EC_{50} and LC_{50} values for various chemicals and test organisms.

determined than, for example, lung cancer. A common method of expressing toxicity is the *LD_{50}*, the medial *lethal dose*, that is, that results in the death of 50% of the test organisms. The LD_{50} is typically presented as the mass of contaminant dosed per mass (body weight) of the test organism, for example, mg/kg. A pesticide with an LD_{50} of 100 mg/kg would result in the death of 50% of a population of rats, each weighing 0.1 kg, if applied at a dose of 10 mg per rat. A dose of 20 mg per rat should result in the death of more than 50% of the population and a dose of 5 mg per rat would result in the death of less than 50%. A series of such experiments at various doses yields a dose-response relationship as depicted in Figure 5-41. A similar term, the median *lethal concentration* (LC_{50}) is typically used in studies of aquatic organisms and represents the ambient aqueous contaminant concentration (as opposed to injected or ingested dose) at which 50% of the test organisms die. More subtle (i.e., behavioral) changes can also reflect a toxic response, but are difficult to assess.

Recall that it is not only the dose, but also the duration that determines exposure and thus toxicity. *Acute* and *chronic* toxicity refer to death (or some other adverse response) resulting from short- (hours to days) and long- (weeks to years) term exposure to a chemical, respectively. Acute effects are experienced at higher contaminant concentrations than are chronic effects. For example, the U.S. Environmental Protection Agency has established acute (2.4 μg/L) and chronic (0.012 μg/L) water-quality criteria for mercury to protect aquatic life from toxic effects. Note that here the acute criterion is 200% of the chronic value. As the duration increases the concentrations that can be tolerated without adverse effect become lower. Acute copper toxicity for rainbow trout decreases from an LC_{50} of 0.39 mg/L at a 12-h duration to 0.13 mg/L at 24 h to 0.08 mg/L at 96 h. The toxicity of a specific chemical may also vary among species. Table 5-9 demonstrates this effect, comparing the 48-h LC_{50} values for 2,4-dichlorophenoxyacetic acid (2,4-D), a common herbicide used on farms and household lawns, for various aquatic organisms.

Table 5-9. Forty-eight-hour LC_{50} Values for 2,4-D

Species	LC_{50}, mg/L
Daphnia magna (zooplankton)	25
Fathead minnow	325
Rainbow trout	358

From Patnaik, 1992.

While the concentrations of 2,4-D listed in Table 5-9 are not likely to be encountered in surface waters (although levels of agricultural chemicals in runoff do increase following spring rains and snow melt), the observed variation in LC_{50} values suggests a scenario where microcrustacean populations would be impacted but fish populations would not. Such a scenario would potentially lead to an alteration and disruption of the food web with ecosystemwide impacts. An understanding of food-web function and the bioaccumulation and toxicity of contaminants (at each trophic level) is necessary to adequately assess the risk posed by the myriad chemical contaminants introduced to our environment.

The species-specific nature of toxicity, shown in Table 5-9, is the fundamental shortcoming in procedures commonly applied for estimating effects on humans, such as criteria based on experiments with test animals. Humans may be substantially more or less susceptible to the toxic effects of a specific compound at a given dose than are laboratory surrogates. When such studies are used to determine standards for human exposure, these "uncertainties" are accommodated through a safety factor that may result in an estimate that is overly conservative by several orders of magnitude—an approach based on a "better safe than sorry" philosophy. In addition, the fact that some wildlife may be more sensitive to toxic chemicals than humans has led to the promulgation of water-quality criteria in which the more stringent of wildlife- or human-health-based standards govern discharge limits. For example, the U.S. Environmental Protection Agency human-health criterion for hexavalent chromium (Cr^{6+}) is 50 μg/L, while the acute criterion for aquatic life is 16 μg/L; here the wildlife standard is approximately one-third the human health-based value.

Sensitive segments of a population must also receive consideration in determining toxicity effects. The embryonic, juvenile, elderly, and/or ill segments of any population are likely to be more susceptible to adverse effects from chemical exposure than are healthy young adults. In some cases, the sex of an individual may also influence its susceptibility. Synergistic toxicity, resulting from the exposure to multiple chemicals, is a phenomenon that is receiving increased attention. For example, consider two compounds with LC_{50} values of 5 and 20 mg/L, respectively. When present together, their combined LC_{50} values might drop to 3 and 10 mg/L, levels that are lower than the individual LC_{50} values. In some cases chemicals may have the opposite (antagonistic) effect, resulting in a combination that is *less* toxic than when present separately. Unfortunately, scientific studies of chronic synergistic effects are lacking, largely due to the countless numbers of chemicals and combinations of chemicals in existence, and to inherent difficulties in performing long-term experiments.

5.8.3 Indicator Organisms

In addition to toxicity from chemicals, there is a concern for protecting humans from *pathogens*, that is, disease-causing organisms. Pathogens causing the waterborne diseases of interest in drinking-water supply and wastewater treatment include viruses (hepatitis, polio, and gastroenteritis), certain bacteria (cholera and typhoid), and protozoa (cryptosporidiosis and giardiasis or backpacker's fever). Standard disinfection practices (e.g., chlorination and ozone) can yield a water supply or wastewater effluent safe from bacterial and viral pathogens. Filtration is a more suitable treatment approach for the elimination of protozoan pathogens, which form cysts that are resistant to chemical or physical attack. The problems posed by protozoan pathogens such as *Giardia lamblia* and *Crytosporidium parvum* have been of increasing interest to engineers following an outbreak of cryptosporidiosis in Milwaukee, Wisconsin, in 1993 that affected approximately one-half of the city's 800,000 water-supply customers.

Wastewater contains a diverse assemblage of microorganisms and, depending on the health of the human population served, may carry numerous pathogens. Indeed, the field of environmental (sanitary) engineering developed in response to concerns over aesthetics and human health relating to water supply and wastewater treatment. Outbreaks of cholera and typhoid were once common worldwide and still occur in underdeveloped countries with poor sanitation. As with chemical toxicity, exposure is the key issue. The pathogens commonly associated with wastewater, such as those responsible for hepatitis and dysentery, may remain viable in wastewater for many days. Again, the risk of infection depends upon the route of exposure and the dose. Combined sewer overflows (CSOs: relief points for sewers carrying a mixture of wastewater and surface runoff) exist in many older cities. During storm events, CSOs discharge the rainwater–wastewater mixture directly to adjoining surface waters in an effort to guard the treatment plant against hydraulic overload. This practice may lead to widespread exposure of the general population to pathogens unless the CSO discharges are collected and disinfected. At the treatment plant, processes such as sedimentation and disinfection serve to remove or destroy most pathogenic organisms.

Drinking water, surface waters that supply drinking water and that are used for contact recreation, and wastewater effluents are routinely monitored to protect the public from waterborne disease. It is logistically impractical and prohibitively expensive to test each water sample for all of the pathogens potentially present. Instead, samples are assayed for those microbes associated with fecal contamination, that is, *indicator organisms*, under the premise that their presence suggests the potential presence of human pathogens. Droste (1997) offers the following as the ideal traits of an organism indicative of fecal contamination:

- That they originate only in the digestive tract of humans and warm-blooded animals;
- That they are easily, rapidly, reliably, and inexpensively identified and enumerated;
- That their survival outside the intestine be longer than for pathogens;
- That they occur in high numbers;
- That they themselves are not pathogenic.

The most commonly used indicator organisms are members of the coliform group, named for the bacterium *Escherichia coli*, a normal inhabitant of the digestive tract of humans and other warm-blooded animals. Tests are made for total coliforms (a group that includes some species widely distributed in the environment) and fecal coliforms (a group that is more specific to fecal contamination). Because the total coliform test potentially includes organisms of nonfecal origin and because the fecal coliform test can produce false-positive and -negative results (Droste, 1997), there is interest in basing bacteriological standards directly on *E. coli*, a species exclusively of fecal origin. Most strains of *E. coli* are not pathogenic; however, some can cause gastroenteritis and occasionally death.

Bacteriological standards for water quality are often set by state and local governments and thus vary somewhat across the nation; some representative values are presented in Table 5-10. As a frame of reference, consider that a gram of human feces may contain on the order of 13,000,000 fecal coliform bacteria (Droste, 1997) and that streams polluted by CSOs may have fecal coliforms in excess of 1,000,000 per 100 mL (Canale et al., 1993).

Fecal and total coliforms are detected in the laboratory by two methods: the membrane-filter test and the multiple-tube fermentation test. The membrane-filter test is the simplest. A known volume of water sample is filtered through a sterilized membrane filter (pore size of 0.45-μm diameter), trapping bacteria on the filter surface. The filter is transferred to a petri dish, growth medium is added to provide nutrients, and the filter incubated at 35°C (98°F, approximate human body temperature) for 24 h. During incubation, a colony of coliform bacteria (exhibiting a metallic green sheen and easily detectable by the naked eye) de-

Table 5-10. Typical Bacteriological Standards for Water Supply, Wastewater Treatment, and Recreation

Water Use	Total Coliforms (cells or CFU/100 mL)
Wastewater effluent	200
Contact recreation	200–1000
Drinking water	<1

velops from each coliform bacterium (colony-forming unit, CFU) initially present. Results are expressed as a CFU/100-mL sample. The multiple-tube fermentation test is based on the fact that coliform bacteria ferment lactose to form carbon dioxide gas. Several dilutions of a water sample are added to sterilized tubes containing lactose broth and are incubated for 48 h at 35°C. The development of gas bubbles suggests the presence of coliforms. The test does not yield a direct count, but rather a most probable number (MPN/100 mL) of coliform bacteria present based on a statistical algorithm (Greenberg et al., 1992).

EXAMPLE 5.24. DETERMINING COLIFORM CONCENTRATIONS

In order to prevent the overgrowth (overlapping) of colonies on the membrane used to detect coliforms, samples expected to have high counts, for example, those from wastewater or polluted surface waters, are diluted prior to filtration.

A dilution is prepared containing 0.01 mL of river water, filtered and prepared for fecal coliform analysis and incubated for 24 h. Following incubation, 15 colonies are counted on the membrane surface. Calculate the fecal coliform density in the original sample and compare that concentration to the standards for contact recreation in Table 5-10.

SOLUTION

The problem is essentially one of unit conversion:

$$\frac{15 \text{ CFU}}{0.01 \text{ mL}} = \frac{1{,}500 \text{ CFU}}{\text{mL}} \quad \text{or} \quad \frac{150{,}000 \text{ CFU}}{100 \text{ mL}}$$

As seen from Table 5-10, this sample is well above allowable limits and even violates the standard for contact recreation.

To place this value into perspective further, consider that clean natural water typically contains 0–100 coliforms/100 mL, polluted surface water contains approximately 10,000 coliforms/100 mL, and raw sewage contains several million coliforms/100 mL. This is a highly contaminated river-water sample and is most likely indicative of untreated sewage pollution.

CHAPTER PROBLEMS

5-1. Mathematical models are used to predict the growth of a population, that is, population size at some future date. The simplest model is that for exponential growth. The calculation requires a knowledge of the organism's maximum specific growth rate (μ_{max}). A value for this coefficient can be obtained from field observations of population size or from laboratory experiments where population size is monitored as a function of time when growing at high substrate concentrations (i.e., $S \gg K_s$):

Time (d)	Biomass (mg/L)
0	50
1	136
2	369
3	1,004
4	2,730
5	7,421

Calculate μ_{max} for this population assuming exponential growth; include appropriate units.

5-2. Once a value for μ_{max} has been obtained, the model may be used to project population size at a future time. Assuming that exponential growth is sustained, what will the population size in Problem 5-1 be after 10 days?

5-3. Exponential growth cannot be sustained forever because of constraints placed on the organism by its environment, that is, the system's carrying capacity. This phenomenon is described using the logistic growth model. (a) Calculate the size of the population in Problem 5-1 after 10 days, assuming that logistic growth is followed and that the carrying capacity is 100,000 mg/L. (b) What percentage of the exponentially growing population size would this be?

5-4. Food limitation of population growth is described using the Monod model. Population growth is characterized by the maximum specific growth rate (μ_{max}) and the half-saturation constant for growth (K_s). (a) Calculate the specific growth rate (μ) of the population in Problem 5-1 growing at a substrate concentration of 25 mg/L according to Monod kinetics if it has a K_s of 50 mg/L. (b) What percentage of the maximum growth rate for the exponentially growing population size would this be?

5-5. The two coefficients defined in Problem 5-4 (μ_{max} and K_s) describe the organism's ability to function in the environment. Populations with a high μ_{max} grow rapidly and take up substrate very fast. Those with a low K_s are able to take up substrate quite efficiently, reducing it to low levels. These characteristics are important when considering the use of microorganisms to clean up pollution from potentially toxic chemicals. Consider two genetically engineered organisms intended for use in a chemical spill cleanup. Organism A has a μ_{max} of 1 day^{-1} and a K_s of 0.1 mg/L. Organism B has a μ_{max} of 5 day^{-1} and a K_s of 5 mg/L. Chemical levels are initially on the order of 100 mg/L; the goal is to reduce concentrations to below 0.1 mg/L. We wish to use the organisms in sequence—first one organism to rapidly reduce chemical levels before they can spread, and second, an organism to reduce chemical levels to the target concentration of 0.1 mg/L. (a) Which organism (A or B) would be most effective in *rapidly reducing levels of pollution*? (b) Which organism (A or B) would be most effective in *reducing the pollutant to trace levels*? Support both of your answers with calculations.

5-6. In wastewater treatment, organism biomass increases as pollutants are taken up and metabolized. This increase is reflected in the amount of sludge generated at the wastewater-treatment plant, a residue that must receive safe disposal. Engineers use the yield coefficient (Y) to calculate biomass (sludge) production. Laboratory studies have shown that microorganisms produce 10 mg/L of biomass in reducing the concentration of a pollutant by 50 mg/L. Calculate the yield coefficient, specifying the units of expression.

5-7. When food supplies have been exhausted, populations die away. This exponential decay is described by a simple modification of the exponential growth model. Engineers use this model to calculate the length of time that a swimming beach must remain closed following pollution with fecal material. For a population of bacteria with an initial biomass of 100 mg/L and a $k_d = 0.4\ \text{day}^{-1}$, calculate the time necessary to reduce the population size to 10 mg/L.

5-8. Organic carbon (C) and ammonia nitrogen (NH_3) are oxidized to carbon dioxide (CO_2) and nitrate (NO_3^-), respectively, by bacteria that are naturally present in wastewater and in natural systems such as lakes and rivers. Both of these reactions consume oxygen and may cause a negative impact on water quality. The amount of oxygen theoretically required to consume a carbonaceous (ThOD) or nitrogenous (NOD) waste may be calculated according to the stoichiometry of the reactions as outlined in this chapter. The production of coke, a fuel produced from coal for use in steel mills, generates a waste stream rich in ammonia, phenol, and naphthalene. Calculate the NOD and ThOD of a waste containing 25 mg/L of ammonia–nitrogen (NH_3-N), 50 mg/L of phenol (C_6H_5OH), and 150 mg/L of naphthalene ($C_{10}H_8$). What is the total theoretical oxygen demand of the waste?

5-9. How much oxygen is *actually* consumed depends on the extent to which the waste is biodegradable, that is, amenable to oxidation by microbes. Ammonia is totally biodegradable, and thus the theoretical nitrogenous oxygen demand (NOD) and the nitrogenous biochemical oxygen demand ultimately exerted (NBOD) are equal. Not so for carbonaceous compounds: the carbonaceous biochemical oxygen demand ultimately exerted (CBOD) may be significantly less than the ThOD if the compound is poorly degradable. To test this, oxygen consumption is measured in the laboratory, inhibiting the oxidation of ammonia so that only carbonaceous oxygen demand is exerted. The measurement is made over a period of five days, yielding the 5-day carbonaceous biochemical oxygen demand ($CBOD_5$). The waste from Problem 5-8 was found to have a $CBOD_5$ of 90 mg/L and a rate constant (k_L) of 0.1 day^{-1}. (a) Calculate the ultimate CBOD of the waste. (b) Is this a readily biodegradable waste? (Why or why not?)

5-10. In order to evaluate the impact of a waste on a receiving water, we must calculate the concentration of the waste as it mixes with the stream and how that concentration changes as the waste moves downstream, degrading as it travels. The first step involves a mixing basin analysis, and the second applies first-order kinetics in a plug flow reactor. A steel mill waste is discharged at a rate of 0.75 m^3/s; the waste has a dissolved oxygen (DO) concentration of 3 mg/L, a CBOD equal to 229 mg/L and an NBOD equal to 114 mg/L. The river receiving the waste has a flow of 6 m^3/s and travels at a velocity of 9 km/day. The river has an ammonia–nitrogen concentration of 0.5 mg NH_3-N/L, a CBOD of 2 mg/L, and a dissolved-oxygen (DO) concentration of 6 mg/L upstream of the discharge. Calculate the DO, NBOD, and CBOD in the river immediately after it becomes mixed with the waste and the NBOD, NH_3-N, CBOD, and $CBOD_5$ 18 km downstream of the point of discharge. Assume a rate constant for ammonia oxidation (k_N) of 0.15 day^{-1} and a rate constant for CBOD exertion of 0.1 day^{-1}.

5-11. As a waste moves downstream, exertion of the oxygen demand removes oxygen from the water and exchange with the air (reaeration) adds oxygen to the water. Initially exertion exceeds reaeration and the dissolved-oxygen (DO) content of the water drops. Later, as the waste becomes stabilized (no longer consuming as much oxygen), the river DO increases. The result is a dissolved-oxygen sag curve. Equations were provided in this chapter to determine the location and value for the minimum point in the DO sag curve. Using data from Problem 5-10 and assuming that there is no NBOD exertion, calculate the oxygen deficit (D_0) in the river after it has mixed completely with the waste discharge, the time and location of the maximum deficit, and the minimum DO concentration. Assume a saturation DO for the river of 10 mg/L, that $k_1 = k_L$, and that the reaeration coefficient (k_2) is 0.3 day^{-1}.

5-12. A population of microorganisms has the growth characteristic described in the table below.

Characteristic	Value
Initial biomass	10 mg DW/L
Maximum specific growth rate	0.3 day^{-1}
Half-saturation constant	1 mg S/L

(a) Write the mass-balance expressions for describing the change in population biomass and substrate with time for growth in a batch reactor (i.e., no inflow or outflow). (b) Would this population ever approach its carrying capacity? (c) Calculate the population biomass after the first 3 days of growth. (d) Calculate the change in substrate concentration over the first 3 days of growth. (e) If the population biomass peaks at 10,000 mg/L when the substrate runs out, calculate the population biomass 10 days after the peak.

5-13. The zebra mussel, an invertebrate organism related to clams and snails, has recently invaded the Great Lakes. They were brought to the United States in the ballast water of ships that originated in Europe. These organisms grow attached at the surface of solid substrates and have been implicated in the clogging of water-intake pipes and other environmental problems. They often inhabit nutrient-saturated environments where renewable resources place no limits on growth. Instead, environmental resistance regulating growth is provided by nonrenewable resources, that is, space. (a) Recognizing that all microorganisms require energy to maintain their metabolic functions (i.e., the cost of doing business) and remembering that Einstein said that we should "keep everything as simple as possible, but no simpler," write mass-balance expressions for the rate of change in population biomass and substrate with time. (b) Draw the expected population-growth curve for these organisms in a nutrient-saturated environment. (c) Calculate the steady-state population biomass if the maximum specific growth rate is 1 day^{-1}, the respiration-rate coefficient is 0.25 day^{-1}, the half-saturation constant is 50 mg/L, and the population carrying capacity is 1,000 mg/L.

5-14. Two populations of microorganisms are growing in a waste-treatment system with characteristics as described in the table below. (a) How many days longer will it take Population B to double its initial biomass than Population A? Think about this problem carefully before you begin to solve it!

Characteristic	Value	Units
Initial biomass	A: 10 B: 10	mg DW/L
Maximum specific growth rate	A: 0.3 B: 0.1	day^{-1}
Half-saturation constant	A: 10 B: 1	mg S/L
Carrying capacity	A: 10,000 B: 100,000	mg DW/L
Respiration-rate coefficient	A: 0.05 B: 0.05	day^{-1}
Substrate concentration	2,000	mg S/L

5-15. A population having a biomass of 2 mg/L at $t = 0$ days reaches a biomass of 139 at $t = 10$ days. Assuming exponential growth, calculate the value of the specific growth coefficient.

5-16. A bacterial population with a maximum specific growth rate of 1 day^{-1} and a respiration-rate coefficient of 0.2 day^{-1} will be applied in cleaning up an organic chemical spill. To quickly attain a low final steady-state chemical concentration, would a higher or lower half-saturation constant be required?

5-17. In the table below, identify the attribute of a microorganism that would maximize the rate of removal of a pollutant in a waste-treatment system.

Variable	Small	Doesn't Matter	Large
Maximum specific growth rate			
Respiration rate			
Yield coefficient			
Half-saturation constant			
Carrying capacity			

5-18. A population doubles its biomass in 10 days. Assuming exponential growth, calculate the value of the specific growth rate.

5-19. Fecal bacteria occupy the guts of warm-blooded animals and do not grow in the natural environment. Their population dynamics in lakes and rivers, that is, following a discharge of raw sewage, can be described as one of exponential decay or death. How many days would it take for a bacteria concentration of 10^6 cell/mL to be reduced to the public health standard of 10^2 cell/mL if the decay coefficient is 2 day^{-1}?

5-20. A consulting firm proposes to perform *in situ* bioremediation of a toxic chemical spill by adding a culture of microorganisms to the subsurface water. The microorganism has a maximum specific growth rate of 0.2 day^{-1}, a half-saturation constant of 5 mg/L, and a respiration-rate coefficient of 0.05 day^{-1}. The remediation goal is to reduce chemical levels to <0.5 mg/L. Do you support or oppose this proposal? Demonstrate your position through calculations.

5-21. You are charged with the cleanup of a subsurface toxic chemical spill. Competing microbiology companies are offering two genetically engineered bacterial species to aid in the cleanup: *Bacillus growsfastus* ($\mu_{max} = 5$ day^{-1}; $K_s = 100$ mg/L) and *Rotobacter eatsallii* ($\mu_{max} = 1$ day^{-1}; $K_s = 1$ mg/L). Your primary objective in the cleanup is to reduce the chemical concentration to <1 mg/L. Select an organism and defend your answer.

5-22. What is the name of the model that relates resource or substrate availability and population growth? For each of the organisms (A and B) in the figure below, use that model to estimate the value for the maximum specific growth rate and the half-saturation constant. Which organism would be most successful in rapidly reducing pollutant concentrations? Which organism would be most successful at reducing pollutant concentrations to low levels?

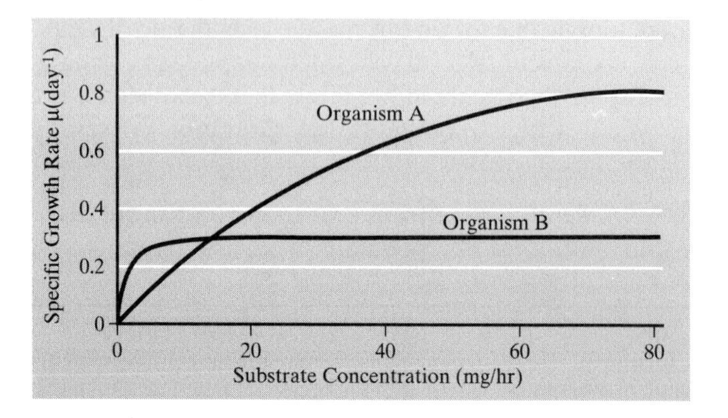

5-23. Figure 5-11 showed world population growth over the past 2,000 years. (a) What population-growth model apparently describes these data? (b) Calculate the specific growth rate of this population. (c) What are the proper units for this coefficient? (d) Why is it unlikely that this mode of growth will be sustained indefinitely? (e) What model would more appropriately describe the future behavior of this population?

5-24. A waste contains 20 mg/L of ammonia (NH_3) and 100 mg/L of ethyl alcohol (C_2H_5OH). What is the total theoretical oxygen demand of the waste?

5-25. A waste contains 100 mg/L of ethylene glycol ($C_2H_6O_2$) and 50 mg/L of NH_3-N. Determine the theoretical carbonaceous and the theoretical nitrogenous oxygen demand of the waste.

5-26. Calculate the NOD and ThOD of a waste containing 100 mg/L of isopropanol (C_3H_7OH) and 100 mg/L of NH_3-N.

5-27. A waste contains 100 mg/L of acetic acid (CH_3COOH) and 50 mg/L of NH_3-N. Determine the theoretical carbonaceous oxygen demand, the theoretical nitrogenous oxygen demand, and the total theoretical oxygen demand of the waste.

5-28. A waste contains 50 mg/L of phenol (C_6H_6O) and 100 mg/L of NH_3-N. Determine the theoretical carbonaceous and the theoretical nitrogenous oxygen demand of the waste.

5-29. The waste-treatment plant for Pine City discharges 1×10^5 m^3/day of treated waste to the Pine River. Immediately upstream of the treatment plant, the Pine River has an ultimate CBOD of 2 mg/L and a flow of 9×10^5 m^3/day. At a distance of 20 km downstream of the treatment plant the Pine River has an ultimate CBOD of 10 mg/L. The state's Department of Environmental Quality (DEQ) has set an ultimate CBOD discharge limit for the treatment plant of 2,000 kg/day. The river has a velocity of 20 km/day. The CBOD decay coefficient is 0.1 day^{-1}. Is the plant in violation of the DEQ discharge limit?

5-30. A waste having an ultimate CBOD of 100 mg/L and a reaction rate coefficient of 0.1 day^{-1} is placed in a vat. Determine the ultimate CBOD and the 5-day CBOD of the waste after it has sat in the vat for 5 days.

5-31. An industry discharges 0.5 m^3/s of a waste with a 5-day CBOD of 500 mg/L to a river with a flow of 2 m^3/s and a 5-day CBOD of 2 mg/L. Calculate the 5-day CBOD of the river after mixing with the waste.

5-32. A waste having an ultimate CBOD of 1,000 mg/L is discharged to a river at a rate of 2 m^3/s. The river has an ultimate CBOD of 10 mg/L and is flowing at a rate of 8 m^3/s. Assuming a reaction-rate coefficient of 0.1/day, calculate the ultimate and 5-day CBOD of the waste at the point of discharge (0 km) and 20 km downstream. The river is flowing at a velocity of 10 km/day.

5-33. A new wastewater-treatment plant proposes a discharge of 5 m^3/s of treated waste to a river. State regulations prohibit discharges that would raise the ultimate CBOD of the river above 10 mg/L. The river has a flow of 5 m^3/s and an ultimate CBOD of 2 mg/L. Calculate the maximum 5-day CBOD that can be discharged without violating state regulations. Assume a CBOD decay coefficient of 0.1 day^{-1} for both the river and the proposed treatment plant.

5-34. A river flowing with a velocity of 20 km/d has an ultimate CBOD of 20 mg/L. If the organic matter has a decay coefficient of 0.2 day^{-1}, what is the ultimate CBOD 40 km downstream?

5-35. A waste has an ultimate CBOD of 1,000 mg/L and a k_L of 0.1 day^{-1}. What is its 5-day CBOD?

5-36. Calculate the ultimate BOD of a waste that has a measured 5-day BOD of 20 mg/L, assuming a BOD rate coefficient of 0.15 day^{-1}.

5-37. A waste has a ThOD of 500 mg/L, a COD of 48 mg/L, and an ultimate CBOD of 5 mg/L. Offer an explanation for the character of this waste.

5-38. A tank truck containing industrial waste was found abandoned on a highway. Witnesses report that the truck had been parked on the highway for 3 days. Laboratory analysis showed the waste in the truck to have a 5-day carbonaceous biochemical oxygen demand of 223 mg/L and a reaction-rate coefficient of 0.1 day^{-1}. The table below gives the ultimate carbonaceous biochemical oxygen demand for waste from five area industries. Which industry in the table is the most likely source of the waste and show the calculations that support your conclusion? If the organic material in the sample is highly biodegradable, estimate what the COD and ThOD are in mg/L.

Paper Mill	Feedlot	Pharmaceutical	Brewery	Fruit Processing
67 mg/L	112 mg/L	223 mg/L	567 mg/L	765 mg/L

5-39. Write the balanced equation for the theoretical oxidation of benzene (C_6H_6) to carbon dioxide and water. What is the ThOD of this chemical if a water sample has 10 mg/L of benzene in it?

5-40. What is the ThOD of the following chemicals? Make sure you show the balanced stoichiometric equation with your work: (a) 5 mg/L C_7H_8; (b) 0.5 mg/L C_6Cl_5OH; (c) $C_{12}H_{10}$

5-41. Your client's wastewater discharge permit states that the wastewater exiting the treatment plant must have a COD less than 25 mg/L. You know the following information about the discharged wastewater; BOD_5/COD = 0.6, BOD of the 60th day = 17.5 mg/L, and $k = 0.10\ \text{day}^{-1}$. (a) Is your client meeting its discharge limit for COD? (b) Your client's wastewater contains only one organic chemical with a chemical formula of C_6H_6. What is the approximate concentration of this organic compound in ppm?

5-42. One individual pours one gallon of water containing 10,000 mg/L of a hazardous chemical in 1,000 gallons of water. A second individual pours 20 gallons of water containing 5,000 mg/L of the same chemical in the water. You learn that the chemical formula of this highly biodegradable chemical is $C_{10}H_{20}$. You then collect a 5-mL sample for your laboratory to analyze. What laboratory results do you expect for (a) the concentration of the chemical after the second individual has added their 20 gallons (in ppm); (b) the ultimate BOD of the 5-mL sample in mg/L; (c) the COD (mg/L) in the 5-mL sample; and (d) the BOD remaining to be exerted after 5 days, provided that the BOD rate constant is $0.1\ \text{day}^{-1}$? List all assumptions used to solve the problem.

5-43. Calculate the dissolved oxygen deficit for a river with a temperature of 30°C and a measured dissolved-oxygen concentration of 3 mg/L. The Henry's law constant at that temperature is 1.125×10^{-3} mole/L-atm and the partial pressure of oxygen is 0.21 atm.

5-44. A waste-treatment plant discharges a waste effluent containing 2 mg/L of dissolved oxygen to a river that has a dissolved-oxygen concentration of 8 mg/L upstream of the discharge. Calculate the dissolved oxygen deficit at the mixing basin if the saturation dissolved oxygen for the river is 9 mg/L. Assume that the river and plant discharge have the same flow rate.

5-45. A river traveling at a velocity of 10 km/day has a dissolved-oxygen content of 5 mg/L and an ultimate CBOD of 25 mg/L at distance $x = 0$ km, that is, immediately downstream of a waste discharge. The waste has a CBOD decay coefficient, k_1, of $0.2\ \text{day}^{-1}$. The stream has a reaeration rate coefficient, k_2, of $0.4\ \text{day}^{-1}$ and a saturation dissolved-oxygen concentration of 9 mg/L. (a) What is the initial dissolved-oxygen deficit? (b) What is the location of the critical point, in time and distance? (c) What is the dissolved-oxygen deficit at the critical point? (d) What is the dissolved-oxygen concentration at the critical point?

5-46. A stream with a temperature of 25°C has a dissolved-oxygen concentration of 4 mg/L. What is the dissolved-oxygen deficit (mg/L)?

5-47. The oxygen concentration of a stream is 4 mg/L. Saturation for the stream is 10 mg/L. What is the oxygen deficit?

5-48. A river traveling at a velocity of 10 km/d has an initial oxygen deficit of 4 mg/L and an ultimate CBOD of 10 mg/L. The CBOD has a decay coefficient of 0.2 day^{-1} and the stream's reaeration coefficient is 0.4 day^{-1}. What is the location of the critical point: (a) in time; (b) in distance?

5-49. A paper mill discharges its waste ($k_L = 0.05\ \text{day}^{-1}$) to a river flowing with a velocity of 20 km/d. After mixing with the waste, the river has an ultimate carbonaceous BOD of 50 mg/L. Calculate the 5-day carbonaceous BOD at that location and the ultimate carbonaceous BOD remaining 10 km downstream.

5-50. For each of the cases below, assuming all other things unchanged, describe the effect of the following parameter variations on the magnitude of the maximum oxygen deficit in a river. Answer: Increase (+), decrease (−), or remain the same (=).

Parameter	Magnitude of the Deficit
Increased initial deficit	
Increased ultimate CBOD @ $x = 0$	
Increased deoxygenation rate	
Increased reaeration rate	
Increased ThOD @ $x = 0$	

5-51. A stream has a flow of 1.8 m^3/s, upstream ultimate BOD of 7.5 mg/L, and dissolved-oxygen concentration of 6.2 ppm. A plant discharges 0.6 m^3/s of wastewater with an ultimate BOD of 5 mg/L and dissolved-oxygen concentration of 0.5 ppm. The stream has a velocity of 0.65 m/s, deoxygenation rate constant of 0.2 day^{-1}, and reaeration constant of 0.37 day^{-1}. The saturated level of dissolved oxygen in the stream is 8 mg/L. Adverse effects on the stream's aquatic life begin to occur if the dissolved-oxygen concentration falls below 4.5 ppm. Is the aquatic life safe in this stream if the only outside source of BOD to the stream is from the plant?

5-52. A lake with a surface area of $1 \times 10^6\ m^2$ receives a water inflow of $1 \times 10^8\ m^3$/yr and a phosphorus input of 1×10^7 gP/yr. Locate the trophic state position for this lake by placing an X on a copy of the Vollenweider plot provided in Figure 5-38. (a) Is the lake currently oligotrophic or eutrophic? (b) Would this lake be more like Lake Erie or Lake Superior? (c) Would this lake have high or low transparency? (d) Nutrient levels? (e) Algal biomass? (f) Bottom water dissolved oxygen? (Explain your answer for parts (c)–(f).) (g) What would be the maximum acceptable phosphorus loading for this lake?

REFERENCES

Becker, D. S. and G. N. Bigham. 1995. Distribution of mercury in the aquatic food web of Onondaga Lake, NY. *Water, Air and Soil Pollution*, 80:563–571.

Budyko, M. I. 1974. Climate and Life. Academic Press, New York.

Canale, R. P., M. T. Auer, E. T. Owens, T. M. Heidtke, and S. W. Effler. 1993. Modeling fecal coliform bacteria—II. Model development and application. *Water Research*, 27:703–714.

Chapra, S. 1997. Surface Water-Quality Modeling. McGraw Hill Company, New York.

Cooke, G. D., E. B. Welch, S. A. Peterson, and P. R. Newroth. 1993. Restoration and Management of Lakes and Reservoirs, 2nd Edition. Lewis Publishers, Boca Raton, FL.

Davis, M. L. and D. A. Cornwell, 1991. Introduction to Environmental Engineering, McGraw-Hill, Inc., New York.

Droste, R. L. 1997. Theory and Practice of Water and Wastewater Treatment. John Wiley & Sons, Inc., New York.

Effler, S. W. 1996. Limnological and Engineering Analysis of a Polluted Urban Lake: Prelude to Environmental Management of Onondaga Lake, New York. Springer-Verlag, New York.

Enger, E. D., J. R. Kormelink, B. F. Smith, R. J. Smith. 1983. Environmental Science. Wm. C. Brown Publishers, Dubuque, IA.

Greenberg, A. E., L. S. Clesceri, A. D. Eaton, 1992. (Eds.) Standard Methods for the Examination of Water and Wastewater, 18th Edition. American Public Health Association, Washington, DC.

Henry, J. G. and G. W. Heinke, 1996. Environmental Science and Engineering, 2nd Edition. Prentice-Hall, Inc., Upper Saddle River, NJ.

Horne, A. J. and C. R. Goldman. 1994. Limnology, 2nd Edition. McGraw-Hill, Inc., New York, 576 pp.

Kupchella, C. and M. Hyland. 1986. Environmental Science. Allyn and Bacon, Inc., Needham Heights, MA.

Jacobson, J. L., S. W. Jacobson, and H. E. B. Humphrey, 1990. Effects of in utero exposure to polychlorinated biphenyls and related contaminants on cognitive functioning in young children. *J. Pediatr.*, 116:38–45.

Lerman, A. 1988. Geochemical Processes: Water and Sediment Environments. Kreiger Publishing, Malabar, FL.

Metcalf & Eddy, Inc., 1989. Wastewater Engineering: Treatment, Disposal, Reuse. McGraw-Hill Book Company, 2nd Edition, New York.

National Science and Technology Council. 1996. The Health and Ecological Effects of Endocrine Disrupting Chemicals: A Framework for Planning, Committee on Environment and Natural Resources.

Nemerow, N. L. 1971. Liquid Waste of Industry: Theories, Practices, and Treatment. Addison-Wesley Publishers, Reading, MA.

Patnaik, P. 1992. A Comprehensive Guide to the Hazardous Properties of Chemical Substances. Van Nostrand Reinhold, New York.

Raven, P. H., L. R. Berg, and G. B. Johnson. 1995. Environment. Saunders College Publishing, Ft. Worth, TX.

ReVelle, P. and C. ReVelle. 1984. The Environment, Willard Grant Press, Boston.

Ricklefs, R. E. 1983. The Economy of Nature, 2nd Edition, Chiron Press, Inc. New York.

Snoeyink, V. L. and D. Jenkins. 1980. Water Chemistry. John Wiley & Sons, Inc. New York.

Stevenson, F. J. 1994. Humus Chemistry: Genesis, Composition, Reactions, 2nd Edition. Wiley & Sons, Inc., New York.

UNESCO, 1978. World Water Balance and Water Resources of the Earth. UNESCO Press, Paris.

Veith, G. D. and P. Kosian, 1983. Estimating bioconcentration potential from octanol/water partition coefficients, in "Physical Behavior of PCBs in the Great Lakes," Mackay, D. (Ed). Ann Arbor Science, Ann Arbor, MI.

Verbrugge, D. A., J. P. Giesy, M. A. Mora, L. L. Williams, R. Rossman, R. A. Moll, and M. Tuchman. 1995. Concentrations of dissolved and particulate polychlorinated biophenyls in water from the Saginaw River, Michigan. *Journal of Great Lakes Research*, 21(2):219–233.

Solutions to Chapter Problems

2-1. (a) i, ii; (b) ii, iii, iv (BOD is not commonly reported in ppm); (c) v; (d) ii, iii, v; (e) vi; (f) i, ii, iii, vi

2-2. (a) 3.40 ppm; (b) because both are in mass units

2-3. (a) 0.41 mg/L; (b) 0.21 mg/L

2-4. (a) 10 ppm; (b) 0.00016 moles/L; (c) 2.26 mg N/L; (d) 10,000 ppb

2-5. 2.8 ppm

2-6. 0.9 coliforms/100 mL, so water is safe

2-7. 11 mg/L and 1.6 mg/L

2-8. (a) 12 mg/L; (b) 12 ppm; (c) 0.0012%

2-9. third well exceeds 10mg NO_3^-—N/L standard

2-10. (a) 0.002 ppb, 2.0 ppt, 3.7×10^{-6} μmoles/L; (b) 0.002 ppm, 2.0 ppb

2-11. (a) 0.005 mg/L; (b) 5 μg/L; (c) 0.005 ppm; (d) 5 ppb

2-12. caffeine, 120 mg/L, so no; TCE, 59 mg/L, so yes

2-13. 0.73 μg/L

2-14. Pb 46 nmoles/L, Cu 31 nmole/L, Mn 156 nmole/L

2-15. for 0.5 mg/L, 1.56×10^{-5} moles/L and 0.5 ppm; for 8 mg/L 2.5×10^{-4} moles/L and 8.0 ppm

2-16. concentration of 3.2×10^{-11} ppm_v equals Pi of 3.2×10^{-17} atm

2-17. (a) 0.088 ppm; (b) 0.0000088%

2-18. ppm = 30; therefore, both are above the standard

2-19. concentration = 860 $\mu g/m^3$, so mass = 0.69 g

2-20. (a) 241 $\mu g/m^3$; (b) 0.125 mole $O_3/10^6$ mole air

2-21. (a) 14.9%; (b) 10.2 L

2-22. 15 ppm

2-23. 26 $\mu g/m^3$ and 5,230 $\mu g/m^3$

2-24. ppm = 57, so yes

2-25. (a) 100 ppb; (b) 100 > 60, so sample may pose a threat

2-26. 0.005 mg/kg

2-27. 5 ppm

2-28. (a) Sum of cations equals 2.92×10^{-3} eqv/L and sum of anions equals 2.90×10^{-3} eqv/L so analysis is correct; (b) total hardness = 89 + 41 = 130 mg/L as $CaCO_3$

2-29. sum of cations equals 3.37×10^{-3} eqv/L and sum of anions equals 3.48×10^{-3} eqv/L so analysis is correct; total hardness = 88 + 79 = 167 mg/L as $CaCO_3$

2-30. (a) 170 mg/L; (b) VSS = mg/L, which is an appreciable amount of organic matter

2-31. (a) 625 mg/L; (b) (solve part c first) 300 mg/L; (c) 325 mg/L; (d) 210 mg/L

3-1. 5.8 g

3-2. (a) 23 days; (b) 46 days; (c) 69 days

3-3. 0.7 days

3-4. (a) first; (b) second

3-5. (a) K = $1.76 \times 10^{-2} = 10^{-1.75}$, so the % PO_4^{3-} = 5%; (b) $\Delta G = -2.55$ kJ so the reaction will proceed as written and under the given conditions; (c) solve for the three activity coefficients. ΔG now equals +0.28; therefore, the reaction will not proceed as written and it will go in the reverse direction until equilibrium is reached.

3-6. (a) first; (b) second; (c) 4×10^{-3} sec

3-7. (a) first; (b) 0.2/day; (c) 3.5 days; (d) 0.3/day

3-8. (a) $\Delta G = -109$ kJ; (b) first order; (c) 0.12/min

3-9. (a) 0.23/min; (b) 3.0 min; (c) 0.61/min

3-10. (a) chloroform = 1,850 yr, TCE = 1.06 yr; (b) chloroform

3-11. k(5) = 0.055/day, k(25) = 0.37/day

3-12. $\Delta G = -1{,}990$ kJ, so yes

3-13. $\Delta G^\circ = +90.3$ kJ, so no

3-14. $\Delta G^\circ = -176.8$ kJ, so yes

3-15. $\Delta G^\circ = -16.86$ kcal, so yes

3-16. (a) $\Delta G^\circ = -213.1$ kJ, so reaction is feasible; (b) ΔG now equals -33.97, so reaction is feasible

3-17. (a) 3.0×10^{-6} moles/L; (b) 3.3×10^{-6} moles/L; (c) (Hint: should activity coefficients be included?) K_{so} increases, so solubility increases

3-18. (a) 3.2 g/m^3; (b) (Hint: how does BP relate to vapor pressure?) DCB

3-19. chloroform

3-20. 4.79×10^{-4} moles/L, 15.3 mg/L

3-21. The measured ratio of aqueous to gas-phase PCB is 13.6 mole/L-atm. Compare to equilibrium ratio of 10 mole/L-atm to determine that PCBs must move from the water into the air to reach equilibrium.

3-22. 0.04 atm-L/mole

3-23. 0.77

3-24. (a) H (TCE) = 0.44, H (PCE) = 1.1, H (dimethylbenzene) = 0.21, H (parathion) = 1.6×10^{-5}; (b) PCE, TCE, dimethylbenzene, parathion

3-25. 96 μg/L

3-26. (a) 3.8; (b) 12; (c) 4.4

3-27. 2.6

3-28. (a) 97%; (b) 76%

3-29. 33%

3-30. (a) 0.001 moles/L, (b) 33 mg/L; (c) 33 ppm

3-31. 1.2×10^{-22} moles/L

3-32. $Q = 10^{-7.8}$; therefore, calcite must continue to precipitate.

3-33. (a) $\Delta G^\circ = -78.4$ kJ/mole; (b) Cd = 2.04×10^3 mg/L, so no, pH must be raised.

3-34. Aqueous phase ferric iron equals 3.9×10^{-14} mg/L, so precipitate equals 0.056 mg/L minus the aqueous phase iron.

3-35. $K_{oc} = 1{,}340$ cm^3/g oc, 95% CI on log $K_{oc} = \pm 0.67$ cm^3/g oc

3-36. 94%

3-37. 30% sorbed and 70% in aqueous phase

3-38. R(TCE) = 30, R(HCB) = 18,000, R(dichloromethane) = 3.1

3-39. (a) 3.16 L/kg; (b) 3,160 mg/kg; (c) $Cr^{3+} = 3.16 \times 10^{-14}$ moles/L; (d) 950 lb Cr^{3+} per tank

3-40. benzene

3-41. 1.19×10^5 yr

3-42. 2.1×10^5 g PAC/day

4-1. 11 mg/L

4-2. (a) the entire lake. Yes, concentrations are changing—the pollutant concentration is dropping with time. Since concentrations are changing with time (the spill is not a continuous source), this is a non-steady-state problem. The chemical is inert; therefore, it is not being destroyed, and there is no information given to suggest that it is being produced. Since the chemical is stated to be inert, it is conservative. (b) The entire atmosphere. Concentrations are changing though the emission rate is constant; carbon dioxide emissions are increasing with time. Since concentrations are changing with time, this is a non-steady-state problem, the chemical is not chemically degraded. The chemical is conservative, because no chemical reaction takes place; (c) the room. Conditions and concentrations are not changing with time. Steady state conditions. The perfume is emitted at a constant rate. The perfume is not being produced or destroyed because of a chemical reaction in the control volume because its decay is slow relative to its time in the control volume. The perfume is conservative, because its first-order decay is much slower relative to the advective flow.

4-3. 49.0 ppb

4-4. (a) 0.90/hr; (b) 1.25 pCi/L

4-5. 6.9 min

4-6. (a) 38 mg/s; (b) 12.7 mg/m^3

4-7. (a) 150 mg/s; (b) 1 hr

4-8. (a) 4.7 mg/L; (b) 4.2 mg/L

4-9. (a) 7.9 coliforms/100 mL, so no, water standard is not being met; (b) not possible, town 1 would need a concentration of -12 coliforms/100 mL.

4-10. (a) 21 mg/L; (b) 17 tons/day

4-11. (a) 10 mg/L; (b) 0.046 days

4-12. (a) Superior 180 yr, Erie 2.6 yr; (b) Superior 414 yr, Erie 6 yr

4-13. 2 hr

4-14. (a) 0.15/min; (b) 30,000 gal; (c) 660,000 gal; (d) discussion; (e) 1,900 g/day

4-15. (a) concentration remaining = 0.005 moles/L; (b) concentration remaining = 3.3×10^{-6} moles/L

4-16. see text

4-17. 63 mg/L

4-18. (a) 0.090/hr; (b) 0.042/hr; (c) 2×10^6 gal

4-19. 12W

4-20. (a) At x = 0.5, 3.5, 4.5 cm, the initial pollutant flux density, J, = 0. At x = 1.5 and 2.5 cm, $J = 10^{-8}$ mg/cm²-s; (b) initial mass flux at x = 0.5, 3.5, and 4.5 cm equals 0, at x = 1.5 and 2.5 cm, $m = 7.1 \times 10^{-8}$ mg/s; (c) graphs; (d) The concentration profile in part (c) is changing due to the random motion of the molecules. The chemical is attempting to reach equilibrium through high concentration areas moving to areas with less concentration.

4-21. (a) 0.028 mg/cm²-s; (b) 0.20 mg/s

4-22. (a) 60,000 m³/day; (b) 0.12 m/s

4-23. (a) 30 cm/s; (b) 96.5 μm

4-24. 13 μm

4-25. 1.18×10^3 cm/s

4-26. (a) figure; (b) $v_t = 1.878 \times 10^7\ D_p$ where v_t = cm/s; (c) 190 cm/s; (d) 0.322 cm/s; (e) 5,840 μm; (f) drift velocity

4-27. (a) 0.06 cm/s; (b) 23 μm

4-28. (a) 0.25 m/day; (b) 0.42 m/day; (c) 238 days

4-29. 62.5 days

5-1. 1.0/day

5-2. 1,101,323 mg/L

5-3. (a) 91,680 mg/L; (b) 8%

5-4. (a) 0.33/day; (b) 33%

5-5. Hint: Determine the specific growth rate of each organism at two different substrate levels. (a) organism B; (b) organism A

5-6. 0.2 mg biomass/mg substrate

5-7. 5.8 days

5-8. 114; 569; 683 mg/L

5-9. (a) 229 mg/L; (b) L_o/ThOD = 0.40

5-10. 5.7, 14.7, 27.2 mg/L; 10.9, 2.4, 22.3, 8.8 mg/L

5-11. 4.3 mg/L, 3.6 days, 32.4 km, 3.7 mg/L

5-12. (a) see text; (b) No (Hint: use values of S, Y, and K to see if K can be reached); (c) 17 mg DW/L; (d) 70 mgS/L; (e) 7,410 mg DW/L

5-13. (c) 1,000 mg/L

5-14. 4.6 days

5-15. 0.42/day

5-16. lower

5-17. discussion

5-18. 0.069/day

5-19. 4.6 days

5-20. Do not support, $\mu = 0.018$/day

5-21. *Rotobacter eatsalli* (Hint: compare μ for each organism at $S = 1$ mg/L

5-22. (a) Monod; (b) A: 0.8/day, 21 mg/L, B: 0.3/day, 2 mg/L; (c) depends on substrate concentration; (d) B

5-23. (a) exponential; (b) 0.0023/yr; (c) time^{-1}; (d) no carrying capacity term; (e) logistic

5-24. 284 mg/L = 75 mg/L + 209 mg/L

5-25. 129, 229 mg/L

5-26. 457, 240, 697 mg/L

5-27. 107, 228, 335 mg/L

5-28. 119, 457 mg/L

5-29. 9,200 kg/day $\gg$ 2,000 kg/day

5-30. 61, 24 mg/L

5-31. 102 mg/L

5-32. 208, 82 mg/L, 170, 67 mg/L

5-33. 7 mg/L

5-34. 13 mg/L

5-35. 390 mg/L

5-36. 38 mg/L

5-37. see solution

5-38. brewery or fruit processor (Hint: determine L_o of waste at time = 0 and 3 days)

5-39. 31 mg/L

5-40. (a) 16 mg/L; (b) 0.27 mg/L; (c) 3.0 mg/L

5-41. (a) 11 mg/L < 25 mg/L; (b) 3.6 ppm

5-42. (a) 108 ppm; (b) 370 mg/L; (c) 370 mg/L; (d) 224 mg/L

5-43. 4.6 mg/L

5-44. 4 mg/L

5-45. (a) 4 mg/L; (b) 2.6 day, 26 km; (c) 7.4 mg/L; (d) 1.6 mg/L

5-46. 4.7 mg/L

5-47. 6 mg/L

5-48. 0.9 day, 9 km

5-49. 11 mg/L, 39 mg/L

5-50. +, +, =, =, + or = (for last part, depends on biodegradability)

5-51. at critical distance, DO = 4.7 mg/L

5-52. (a) $W' = 10\ gP/m^2$-yr and $Q' = 100$ m/yr; (b) Erie; (c) low; (d) high; (e) high; (f) low; (g) 1×10^6 gP/yr

Index